# Carbon

Tapan Gupta

# Carbon

## The Black, the Gray and the Transparent

Springer

Tapan Gupta
La Mesa, CA
USA

ISBN 978-3-319-66404-0 ISBN 978-3-319-66405-7 (eBook)
DOI 10.1007/978-3-319-66405-7

Library of Congress Control Number: 2017950211



Printed on acid-free paper

*Photo credit line:* Graphite, Diamond, Coal, Graphene, DNA, Photosynthesis, Buckyball (C60), CNT, Protein (Clockwise)

This Springer imprint is published by Springer Nature
The registered company is Springer International Publishing AG
The registered company address is: Gewerbestrasse 11, 6330 Cham, Switzerland

*The book is dedicated to the memory*
*of my father Dr. Gopeswar Gupta*
*and my mother Dr. Kanak Lata Gupta.*

# Preface

According to N.R. Pace, "All living things contain carbon in some form and it is the primary component of macromolecules including proteins, lipids, nucleic acids (RNA and DNA), and carbohydrates." Indeed, it exists in all living animals. As a matter of fact, it is the backbone of all organic (chemistry) compounds.

Organic chemistry is the elementary description of life because the chemical reactions required for the faithful propagation of a living organism require a high degree of specificity in the interactions of the molecules. Only two of the natural atoms, carbon and silicon, are known to serve as the backbones of molecules large enough to carry biological information; however, opinion is divided on whether silicon-based life is a realistic prospect.

Carbon is a long-studied element, but there is always more to discover. Indeed, in 1985, Richard Smalley and Robert Curl of Rice University, and their colleagues discovered a new form of carbon, the fullerene, also known as "buckyballs." Since then, "buckyballs" and "buckytubes" have been the subject of intense research both for their unique chemistry and technological applications in material science, electronics, and nanotechnology.

In 2010, Andre Geim and Konstantin Novoselov of the University Of Manchester, England, won the Nobel Prize in Physics for their groundbreaking experiments regarding the two-dimensional (2-D) material graphene—an allotrope of carbon. The global market for graphene reached 9 million dollars in 2014. In the year 2010, scientists reported the invention of flexible conductive textiles dipped in carbon nanotube (CNT) ink that could be used to store energy, perhaps paving the way of wearable batteries, solar cells, and other electronics.

The next important chemical which belongs to the carbon family is polymer. The term "polymer" was first introduced in 1833 by the Swedish chemist Jöns Jacob Berzelius. Many polymers such as silk, cotton, starch, sand, asbestos, as well as complex polymers including RNA (ribonucleic acid) and DNA (deoxyribonucleic acid), which hold genetic codes, exist in nature. John Steele Gordon wrote: "It has been a curious silent revolution. . .when we think of scientific triumphs of twentieth Century, we think of nuclear physics, medicine, space exploration, and the computers. But all these developments would have been much impeded, in some cases impossible, without ... plastics ... most of them are made of polymers." Indeed, Moore's law, which dictates the size of a transistor, depends mostly on the magic polymer—the photoresist.

I was fascinated with and excited by all the discoveries, as the wonder nacre—carbon—can exist in many forms and do wonderful things. Indeed, one of its allotropes, the diamond, is exquisitely beautiful. All the characters of carbon that have been explored so far led me to write a small scientific history of carbon. This book *Carbon: The Black, the Gray, and the Transparent* is not a complete scientific history, but focuses on some of the wonderful discoveries about this old faithful element that was known to prehistoric humans in the form of charcoal.

This book is not intended strictly as a textbook, but should serve as a reference book, which will satisfy the expressed needs that brought it about. I strongly believe that it will be an interesting document to all polymathic scholars. I realize it is difficult to provide a complete picture and the history of carbon, but that is the mission of this book.

Chapter 1 attempts to illustrate a complete picture about the existence of carbon. Carbon's discovery is said to be a lost one. According to the Swinburne Center for Astrophysics and Supercomputing, it forms in the belly of stars in a reaction called the triple-alpha process.

Chapter 2 is dedicated to activated carbon (AC) which is used in our everyday lives, starting with water filtration to medical science as an absorbent of poisonous materials. I also discuss the kinetic theory related to its absorption and desorption properties.

Chapter 3 is devoted to some of the important composites of carbon, like carbon fibers, and metal matrix composites that are being used in so many commercial products like airplanes, cars, building walls, footbridges, diffusion coatings, electroplating, and in heat sinks. It has been a material of great interest to the Air Force and NASA because of its high modulus or resistance to stretching. The inelasticity of carbon composites plays an important role in reinforcing rigid structures such as the nose cones in hypersonic aircraft.

Chapter 4 discusses the miracle material polymer. It has been said that all the twentieth centuries' developments would have been much impeded, in some cases impossible, without plastics, and almost all of them are made of polymers. Polymers are also a part of the nano-world, and with the development of a miracle polymer known as photoresist, it is possible to fabricate a transistor having a structure of ~22 nanometer (which has raised a dilemma in tunneling physicists regarding George Gamow's theory).

In Chap. 5, some of the important characters of black carbon are presented. According to physics, it is difficult to distinguish a material which is black and which is not. The black carbon coal discussed in Chap. 5 is not exactly black, but to ordinary people it is black. The highest rank coal, anthracite, followed by bituminous and lignite are discussed along with their use as a source of energy in our daily lives. Products from the coal, mainly coal tar and its derivatives, are also discussed.

Chapter 6 is all about graphite—the gray carbon. The highest grade of black coal is called carbon gray, which is also known as meta-anthracite. It has been used extensively by the manufacturers of lead pencils and of lead-bismuth alloy (due to its relatively low melting points and low neutron absorption). It has also been used as a coolant in some nuclear reactors. Lead has been proposed as a coolant in fast reactors, as part of the Generation IV reactor initiative. Its physical characteristics, along with its electrical properties, are discussed in this chapter. Graphite mining, beneficiation, and milling are also discussed.

Chapter 7 deals with the two-dimensional (2-D) carbonaceous material called graphene. It is an allotrope of carbon, and the groundbreaking experiments to produce the 2-D graphene are discussed in this chapter. According to Connor, the creation of graphene, a wonder material and the result of brilliant serendipity, stands alongside the accidental discovery of penicillin by Alexander Fleming. The electrical and the optical properties of this remarkable material are also discussed.

Chapter 8 is all about the carbon nanotube (CNT). Everything, when miniaturized to the sub-100-nanometer scale has new properties. CNTs are no exception to this. Indeed, CNTs are used in materials from the mundane to the lifesaving, from sunscreen to chemical catalysis to antibacterial agents. In this chapter, its fabrication along with its different characteristics is discussed.

Chapter 9 describes briefly the transparent carbon diamond, an allotrope of carbon. Its splendid colors, transparency, and physical properties have made diamond not only a precious gemstone, but have made it the material of choice for radiation detection in nuclear medicine. Different experimental methods for synthesizing synthetic diamond, nano-diamonds, and diamond-like carbon (DLC) are presented along with their properties and uses.

Chapter 10 addresses another important material—the buckminsterfullerene. It is said that fullerenes are beautiful both physically and intellectually. Their qualities, and even some of their properties, can be appreciated instantly and intuitively by non-scientists. This chapter discusses the synthesis, properties, separation, and purification of fullerenes and the effects of dopants on the original material.

## Acknowledgment

The book *Carbon: The Black, the Gray and the Transparent* would not have been possible to write without direct and indirect help of so many scientists. It is not possible to mention all the names of the scientists, but I shall mention some of the leading scientists in this area who have helped me the most. First of all I want to mention Drs. Andre Geim and Konstantin Novoselov, Nobel Laureates in Physics (2010), at the University of Manchester, England. Professor Geim has kindly given me some of his valuable publications to include in my book. I gratefully acknowledge the help. Professor James Hone of Columbia University of New York and his team helped me in my endeavor to write the book. The other professors and the researchers that are to be well mentioned for their help are Drs. Robert Young, University of Manchester, England; R.B. Hallick, University of Arizona, USA; J. W. Lyding, University of Illinois at Chicago, ILL,USA; J. P. Gore, Purdue University, IN, USA; R. Rao, Pennsylvania State University, PA, USA; J.M. Tour, Rice University, TX, USA; R. Santoro, Cornell University, NY, USA; and J.B. Vander Sande, Massachusetts Institute of Technology (MIT), MA, USA. I gratefully acknowledge the comments of the reviewers, especially one of them who thoroughly gone through each and every page of the book and made corrections and suggestions for improvement.

The publishers and the researchers whom I owe much are Wikipedia (Wiki), Elsevier, Wiley, Springer, and Academic Press. Most of the referred pictures and drawings are from Wiki. It is my pleasure to gratefully acknowledge the help of Wikipedia (Wiki). I sincerely apologize if I have missed to mention the help of any one of the scientists, research scholars, and the publishers.

Lastly, I want to acknowledge my gratitude to my wife, Mrs. Arundhati Gupta, and my daughter, Ms. Atreyee Gupta, for their steadfast support, understanding, and help throughout the several years required for the completion of the book.

La Mesa, CA, USA Tapan Gupta

# Contents

# Chapter 1
# Carbon (C) the Nacre and Its Allotropes

Carbon: ***Carbon*** (identified by a symbol ***C*** in the periodic table of elements, Fig. 1.3) is an interesting element; you cannot see (C, pronounced as *see*) the ***C*** all the time, but the existence of the element (***C***) is almost everywhere. It is in your *clothes*, in your *backpack*, in your *food*, and in your *body*. It provides the framework for all *tissues* of *plants* and *animals* (rational and irrational). These *tissues* are built of elements grouped around *chains* or *rings* made of ***carbon*** atoms. Indeed, most of the structures that make up animals, plants, and microbes are made from three basic classes of molecules: *amino acids*, *carbohydrates*, and *lipids* (often called fatty acids), and all of these materials contain carbon as their backbone [1, 2]. As these molecules are vital for life, *metabolic reactions* either focus on making these molecules during the construction of cells and tissues, or breaking them down and using them as a source of energy during digestion [3]. As a matter of fact, carbon (C) occurs in nature as the sixth most abundant element in the universe and the 19th element in order of mass in the Earth's crust.

Indeed, *metabolism* (from Greek: *change*) is the set of life-sustaining chemical transformations within the cells of living organisms [4]. The core of the *cellular metabolism* is the *central carbon metabolism* (CCM). Indeed, CCM is a series of processes that convert nutrients from environment to *power cell metabolism*. CCM is recognized as a set of biochemical pathways for transport and oxidation of main carbon sources in the cell [5]. Generally, it consists of the *phosphotransferase* system, *glycolysis*, *gluconeogenesis*, *pentosemonophosphate* bypass with *Entner-Doudoroff* pathway, *Krebs cycle* with *glyoxylate* bypass and the respiration chain. It was known since early biochemical studies on bacterial cells that prokaryotes prefer glucose over other carbon sources, and it is reflected in specific regulations of expression of gene coding for carbolic enzymes [6, 7]. In *mycobacterium tuberculosis* (Mtb), CCM appears to have evolved to serve independent physiological and pathogenic roles [8]. A genomically annotated inventory of known or homology-predicted CCM enzymes in Mtb is presented in Fig. 1.1.

The construction of the human cells and tissues are through three basic classes of molecules: *amino acids*, *carbohydrates*, and *lipids* (Table 1.1). These biochemicals


T. Gupta, *Carbon*, DOI 10.1007/978-3-319-66405-7_1

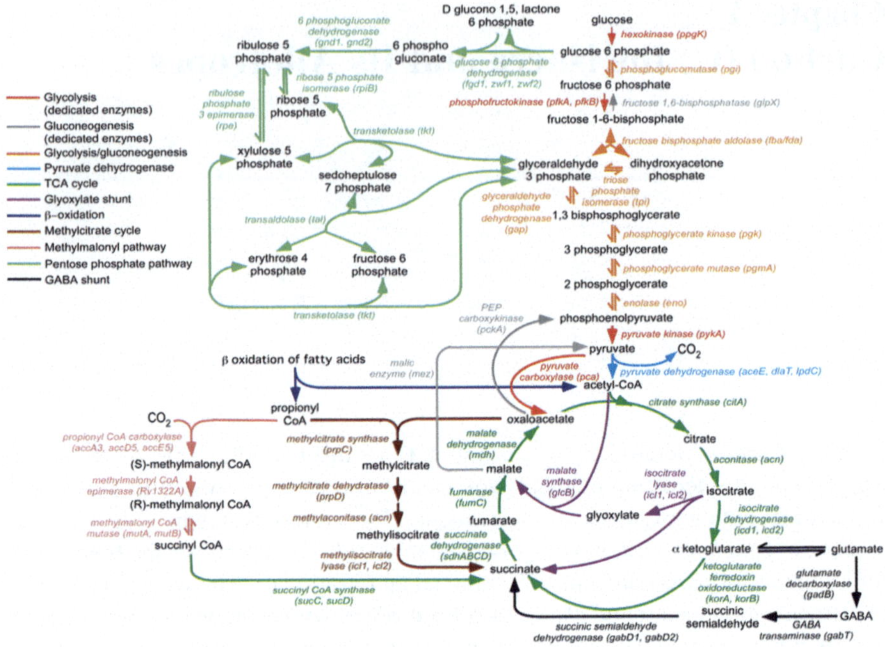


**Fig. 1.1** A genomically annotated inventory of known or homology-predicted CCM enzymes in Mtb (Courtesy, ncbi.nih.gov.USA, reproduced with kind permission of Elsevier Sci. Pub.)

**Table 1.1** Life-forming macromolecules and the polymers[a]

| Type of molecule | Name of monomer forms | Name of polymer forms | Example of polymer forms |
|---|---|---|---|
| Amino acids | Amino acids | Proteins (also called polypeptides) | Fibrous proteins and globular proteins |
| Carbohydrates | Monosaccharides | Polysaccharides | Starch, glycogen, and cellulose |
| Nucleic acid | Nucleotides | Polynucleotides | DNA and RNA |

[a]Adapted from Wikipedia

can be joined together to make polymers such as DNA and proteins – the essential macromolecules of life [9]. Indeed, all the biochemicals and the other materials that have been discussed so far belong to the *carbon family*.

Thus, it will not be an exaggeration to say

*Carbon, carbon, carbon everywhere*
*and without the element carbon (C),*
*life as we know would be no where.*

Indeed, according to poet Janet Kuypers [10],

**Fig. 1.2** (**a**) Picture of carbon and (**b**) spectral lines of carbon using 600lpmm grating (Courtesy: teravolt, WIKI)

*It's very possible*
*that throughout the early history of earth,*
*asteroids collided with this planet,*
*leaving their* ***carbon****-rich* ***DNA*** *structures behind*
*to help start life, and populate the earth.*

Carbon: (*C*) compounds form the basis of all known life on earth, and the carbon–nitrogen cycle provides some energy produced by Sun and other stars. Thus, we can see that carbon is an interesting and fascinating material, which has revolutionized material science (since 3750 BC by Egyptians) [11], and studies of its origin and its allotropes and how it has shaped the lives of different living organisms will be worth and meaningful.

Figure 1.2a shows the picture of *carbon particles* and Fig. 1.2b shows the emission line spectrum of neutral and ionized carbon excited in electric discharge. Carbon (from Latin word *carbo*, meaning *coal*) is a chemical element placed in periodic table with a symbol ***C***. In 3650 BC, *Egyptians* used *carbon* in the form of *coal*. It was used to absorb malodorous vapors from wound. During World War I, the absorbent properties of carbon were used in gas masks and in other ways to defend against poisonous gas attacks. During the nineteenth century, carbon in the form of charcoal became important for decolorizing sugar syrups. Another important material derived from carbon (*C*) *is called cellulose*. It is a *carbon polymer* used to make cotton, linen, and hemp material. This is not all. The *plastics*, one of the most important materials in our daily life, are made from hydrocarbon carbyne ($H_2C$:), which contain carbon atom [12, 13].

Figure 1.3 is taken from the periodic table of elements showing the element carbon (C), with atomic number (Z) 6 and atomic weight 12.

The classification scheme of carbon is based on the types of chemical bonds in carbon, with each valence state corresponding to a certain form of simple substance. Elemental carbon can exist in three bonding states corresponding to $sp^3$, $sp^2$, and $sp^1$ hybridization of the atomic orbitals, and the corresponding three carbon

**Fig. 1.3** The element carbon (C), existing in the periodic table, showing atomic number (Z) in the upper left-hand side and the atomic weight at the bottom

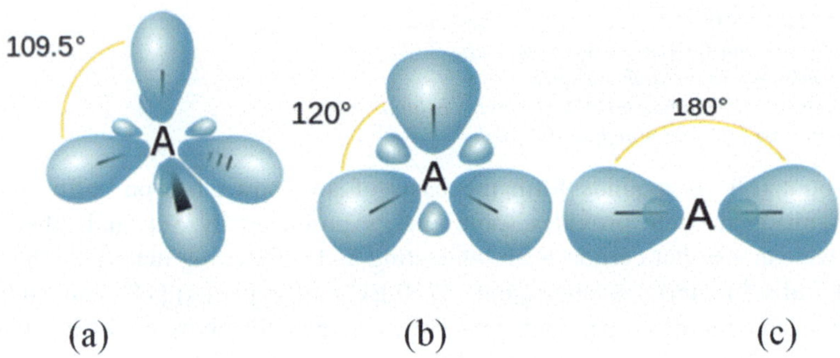


**Fig. 1.4** Orbital hybridization of carbon (**a**) Four $sp^3$ orbitals, (**b**) Three $sp^2$ orbitals and (**c**) Two sp. orbitals (Courtesy: J. Fernando, WIKI)

alloptropes with an integer degree of carbon bond hybridization are *diamond*, *graphite*, and *carbyne*. In 1931, chemist Linus Pauling first developed the hybridization theory to explain the structure of the simple molecules like methane ($CH_4$) using atomic orbitals [14, 15] (Fig. 1.4).

All other carbon forms constitute the so-called transitional forms that can be divided into two big groups. The first group comprises mixed short-range order carbon forms, and the second group includes intermediate carbon forms with a noninteger degree of carbon bond hybridization, $sp^n$. When the value of n is in between 1 and 2 ($1 < n < 2$), it forms a monocyclic carbon. On the other hand, when n lies between 2 and 3 ($2 < n < 3$), it forms closed-shell carbon structures such as *fullerene*, *carbon onions*, and *naonotubes*. A tentative ternary carbon allotropy diagram based on carbon valence bond hybridization is shown in Fig. 1.5 [16].

Carbon has three naturally occurring *isotopes*, $^{12}C$, $^{13}C$, and $^{14}C$. The isotope $^{13}C$ is stable, while $^{14}C$ is radioactive, decaying with half life of about 5730 years. It is a first row element with a small ($1s^2$) core having the valence electron 4. This means carbon has four available atoms for covalent bonding.

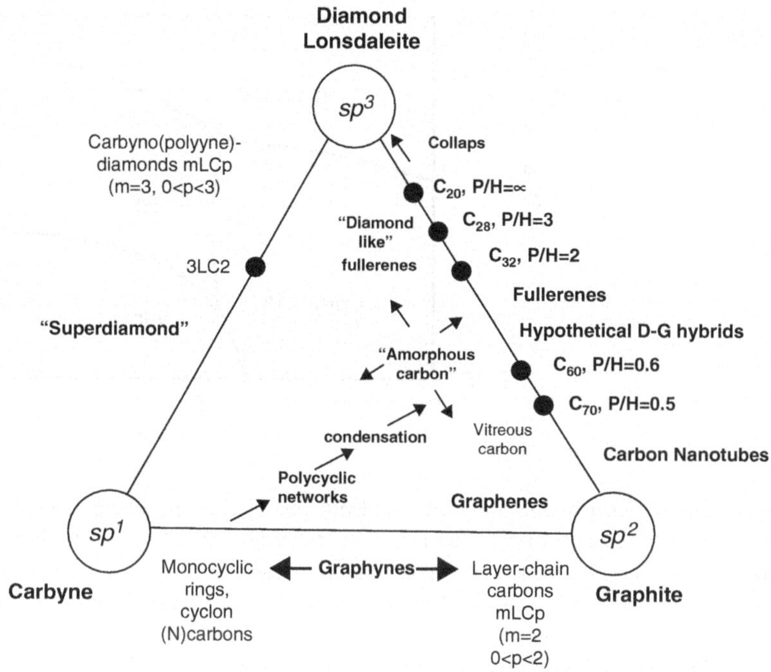


**Fig. 1.5** Ternary phase diagram of carbon allotropes (Reproduced with permission, Elsevier science, Ref. [17])

The overlap between p orbitals perpendicular to C–C bond direction is very sensitive to bond length and the ability of two carbons to approach each other closely enables π bonds to form. It is this fact which gives carbon its unique energetic *degeneracy* in *allotropes* with radically different bonding topologies (sp $^3$/tetrahedral of *diamond phase* and sp$^2$/planar and *curved graphite nanotubes,* or *fullerene-like phases*) (Fig. 1.5). As a result, the unique *degeneracy* coupled with the tight core of the carbon atom presents great challenges to computer modeling [18, 19].

However, these challenges have been met with different approaches, mostly by either *Density Functional Theory (DFT)* or *Hartree-Fock theory (HFT)*, depending on the scale of phenomenon under investigation and computing facilities available [20–22]. The first approach (*DFT*) could give rise to an exact *first principles* or *parameter-free* method while the second one (*HFT)* is referred to the Latin term *ab initio*-derived calculations. Both of these two methods yield internal energy of a *cluster* of *upper cell* of atoms as a function of nuclear coordinates, generating an energy hypersurface.

The exemplar of *cluster codes* we will apply here is AIMPRO [23] based on *DFT* within spin density approximation (SDA) for exchange and correlation [24] and norm-conserving pseudopotentials [25] obviating the numerically difficult

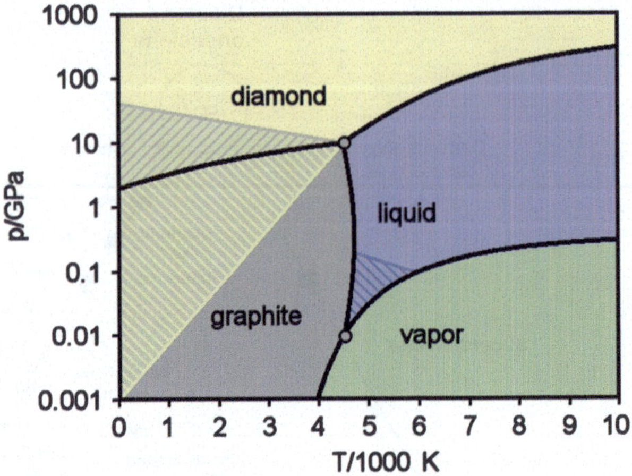


**Fig. 1.6** Theoretically predicted phase diagram of carbon (C) (Courtesy: WIKI)

treatment of core electrons. Among all the allotropes of carbon, *graphite* is perhaps the worst possible case and best illustration of the problems of size calculation.

Figure 1.6 shows theoretically predicted phase diagram of carbon (C). From Fig. 1.6, we can see that carbon has the highest sublimation point of all the elements. At atmospheric pressure, it has no melting point as its triple point is at 10.8 ± 0.2 MPa and 4600 ± 300 k (4330 °C or 7820 °F) [26, 27].

## 1.1 Coal

*Coal*, the other most important material which shaped the prehistoric and modern life, is also a form of *carbon*. We need coal for generating electricity, to run steam engine, and the most important of all that we need coal as a cooking fuel. According to the US Energy Information Administration (EIA), 40% of American electricity in 2030 is expected to come from coal. In the winter of 2013, in New England area of the United States, the cold blast of arctic due to *polar vortex effect* pushes many utilities to burn coal reserves to meet the surge in electricity demand. Figure 1.7 shows the picture of one of the many varieties of coal – the *anthracite*.

The word *coal* in English is a *fossilized carbon mineral*, and *anthracite coal* is one of the varieties of *coal* that burns slowly with nonluminous flame giving out intense heat. It is mined from the oldest geological formations and therefore has spent longest time underground. As a matter of fact, anthracite coal has become such a part of Scranton, Pennsylvania area's history that they established an entire museum devoted to it (*the Pennsylvania Anthracite Heritage Museum*).

**Fig. 1.7** Picture of anthracite coal (Courtesy: Theodore Gray)

## 1.2 Graphite

There are several *allotropes* of carbon of which the best known are *graphite*, *diamond*, and *amorphous carbon* [28]. *Graphite*, an *allotrope of carbon*, is used as an electrode. It is found in pencil, in lubricants, and it is also used to make rods to control nuclear reactors. However, according to some experts, amorphous carbon is not generally called a third allotrope, because it is a form of graphite consisting of microscopic crystal and can be obtained by heating any kind of a variety of carbon-rich materials.

Figure 1.8 shows the picture of a piece of graphite procured from graphite ore. Deposits of graphite were first found in Borrow dale, Cumberland, UK and England, and they were having sufficient size and purity. Because of the purity of the graphite ore, until nineteenth century, pencils were made simply by sawing blocks of natural graphite into strips before encasting the strips in wood [29].

## 1.3 Fullerene

To understand the mechanisms by which long-chain carbon molecules are formed in interstellar space and *circumstellar* shells, *graphite* has been vaporized by laser irradiation, producing a remarkably stable cluster consisting of carbon-60 ($C_{60}$) atoms [30]. The carbon-60 molecules are also called *fullerenes*. Since its discovery in September 1985 by Harold Kroto and coworkers, *fullerene*, the *buckyball*, became a first priority research in carbon-based nanoscience materials. $C_{60}$

**Fig. 1.8** Picture of a piece of graphite procured from graphite ore (Courtesy: US Geological Survey, WIKI)

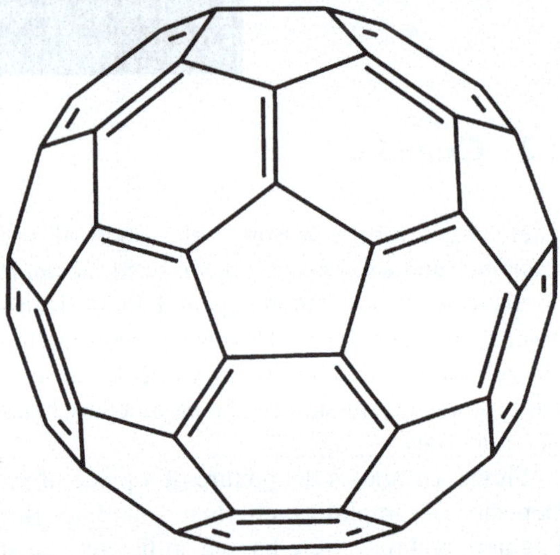

**Fig. 1.9** Schematic of a geodesic dome structure of buckminsterfullerene

molecule is very stable and can withstand high temperatures and pressures. The hollow structure of *buckyball* or *buckminster fullerene* can encase other atoms and molecules.

Figure 1.9 shows the schematic of a geodesic dome structure of *buckminsterfullerine*. The dome is composed of interlocking hexagons or pentagons. The *buckminsterfullerene* is often shortened to the term *bckyball*. Sooner, the carbon-60 fullerene became an interesting field of study, and the wave–particle duality of *buckyball* created an excitement in the research community.

Of all the various alloptropic forms of carbon, the discovery of fullerenes, which is a special class of *diamond* with a cubic lattice structure, has recently attracted the attention of the scientists (Fig. 1.10).

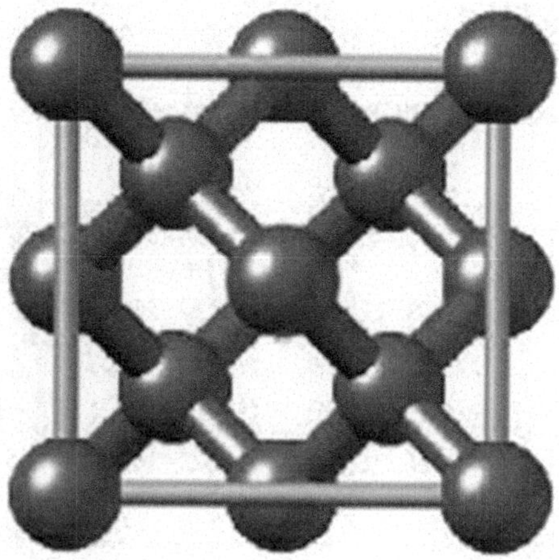

**Fig. 1.10** Cubic lattice of diamond (Courtesy: Marian vladivostop, WIKI)

## 1.4 Diamond

Diamond forms a uniquely stable structure in which each atom has four covalent bonds directed to the vertices of a rectangular tetrahedron. These arise from bonding between $sp^3$ hybrid orbitals of neighboring carbon atoms which form a cubic lattice (*the diamond lattice*) of great rigidity and strength, with a bond length of 0.1545 nm.

Diamond has remarkable physical properties stemming from the rigidity of lattice, to the highest Debye temperature of any material (2230 K). Diamond, the hardest substance known is due to the consequence of its rigidity. At room temperature, diamond is considerably a better thermal conductor than copper and aluminum, because in diamond thermal conduction takes place through phonon transport instead of electron transport as can be seen in case of copper and aluminum.

The electrical properties of diamond are largely determined by its bandgap (5.45 eV), and at room temperature the electrical resistivity is in the range of $10^{13}$ $-10^{16}$ Ω-cm, and the refractive index of diamond is 2.419 at the wavelength of sodium light, which is responsible for its beauty as a gem stone.

Figure 1.11 shows the picture of a cut and shaped diamond gem stone. Natural diamond, generated by geological processes, has been the object of intense interest to mankind for many centuries. In modern era, the experimental particle physics and medical physics are using diamonds as radiation detectors. To make it possible to contemplate the use of diamond in such detectors, the development of a process known as chemical vapor deposition (CVD) has been very successful. However, the development of the CVD process and the potential for industrial scale diamonds (especially for gem) have been driven by the requirements. But for particle physics or medical science research the potential market is relatively small. We will explore all these in a separate chapter.

**Fig. 1.11** Picture of a cut and shaped diamond gem stone (Courtesy: Mario Sarto, GNU, WIKI)

The studies of the *electrical resistivity*, *magnetoresistance*, and *thermal conductivity* of carbon will reveal new information about *static lattice defects*, *microstructure*, and *in-plane coherence length*. However, the electrical resistivity measurements in carbon and graphite are very delicate [31, 32]. In principle, *magnetoresistance* measurements probe the *mobility*, and thus are essentially sensitive to *scattering* mechanism. The analysis of *thermal conductivity* measurements yields information about *lattice defects* and other *phonon-scattering events*. The *electrical resistivity* on the other hand is much dependent on the impurities or dopants concentration within the host material, which are electrically active and at low temperatures affect the magnetic impurities. Thus, intercalation by means of acceptor or donor impurities may increase drastically the electrical resistivity [33, 34].

The most common and very frequently used form of black *carbon* (C) is *coal*. It is interesting to note that in spite of scientific evidence, common people like to call *coal* – one of the allotropes of *carbon* (C) as *black coal*. As a result, coal is very often referred to compare a near black object (*a black body is an idealized physical body with emissivity (ε) = 1. A body which has a lower value than this value is called a gray body. Carbon has emissivity ~ 0.8. The advent of universality is reviewed by Pierre Previst* et al. *These writings illustrate that black body radiation, as experimentally produced in cavities and as discussed theoretically, has remained dependent on thermal equilibrium with at least the smallest carbon particle*) [34].

To give an example, we want to show the picture of a new planet identified by NASA's Kepler spacecraft as *a coal black* alien planet. So far, it is the darkest planet ever known. The planet is as big as Jupiter and is lurking around the yellow sun-like star GSC 03549–02811 some 750 light-years away in the direction of the constellation *Draco*.

The distant exoplanet TrES-2b, shown in Fig. 1.12, is the artist's conception, and the planet is darker than the *blackest coal* [35].

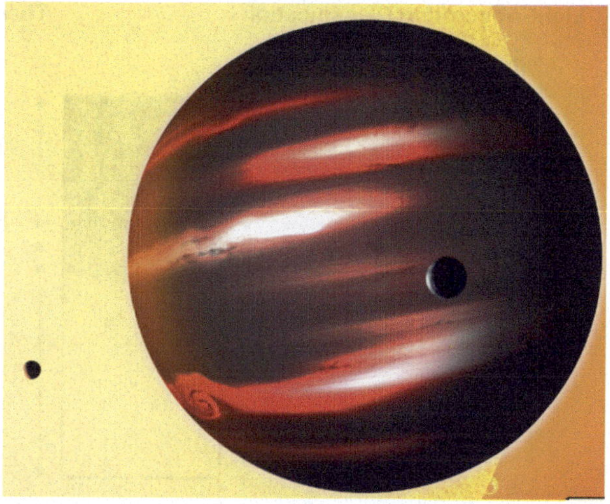

**Fig. 1.12** Artist's conception of the exoplanet TrES-2b (Photo courtesy: NASA)

The researchers found that this gas giant reflects less than 1% of the sunlight falling on it (*lampblack reflects less than 2% of the radiation falling on it and crudely approximates an ideal black body*), making it darker than any planet or moon seen until now. As a result, scientists think that the planet can be compared to a *black body* comparing its absorption of radiated light (Fig. 1.12). To *a physicist*, a *black body* is a hypothetical body that completely *absorbs all wavelengths of thermal radiation incident on it*. Such a body does not *reflect light*, and therefore appear *black* if their temperatures are low enough so as not to be self-luminous.

## 1.5 Soot and Light-Absorbing Carbon (LAC)

The *black coal* (the allotrope of carbon) when burnt produces *carbon mono/dioxide* ($CO/CO_2$) gas, and mostly *black soot*, which is responsible for absorbing light from the atmosphere. As a result, the *black soot* is also called *light-absorbing carbon* (LAC). The *refractive index of* LAC (light-absorbing carbon), together with size and density are key parameters in the Lorentz-Mie calculation of the *mass absorption efficiency* ($\alpha_{abs}$) of a single soot particle.

Figure 1.13 shows the *refractive index of* LAC (light-absorbing carbon). The size and density of LAC are the key parameters in the Lorentz-Mie calculation of the mass absorption efficiency ($\alpha_{abs}$) of a single soot particle [36, 37].

For atmospheric soot particles, Bond and Bergstorm propose a value of ($\alpha_{abs}$) $=7.5 \pm 1.2$ m$^2$/g [38]. The absorption of light (due to the presence of atmospheric carbon, black/brown?) depends on the morphological and chemical properties of the carbon particles and their origin. As a result, the measurements and the data provided by so many scientists are some time very confusing, complex, and

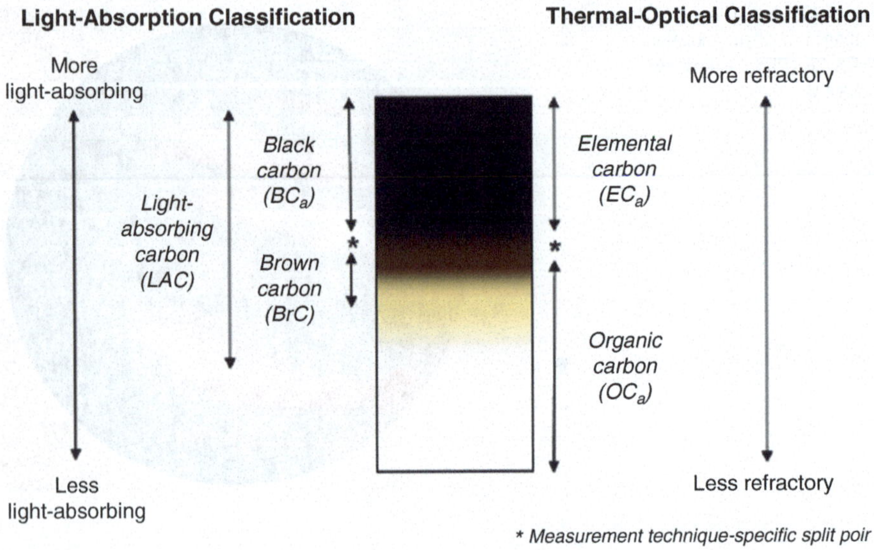


**Fig. 1.13** *Refractive index of* LAC (Light-absorbing carbon)

contradictory. Since the main goal of this book is not to be limited on this, we will have to leave the matter here.

Starting from the *fuels* like *coal*, *coke*, *oil*, *gasoline*, and *natural gas* to the most common household items like *sugar*, *starch*, and *paper* are compounds of *carbon* with *hydrogen* and *oxygen*. Last, but not the least, *proteins*, such as *hair*, *meat*, and *silk* contain *carbon* and other elements like *phosphorus, nitrogen,* and *sulfur*.

The lifeline of living organisms is nothing but the extended networks of *carbon* (C). The networks of carbon chains form the skeletons of most organic molecules. The skeletons of the organic molecules may vary in length and from one organic material to another. It may be straight, branched, or arranged in closed ring. As a matter of fact, organic compounds contain carbon and are prevalent in living systems [39].

Carbon moves from living things to the atmosphere. During exhalation, living things release carbon dioxide gas ($CO_2$) into the atmosphere, which moves from the atmosphere to the plants, and then to the ground. From the ground over millions of years, weathering of rocks on land can add carbon to the surface water which eventually runs off to the ocean. Over long timescales, carbon is removed from seawater when the shells and bones of marine animals and planktons collect on the seafloor.

**Fig. 1.14** Picture of a seashell collected from seashore of Florida (US) (Photo courtesy: Theodore Gray)

## 1.6 Carbon Cycle

Figure 1.14 shows the picture of a seashell collected from the seashore of Florida. The colorful design on the shell is believed to be the joint work of the special organ *mantle* and pigment glands of the seashell. These shells and bones are made of limestone (mostly calcium carbonate, $CaCO_3$), which contains carbon. The carbon then can be released back to the atmosphere if the limestone melts or is metamorphosed in a subduction zone completing the *carbon cycle*, which is shown in Fig. 1.15. Under terrestrial conditions, conversion of one element to another is very rare. Therefore, the amount of carbon on earth is effectively constant. Thus, processes that use carbon must obtain it from somewhere else. The element carbon follows a path in environment, which makes up the carbon cycle (Fig. 1.15) [40, 41].

## 1.7 Photosynthesis

One of the most important activities in the plant kingdom is *photosynthesis* which is another way of carbon fixation in nature. *Photosynthesis* is the synthesis of carbohydrates from sunlight and carbon dioxide ($CO_2$). Carbon dioxide ($CO_2$) and water, used in *photosynthesis*, can be traced through various intermediate substances produced during the reactions [42]:

$$6CO_2 + 6H_2O = C_6H_{12}O_6 + 6O_2 \tag{1.1}$$

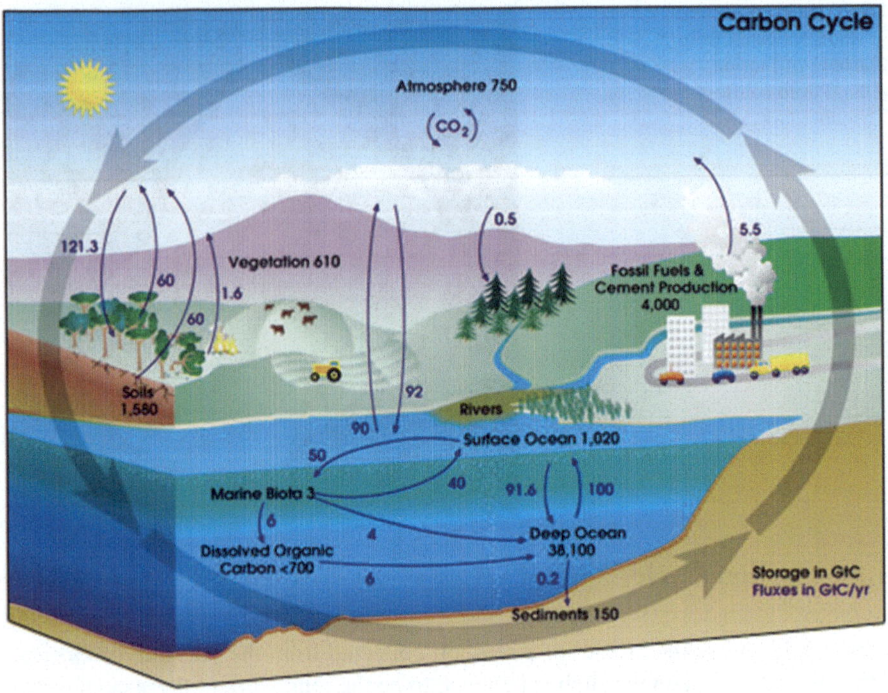


**Fig. 1.15** Carbon cycle. (Courtesy: NASA, USA)

Fraction of the chlorophyll molecules in the chloroplast absorbs photons in the red part of the visible spectrum and passes this energy of excitation on to other species in the reaction chain. Indeed, the ability to absorb light is due basically to the conjugated polyene structure of the porphine ring system. The role of magnesium (in Fig. 1.16) ion is, at least, twofold: (i) it helps to make the entire molecule rigid and (ii) enhances the rate at which the short-lived singlet-excited state initially formed by photon absorption. It is then transformed into the corresponding triplet state, which has longer lifetime and the ability to transfer its excitation energy into redox chain [43].

In *photosynthetic* prokaryotes, the mechanisms of carbon fixation are more diverse, because it can be fixed either by the Calvin–Benson cycle, a reversed citric acid cycle [44], or by the carboxylation of acetyl (CoA) [45]. Figure 1.17 shows the bacteria clustering of the parts of an algal filament exposed to red light in the wavelength between 650 and 680 nm, which is most effective in the photosynthesis activity in plants [46].

In plants, bacteria, and algae, oxygenic *photosynthesis* splits water, with oxygen produced as the waste product. This process uses ATP (adenosine triphosphate), which is a *coenzyme to some reactions*, and NADPH (nicotinamide adenine dinucleotide), a derivative of vitamin $B_3$ (niacin), also an important coenzyme [47, 48]

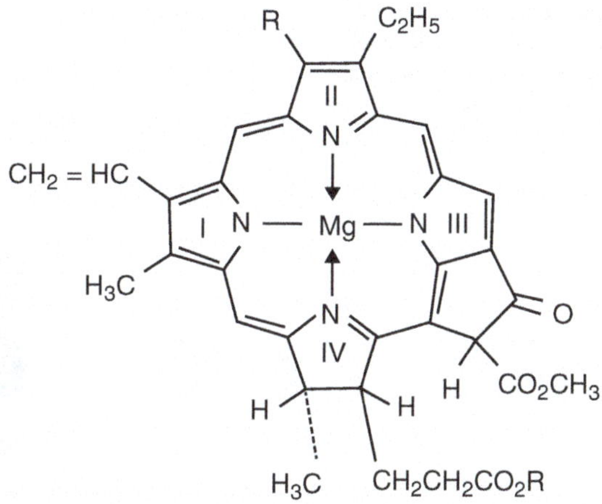


**Fig. 1.16** One of the chlorophyll molecules (Reproduced with permission of Wiley)

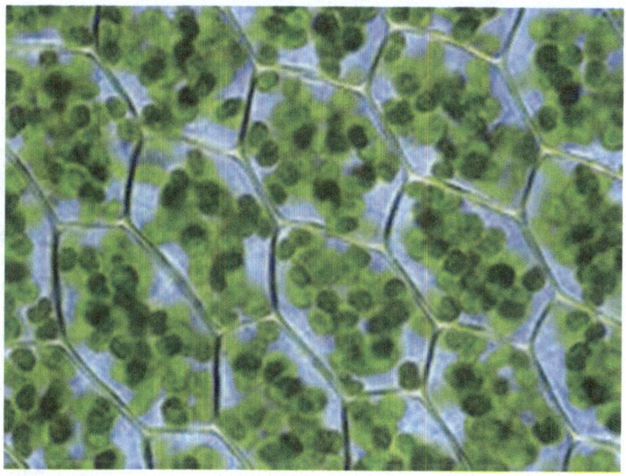

**Fig. 1.17** Bacteria clustering of the parts of an algal filament exposed to red light, the most effective wavelength in photosynthesis (Photo courtesy: Kristin Peters, GNU,WIKI)

produced by the photosynthetic reaction centers, to convert $CO_2$ into glycerate 3-phosphate, which can then be converted into glucose.

Figure 1.18 shows the space-filling model of the coenzyme NADH. Coenzymes transport chemical groups from one *enzyme* to another.

## 1.8 Enzymes

*Enzymes* are large biological molecules and are responsible for the thousands of metabolic processes that sustain life [49].

Figure 1.19 shows human glyoxalase I. Two zinc ions that are needed for the enzyme to catalyze its reaction are shown in purple spheres, and an enzyme inhibitor called s-hexylglutathione is shown as a space-filling model, filling the

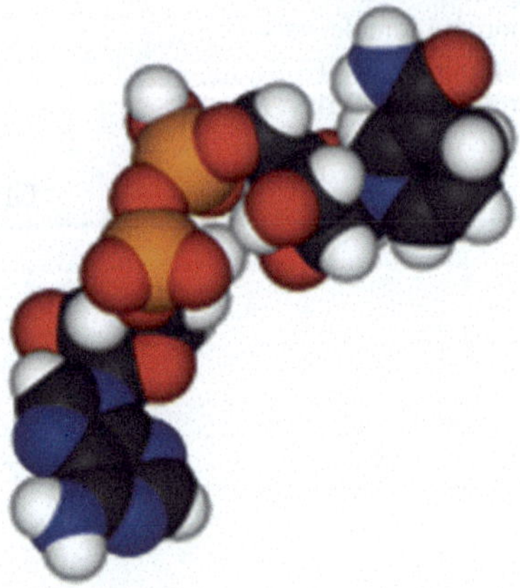

**Fig. 1.18** Space-filling model of the coenzyme NADH (Courtesy: Benjah-bmm27, WIKI)

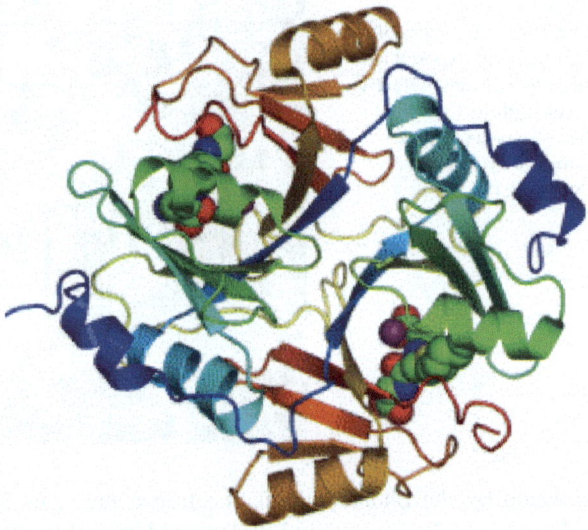

**Fig. 1.19** Human glyoxalase I. (Photo courtesy: W. Willow, GNU, WIKI)

two active sites. *Enzymes* are very specific, and it was suggested by the Nobel laureate chemist Emil Fischer in 1894 that both the enzyme and the substrate possess specific complimentary geometric shapes that fit exactly into one another. Like all proteins, enzymes are long, linear chains of amino acids that fold to produce a three-dimensional product.

Figure 1.20 shows *carbon fixation* reaction, which is carried out by the *enzyme* RuBisCO as part of the Calvin—Benson cycle [50].

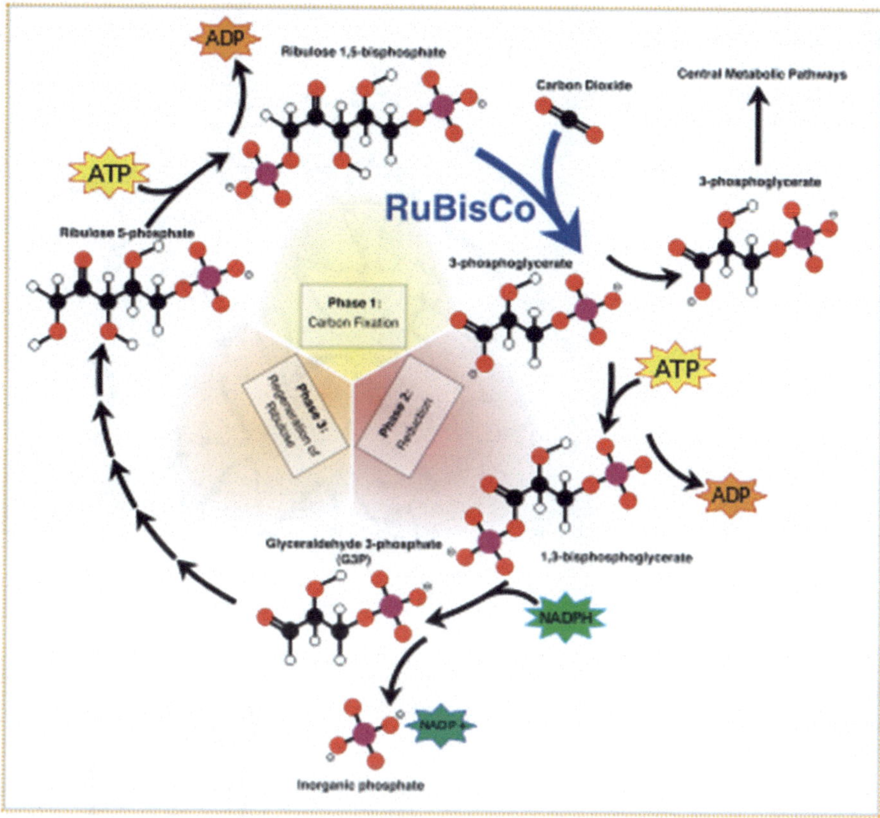


**Fig. 1.20** *Carbon fixation* reaction. It is carried out by the enzyme RuBisCO as part of the Calvin—Benson cycle (Photo courtesy: Mike Jones,WIKI)

## 1.9 Bio-organic–Inorganic Materials

*Carbon* (C), one of the constituents of almost all the organic materials and the essential lifeline of all living organisms, extends its wing to another important field of chemistry – the *biochemistry*. *Biochemistry* is the chemistry of living organisms, involving the essential and indispensable way of shaping life, where inorganic metal chemistry is married to organic chemistry to give support to sustain life. The metals like iron (Fe), cobalt (Co), zinc (Zn), copper (Cu), manganese (Mn), and molybdenum (Mo) are inorganic metals and are essential parts of living beings. Here we will survey the major aspects of metals in *biological systems* that form the *bioorganic–inorganic chemistry (BOIC)* [51–53].

The *BOIC* of iron (Fe) is certainly the most widespread history of organic *heme protein*, and the inorganic transition metal transports oxygen and mediates in electron transfer chains. One of the main *heme proteins* called *hemoglobin* binds

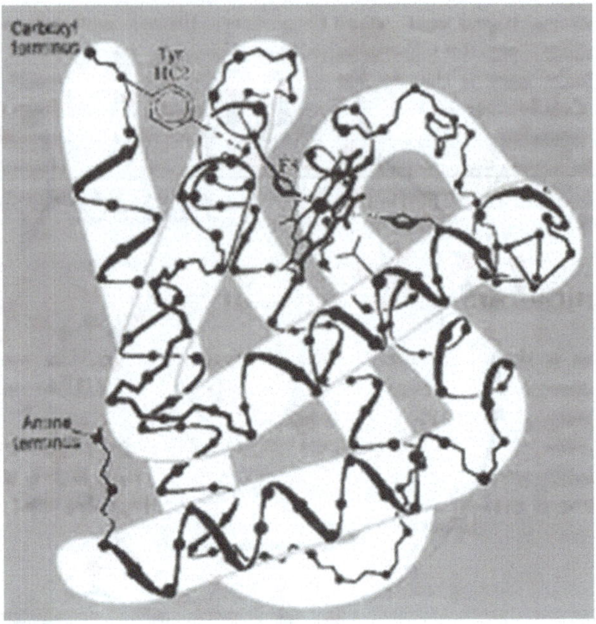

**Fig. 1.21** Schematic representation of one of the four subunits of hemoglobin (Reproduced with permission from Wiley)

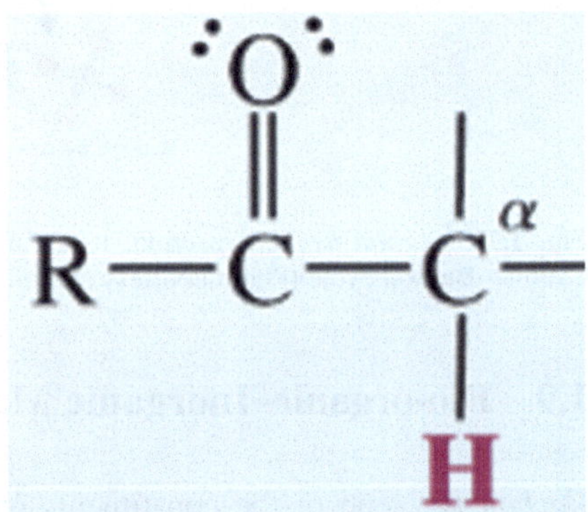


**Fig. 1.22** α-carbon attached next to the carbonyl carbon

oxygen molecules to its iron atoms and transports them from lungs to muscles where they are delivered to another *heme protein* called *myoglobins* [54].

Figure 1.21 shows a schematic representation of one of the four subunits of *hemoglobin*. The continuous black band represents the *peptide chain*. Dots on the chain represent the *α-carbon* atoms (Fig. 1.22).

Figure 1.22 shows the α-carbon attached next to the carbonyl carbon and refered to as the first carbon atom that attaches to a functional group carbonyl.

It is said that hemoglobin is not simply a passive container of oxygen but an *intricate molecular machine* [55]. Another contribution of iron (Fe) in shaping the organic–inorganic chemistry of biochemistry is rubredoxin, the substance that

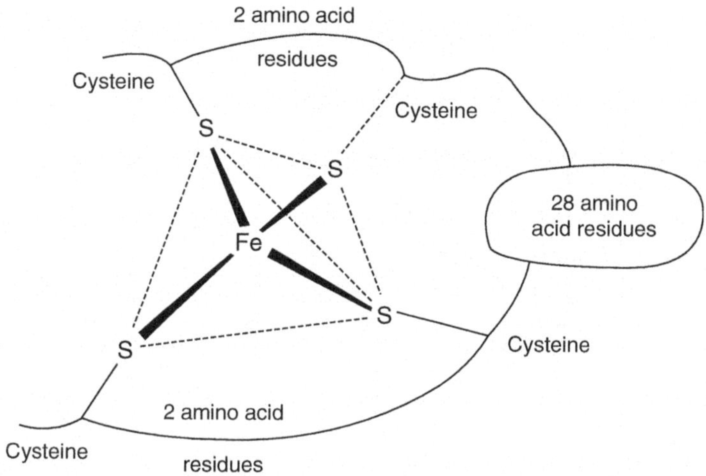


**Fig. 1.23** Location of the iron atom in rubredoxin (Reproduced with permission from Wiley)

participates in a number of biological redox reactions, especially in *anaerobic bacteria*. An X-ray crystallographic study has shown how the iron (Fe) atom in the ferric state coordinated with four sulfur atoms of cysteine ($C_3H_7NO_2S$) residues (Fig. 1.23). The coordination is roughly tetrahedral with several distortions.

In microorganisms, iron is transported by substances called ferrichromes (Fig. 1.24a) and ferrioxamines (Fig. 1.24b) [56]. The former are trihydroxamic acids in which the three hydroxamate groups are on the three side chains of a cyclic hexapeptide. The latter have the three hydroxamate groups as part of the peptide chain, which may be cyclic or acyclic. Typical structures of ferrichromes and ferrioxamines are shown in Fig. 1.24. The importance of these compounds is derived from their exceptional ability to chelate Fe (III) and then pass through cell membranes, carrying iron from inorganic sources.

The biochemistry of cobalt (Co), on the other hand, is its intimate involvement in the coenzymes related to vitamin $B_{12}$. The term, vitamin $B_{12}$, refers to cyanocobalamin, which has cobalt in the +3 oxidation state and $CN^-$ as the ligand X. The basic structure is shown in Fig. 1.25. Here, Co atom is coordinated by four nitrogen atoms neglecting the ligand X, which is called cobalamin. A macrocyclic ligand called corrin ring, shown in bold lines, resembles the phosphine ring, but differs in various ways, notably in the absence of one methane = CH- bridge between a pair of pyrrole rings [57, 58].

There are several other metals (Cu, Zn, Mn, Mo, etc.) that help to shape the daily lives of living creatures. Finally, we will discuss the contribution of metal zinc (Zn). There are at least 20 known zinc metalloenzymes. Two of them that are studied mostly are carbonic anhydrase and carboxypeptidase. The enzyme carboxypeptidase is present in the pancreas of mammals and it catalyzes the hydrolysis of the peptide bound at the carboxyl end of a peptide chain (Fig. 1.26). In Fig. 1.26a, we

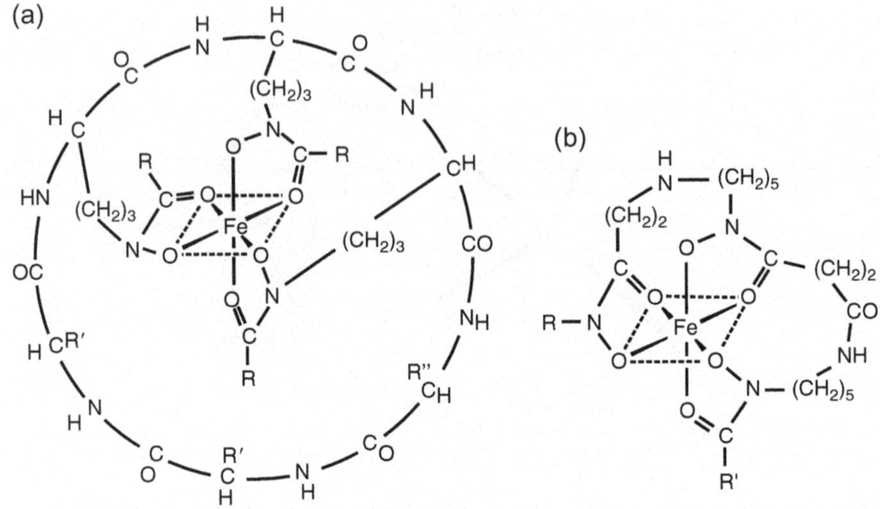


**Fig. 1.24** (**a**) Schematic of the structure of ferrichrome and (**b**) a typical structure of an acyclic ferrioxamine (Reproduced with permission from Wiley)

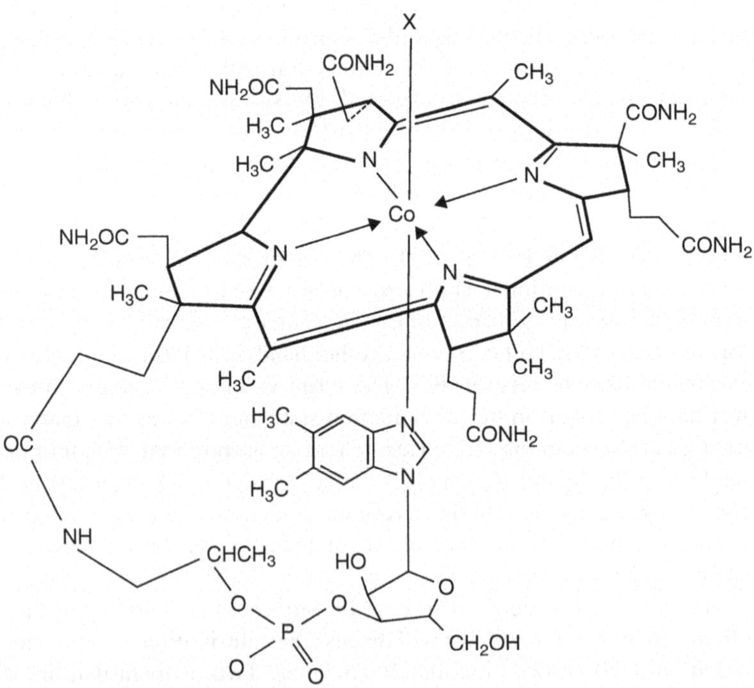


**Fig. 1.25** Structure of cobalamin. The corrin ring is shown in heavy lines (Reproduced with permission from Wiley)

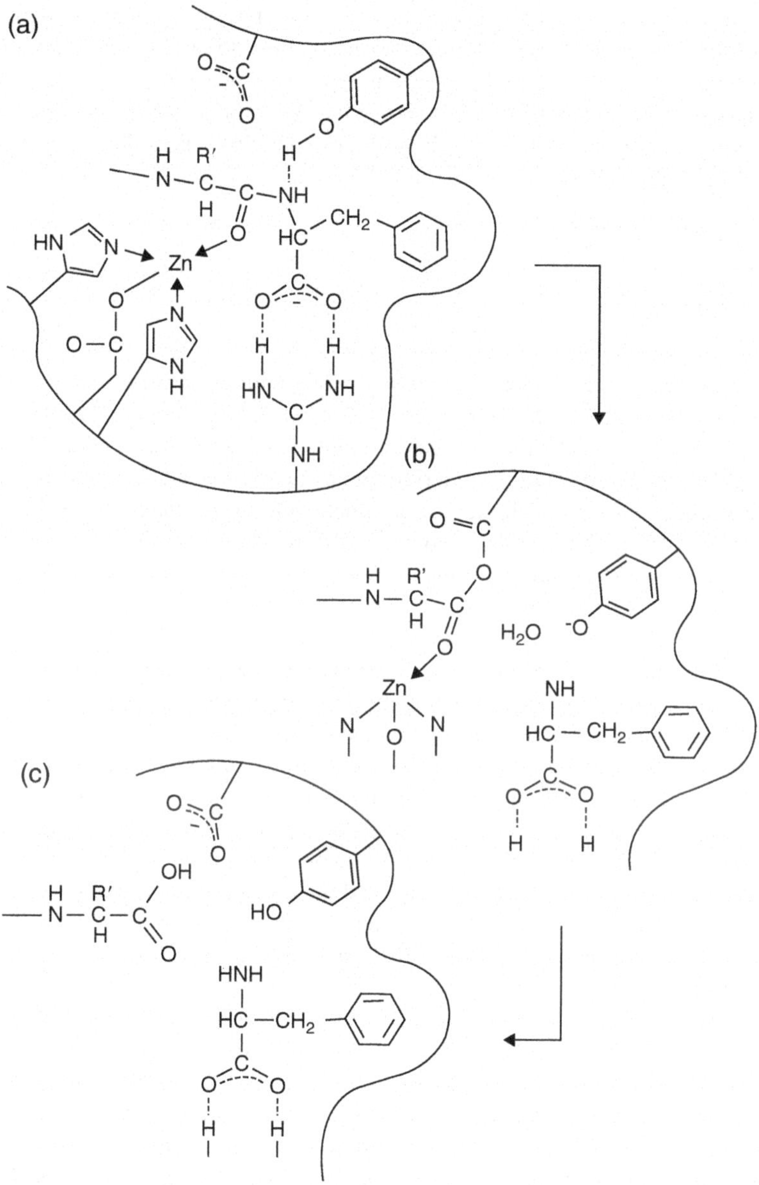


**Fig. 1.26** (**a**) A proposed mode of binding of the substrate in carboxypeptidase. The substrate is shown in heavy type and lines. The curved lines schematically define the surface of the enzyme molecule. (**b**) A possible first step in the mechanism, wherein a carboxyl side chain attacks the carbonyl carbon atom, forming an anhydride. (**c**) The next steps would be hydrolysis of the intermediate anhydride and dissociation of the products from the active site (Reproduced with permission from Wiley)

can see that the zinc (Zn) ion is bound in a distorted tetrahedral environment, with two histidine–nitrogen atoms. On the other hand, the carbonyl oxygen atom of the peptide linkage (which is to be broken) has replaced water molecule in the coordination sphere of the zinc ion, which is shown in Fig. 1.26b. Once the peptide bond is broken, with the formation of acid anhydride, rapid hydrolysis of the acid anhydride would occur (Fig. 1.26c).

## 1.10 Polymers

Carbon provides us colors, tastes, and odors, along with the abilities to perceive and respond to our environment. In other words, carbon and its compounds are essential to our existence. It is in the food we eat, the clothes we wear, the cosmetics we use, and the gasoline that fuels our car. In other words, carbon is the sixth most common element in the universe which plays a dominant role in the *chemistry of life*.

The reason that carbon can form so many different compounds is that carbon can link together with other carbon atoms in different arrangements – chains and rings. Long chains of carbon compounds are called *polymers*, and they can form extremely large molecules. In the presence of free radical initiators such as peroxides or AIBNs (azoisobutyronitriles), many alkines react to form polymers, which are large molecules composed of repeating units. In polymerization reaction, small molecules known as monomers react to form *polymer* [59].

The polymerization of carbon can also form double or triple bonds in addition to single bonds. Carbon's ability to combine with itself and with other elements explains why such a large number of carbon compounds exist. The polymerization chemistry for the design of photoresists that has been successfully used in microelectronics is a task, which usually has succumbed to the pressures of very competitive environments of high technology and has therefore progressed very rapidly to selected directions corresponding to the most acute needs of the *microelectronics industry* [60, 61].

Figure 1.27 shows the Novolak resin and the photopolymerization. During exposure to UV light, the fundamental structure of resist changes. As for example, at 248 nm UV exposure, i-line's Novolak-based resist's structure changes. Similar effects are observed in chemically amplified (CA) hydroxystyrene-based resist, and cyclic olefin or acrylate-based resist at 248 and 193 nm respectively. Even as-yet undetermined polymer family shows a structural change when exposed to 157 nm UV light [62, 63]. Novolaks are phenol–formaldehyde type polymers generally synthesized using cresol rather than phenol, with the polycondensation reaction halted before the polymer becomes cross-linked.

Other two important polymers produced on an industrial scale are *nylon* and *polyesters*. Both of these products are the examples of *condensation polymers* [64, 65]. *Nylon* is a thermoplastic, silky material. It has been commercially used first in 1938 as a nylon bristled toothbrush, and became very popular because of its

**Fig. 1.27** Novolak resin and the photopolymerization

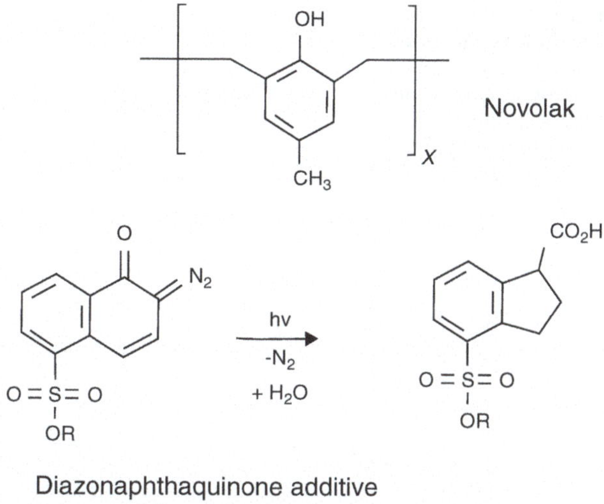


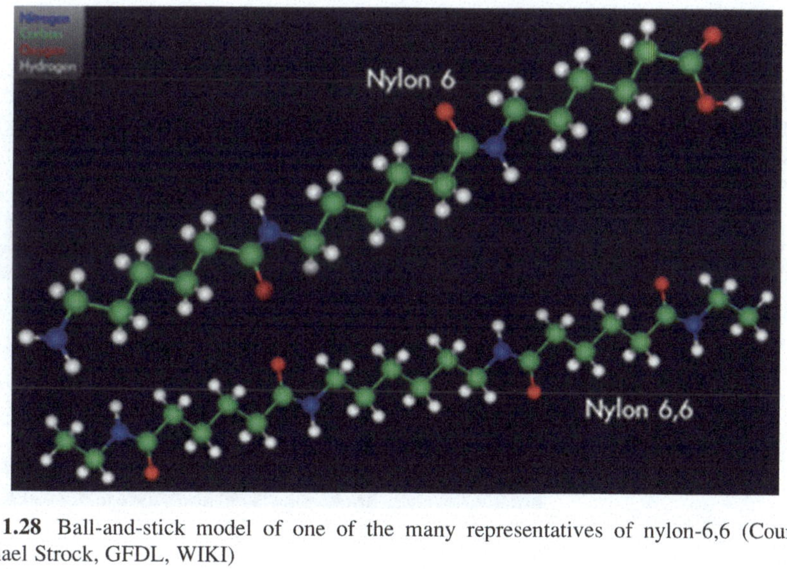


**Fig. 1.28** Ball-and-stick model of one of the many representatives of nylon-6,6 (Courtesy: Michael Strock, GFDL, WIKI)

different colorful appearance. Nylon is made of repeating units linked by amide bonds and frequently referred to as polyamide (PA) [66, 67].

Figure 1.28 shows one of the many representatives of nylon-6.6. After World War II, DuPont and other corporations were able to make some fine garments out of nylon to replace Asian silk. Soon, the fabrics made out of nylon became very popular and by the end of 1945 manufactured nylon fibers had taken a market share of 25% at the expense of cotton [68].

**Fig. 1.29** Picture of a woman with blue nylon fabric ball gown (Picture courtesy: Emma Domb Chemical Heritage Foundation, WIKI)

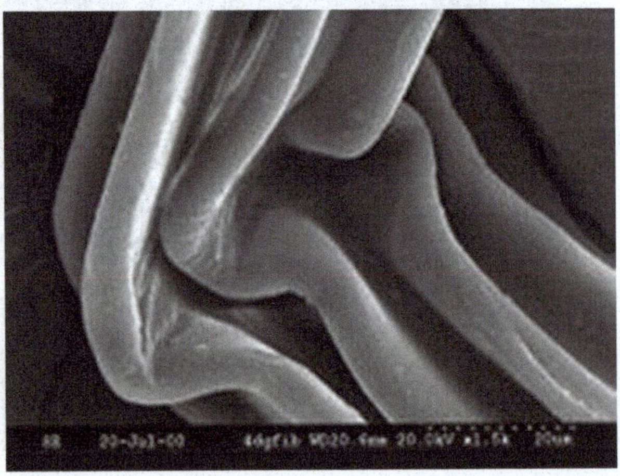

**Fig. 1.30** Scanning electron micrograph (SEM) of high surface area polyester fiber (Courtesy: Pschemp, GNU, WIKI)

Polyester is also a form of thermoplastics that contain the ester functional group in their main chain [69]. Synthesis of polyesters is generally achieved by polycondensation reaction. Polyesters are used to make bottles, films, tarpaulin, canoes, liquid crystal display (LCD), dielectric film for capacitors, and insulating tapes. Figure 1.29 shows the picture of a woman with light blue nylon fabric ball gown. There is a kind of fabric sold in the market as “Silk”. It is actually a 100% polyester fiber woven to resemble the sheet and durability of insect derived silk. Figure 1.30 shows the scanning electron micrograph (SEM) of a high surface area polyester fiber.

**Fig. 1.31** Picture of a polyester shirt (Courtesy: Damieng at en., GNU, WIKI)

Synthetic fiber used to manufacture silk often provides improved wrinkle resistance, durability, and high color retention compared to natural silk. Figure 1.31 shows a close-up picture of a polyester shirt.

## 1.11 Carbon Nanotubes (CNTs)

Another example of an allotrope of carbon is carbon nanotube (CNT). As a matter of fact, the whole nanotechnology – a very fascinating branch of science is based on the study of *single-walled carbon nanotube* (SCNT) which owes very much to the polymer science [70, 71]. All these properties of carbon allow the element to form many types of simple compounds and chains that form very complex molecules.

Figure 1.32 shows a network model of a single-walled carbon nanotube (CNT). Nanotubes are constructed with length to diameter ratio up to 132,000,000 [72]. Nanotubes are the members of fullerene structural family. From the above discussions and illustrations of so many members of the carbon (C) family and their fascinating behaviors, it will be worth to study the chemistry, physics, and materials engineering of carbon [73].

Carbon has four electrons in its outermost orbital. The valence of 4 allows each carbon atom to form stable associations with up to four other atoms, including other carbon atoms. It can form single, double, or triple bonds. Figure 1.33a shows some of the organic compounds that contain *carbon* and Fig. 1.33b shows different configurations [74].

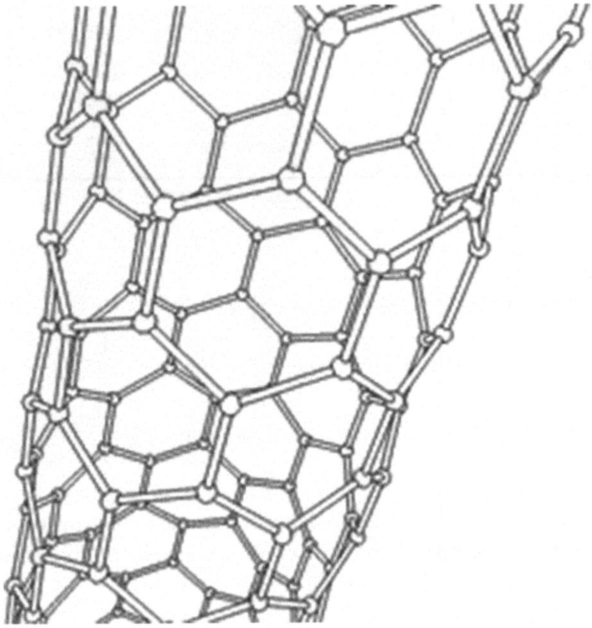

**Fig. 1.32** Network model of single-walled carbon nanotube (CNT) (Courtesy: Hochgeleden van Schwarzm, GNU, WIKI)

## 1.12 Organic Molecules

Figure 1.33 shows some of the common organic compounds used in our daily life. All organic compounds share similar characteristics. Most importantly, they all contain carbon and pertain to living things. However, there are different subgroups of organic compounds that share additional characteristics. These subgroups include *hydrocarbons*, *alcohols*, and *organic acids* (Fig.1.33a).

*Hydrocarbons ($C_mH_n$)* are organic compounds that contain only hydrogen and carbon atoms [75]. Many hydrocarbons are used as fuel for cooking, not to speak of the heaters in hot-air balloons. Examples of hydrocarbons include the compounds, methane ($CH_4$), propane ($C_3H_8$), butane ($C_4H_{10}$), and ethane ($C_2H_6$) (Fig.1.33b. Among the alcohols, isopropanol is a common type of alcohol used as rubbing alcohol, and ethanol ($C_2H_5OH$) is used in beer and wine. Some of the organic acids like *citric acid*, *lactic acid*, and *acetic acid* have very similar structure like previously mentioned alcohols. Figure 1.34a represents the structure of citric acid and Fig. 1.34b shows the picture of a lemon/lime which is one of the sources of citric acid.

*Citric acid* is found in fruits such as oranges and lemons, while *lactic acid* is caused by a type of fermentation. In industry, lactic acid fermentation is performed by lactic acid bacteria, which convert glucose and sucrose to lactic acid. Figure 1.35a shows the structural formula of lactic acid ($C_3H_6O_3$) and Fig. 1.35b shows the ball-and-stick model of lactic acid. The *acetic acid* on the other hand is a weak acid, meaning the free energy of the conjugate base carboxylate ion is higher

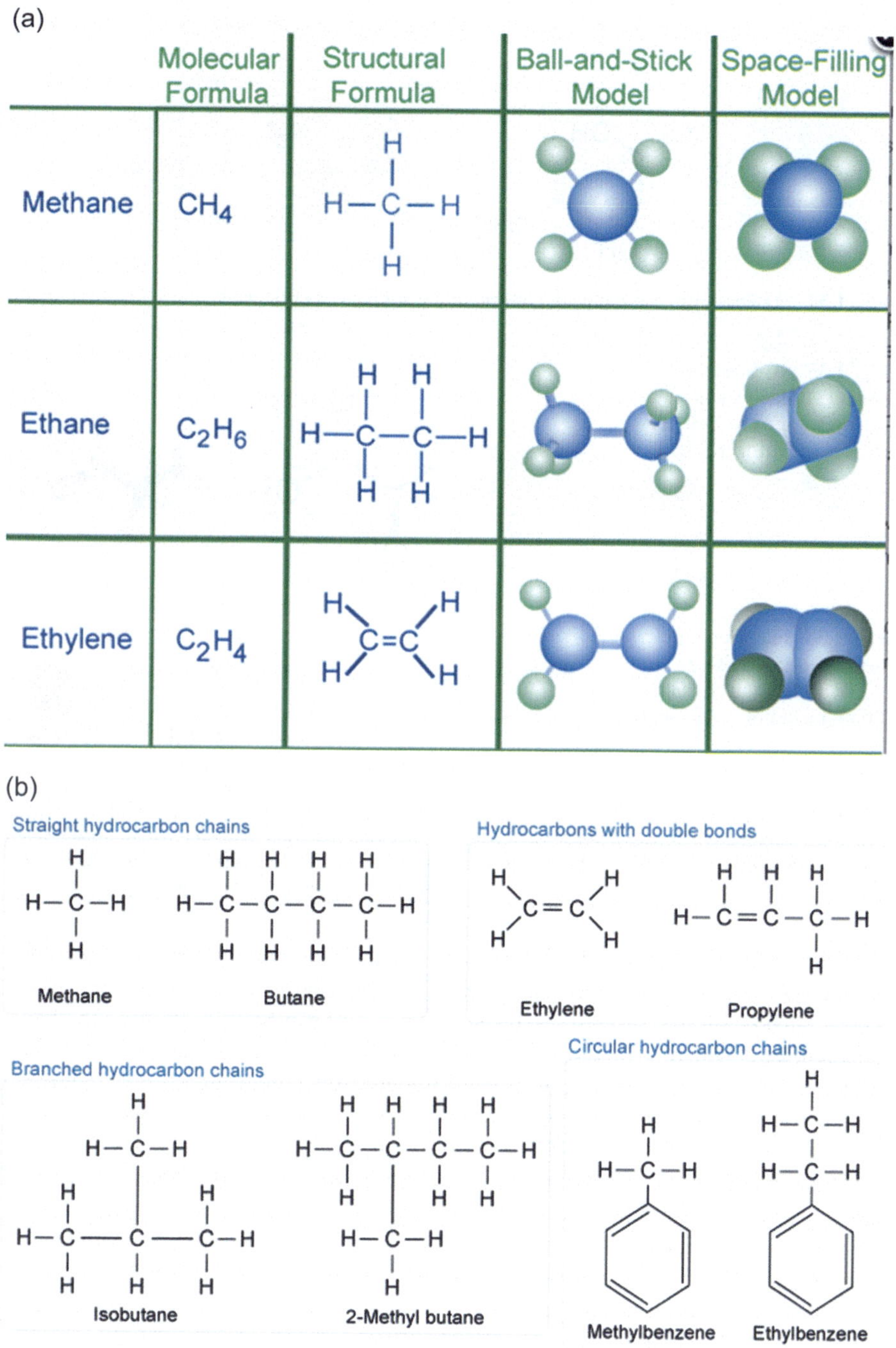


**Fig. 1.33** (**a**) Organic compounds containing carbon; (**b**) Different configurations of organic compounds. (Courtesy: Penn State Univ. Bios, and WIKI)

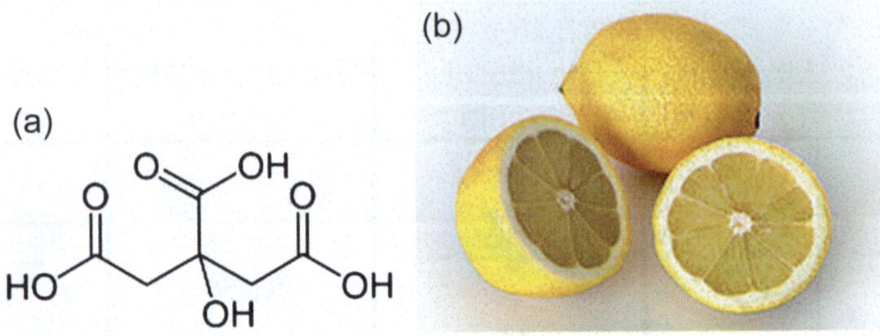


**Fig. 1.34** (**a**) Structure of citric acid; (**b**) one of the sources of citric acid

**Fig. 1.35** (**a**) Structural formula of lactic acid ($C_3H_6O_3$) and (**b**) Ball-and-stick model of lactic acid

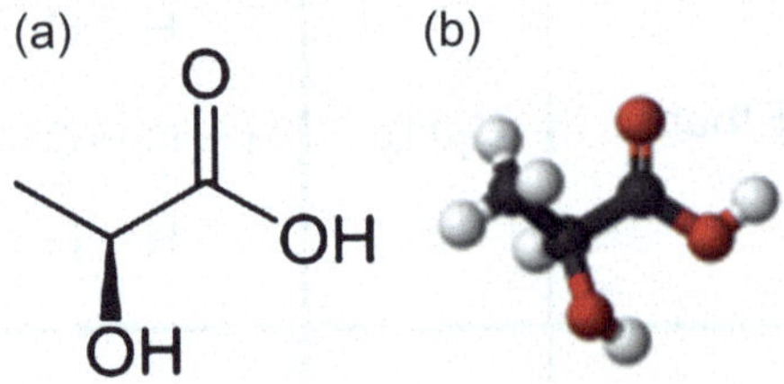


**Fig. 1.36** (**a**) Structural formula of acetic acid ($CH_3COOH$) and (**b**) space-filling model of acetic acid

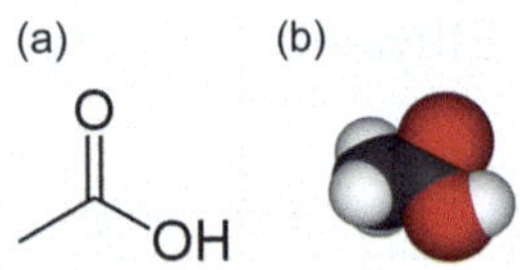


(when the ion is not very stable, the $pK_a$ – the scale of acidity – which is equivalent to $-logK_a$ is higher, and the acid is less acidic) [76]. Figure 1.36a shows the structural formula of acetic acid ($CH_3COOH$) and Fig. 1.36b shows the space-filling model of acetic acid.

Carbon (C) forms four classes of large organic compounds (macromolecules): *carbohydrates, lipids, proteins*, and *nucleic acids*. The three of these macromolecules, *carbohydrates, proteins,* and the *nucleic acid* can be formed through chemical synthesis which is very similar to identical building blocks (monomers) linked to polymers. The general rule to *form* a *polymer* by joining *monomers* is termed *anabolism*. Conversely, the process of *breaking* a *polymer* into *monomers* is called *catabolism*, which also can involve multiple steps (Fig. 1.37).

Organic molecules of carbon, together with hydrogen and oxygen, are called *carbohydrates*, in which there are two atoms of hydrogen for every one atom of oxygen. In general, *carbohydrates* include types of *sugars* and *starches* (*polymer of glucose*, $C_6H_{12}O_6$). The *sugars* are called *monosaccharides* – the building blocks for all other monosaccharides and are identified with the prefixes *D* and *L* (Fig. 1.39) according to the senses of rotation in a linearly polarized light (rotation of the axis of the polarized light, clockwise or anticlockwise) that will distinguish two particular stereoisomers that are mirror images of each other. When one

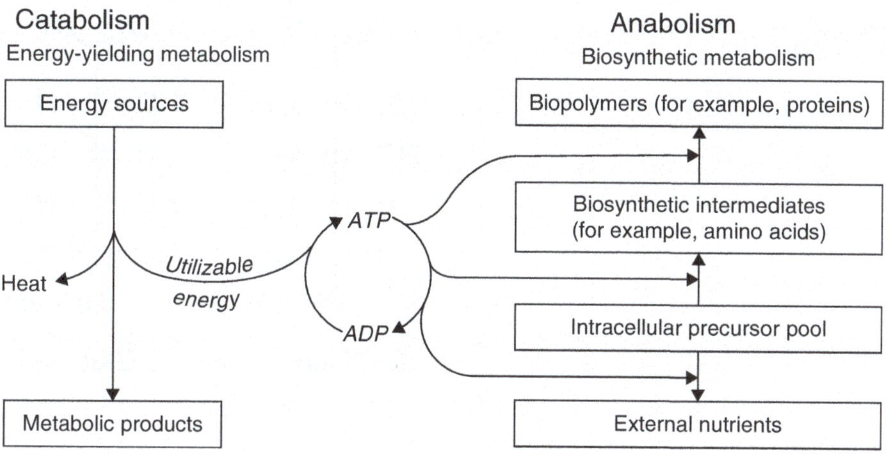


**Fig. 1.37** Processes of catabolism and anabolism (Courtesy: WIKI)

**Fig. 1.38** Some of the food products that are rich in carbohydrates (Courtesy: US Govt.WIKI)

monosaccharide binds another monosaccharide, it forms a disaccharide. Figure 1.38 shows some of the food products that are rich in carbohydrates.

On the other hand, when two fructose molecules bind together, they form disaccharide (complex sugar) *sucrose* ($C_{12}H_{24}O_{12}$), which is derived from the plant sugarcane. Monosaccharides and disaccharides are both sugars, and the starches are *polysaccharides* – the long chains of sugar molecules. *Glucose*, *sucrose* and *amylase* (starch) are examples of *mono*, *di*, and *polysaccharides*, respectively.

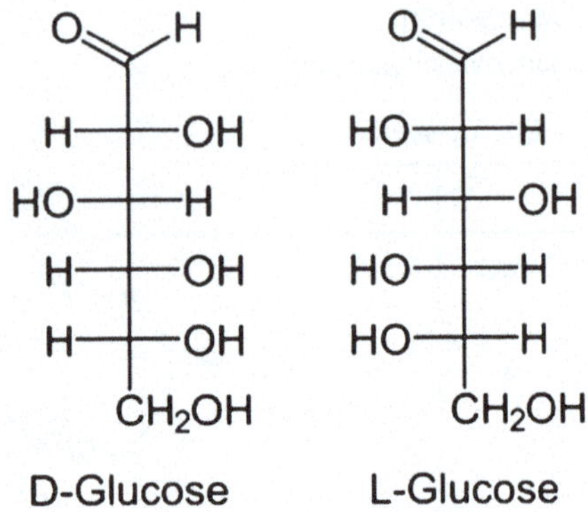


**Fig. 1.39** Two structures of monosaccharide

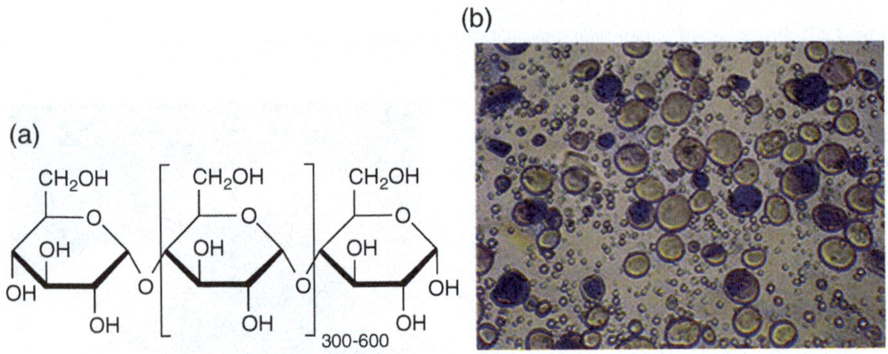


**Fig. 1.40** (**a**) Long-chain structure of amylose, the starch and (**b**) granules of wheat starch stained with iodine (as seen under microscope) (Courtesy: (**a**) NEUROtiker and (**b**) Kiselov Yuri,WIKI)

*Starches*, on the other hand, are *polysaccharides*, having long chain of simple sugar molecules. Figure 1.39 shows the two structures of monosaccharide [77].

One common form of starch that is found in human body (liver and muscle cells) is called glycogen. Glycogen provides high energy and long-lasting fuel in human body. So we can see that carbohydrates, the organic compounds containing carbon, are the main source of energy for living beings. Figure 1.40 (a) shows the long-chain structure of amylase – the starch and (b) shows granules of wheat starch stained with iodine [78, 79].

*Cellulose* $(C_6H_{10}O_5)_n$ is also a *polysaccharide* consisting of a linear chain and the principal structural components of plants. It is the most abundant organic compound on earth. *Cotton* is almost a pure *cellulose*, whereas wood is a *cellulose* combined with a *polymer* called *lignin* [80]. Figure 1.41a shows the linear chain of cellulose, and Fig. 1.41b shows the long-chain structure of cellulose.

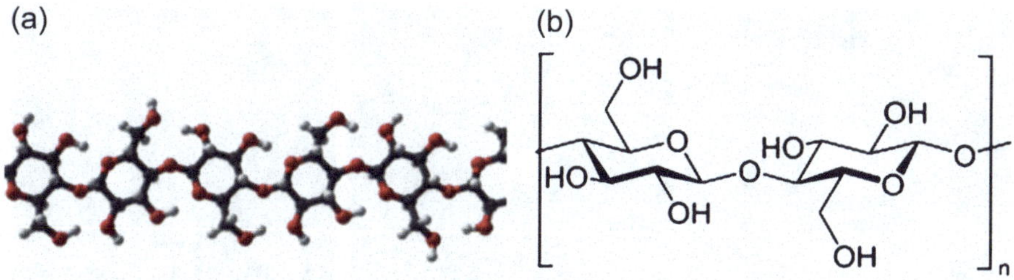


**Fig. 1.41** (**a**) Linear chain of cellulose and (**b**) long chain structure of cellulose (Courtesy: (**a**) Ben Mills and (**b**) NEUROtiker, WIKI)

## 1.13 Proteins

Figure 1.42 shows proteins in different cellular compartments and structures tagged with green fluorescent protein (here, white) [81]. *Proteins* are condensation polymers of amino acids and considered as very large *peptides*. It is the second most important material essential for life [82]. They are large complex molecules that consist of *amino acids*, containing the elements carbon (C), hydrogen (H), oxygen (O), and nitrogen (N).

Figure 1.43 shows a secondary amine (NH) group. It has been detected that spatial orientation of the L-form of amino acids can form when walking along the C-C-N bridge from the C-O end to the $NH_2$ end. The L-isomers of amino acids are found in nature (Fig. 1.43).

A large number of proteins exist in nature. Human body contains an estimated 1000, 000 different proteins. An example of a protein essential to human blood cells is called hemoglobin. Without the hemoglobin protein, human body will not be able to deliver essential oxygen to function efficiently. Hemoglobin is a complex protein molecule that consists of a chain of about 300 amino acids [83]. Analysis of amino acids and their constituents helps us to predict secondary, tertiary, and quaternary protein structures, and also helps us to predict about the complex hemoglobin that contains *heme* units (Fig. 1.44) [84].

When condensation reaction occurs, one molecule of water is eliminated from the carboxylic acid of one amino acid and the amino group of another. The result is a peptide bond (called an amide group in simpler molecules), and the new molecule is called a dipeptide. When two different amino acids form dipeptides, two different combinations are possible, depending on which amine group reacts with which acid group.

Figure 1.45 shows the peptide bond in natural opiates [85]. Many short-chain peptides are biochemicals like enkephalins and endorphins, and are referred to as natural opiates because they moderate pain. In 1973, Solomon Snyder and coworkers at John Hopkins University discovered that the brain and the spinal cord contains specific bonding that opiate molecules fit as a key fits into a lock [86].

Another type of structure that consists of protein is called *enzyme*. The biological processes of nature involve thousands of chemical reactions, most of which have

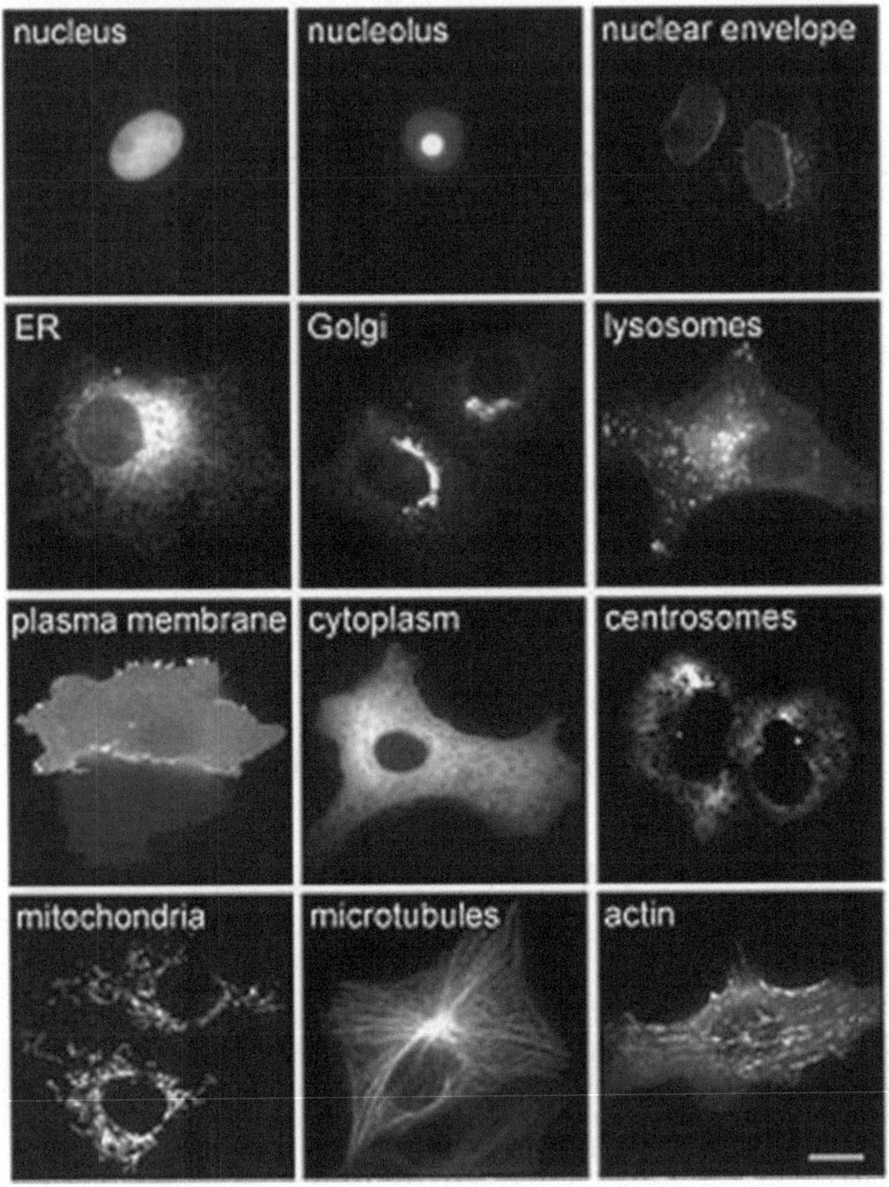


**Fig. 1.42** Proteins in different cellular compartments and structures tagged with green fluorescent protein (here, white) (Courtesy: J. Simpson and R. Pepperkok, WIKI)

their own unique naturally occurring catalysis. These biological catalysts are known as enzymes. The enzyme catalysts are important not only in nature, but also in industry and laboratory. An example of an important enzyme-catalyzed addition to an alkene is the hydration of fumarate ion to malate ion. Figure 1.46

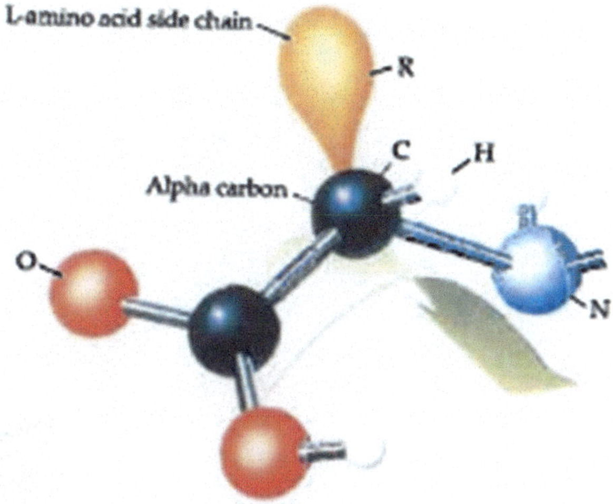


Fig. 1.43 Orientation of the L-form of amino acid

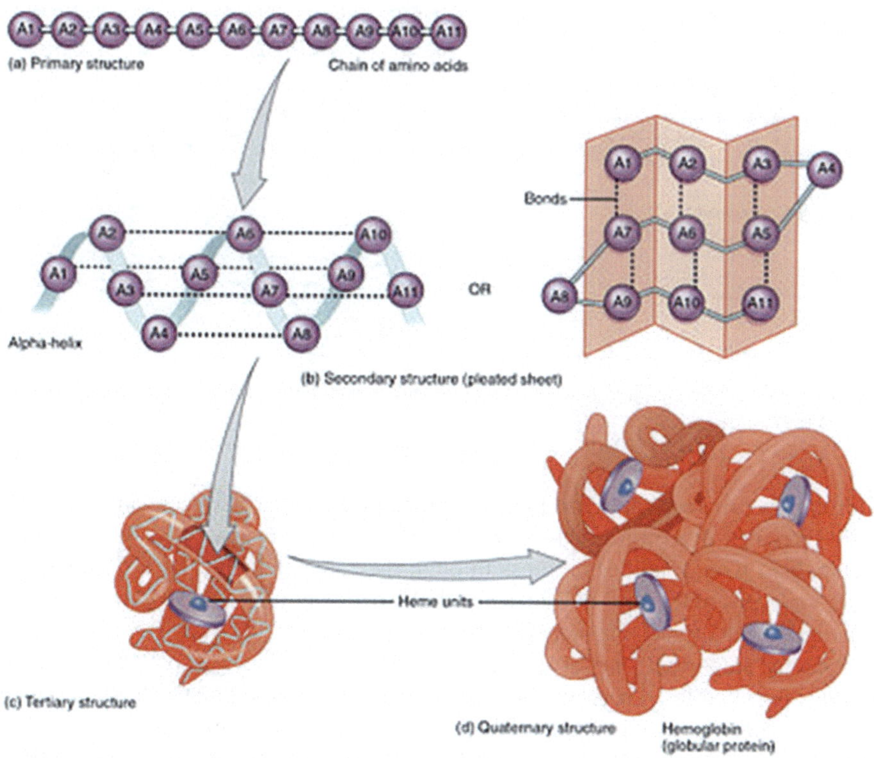


**Fig. 1.44** Constituent amino acids can be analyzed to predict secondary, tertiary, and quaternary protein structures (Courtesy: Open Stax college, WIKI)

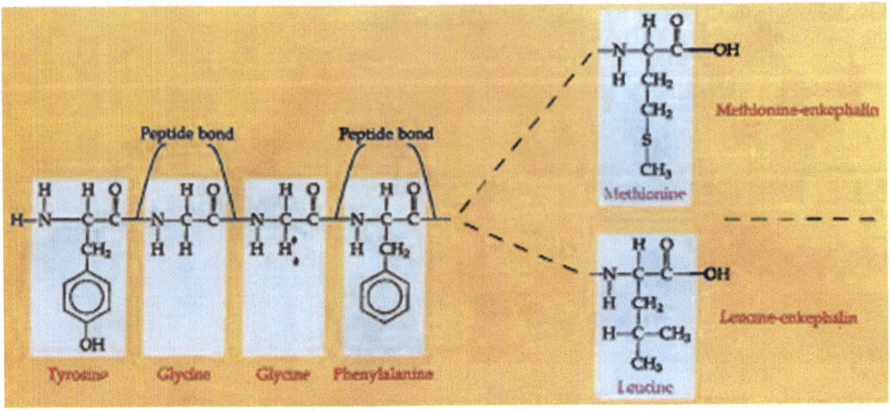


**Fig. 1.45** Peptide bond in natural opiates

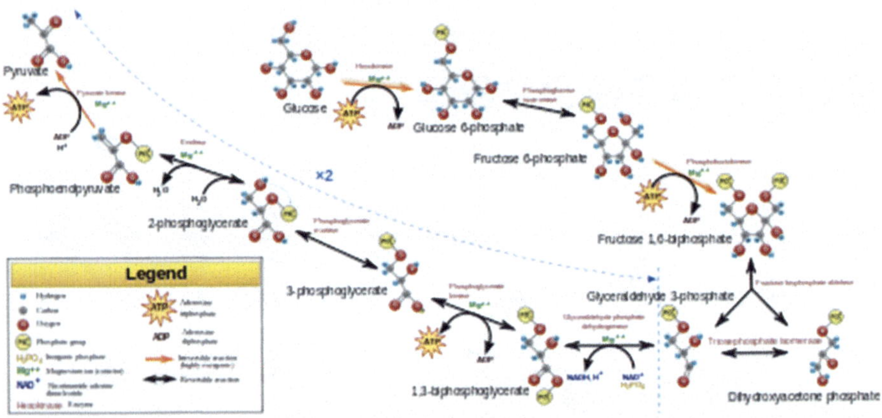


**Fig. 1.46** Glycolytic enzymes and their functions in the metabolic pathway of glycolysis (Courtesy: WIKI)

shows glycolytic enzymes and their functions in the metabolic pathway of *glycolysis*. Enzymes determine what steps occur in the pathways. Without enzymes, metabolism would neither progress through the same steps nor be fast enough to serve the needs of the cell. Indeed, a metabolic pathway such as *glycolysis* could not exist independently of enzymes [87, 88].

Another important chemical ingredient which contains carbon and found in animals is called fat. There are two kinds of fat found in animals. They are called *saturated fat* and *unsaturated fat*. The first one consists of triglycerides containing only saturated fatty acids without any double bonds between the individual carbon atoms of the fatty acid chain. Most common sources of fatty acids are meat, eggs, chocolates, and nuts that primarily contain triglycerides of palmitic and stearic acids. Unsaturated fatty acids, on the other hand, contain at least one double bond

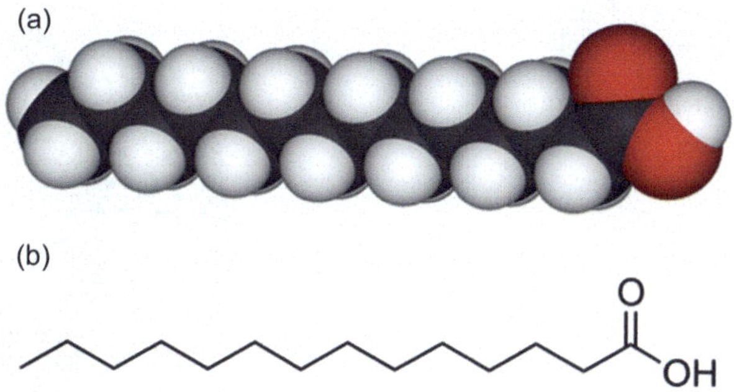


**Fig. 1.47** (**a**) Space-filling model of the saturated fatty acid; (**b**) two-dimensional representation of the fatty acids (Courtesy: (**a**) Benjah-bmm27 and (**b**) LadyofHat, WIKI)

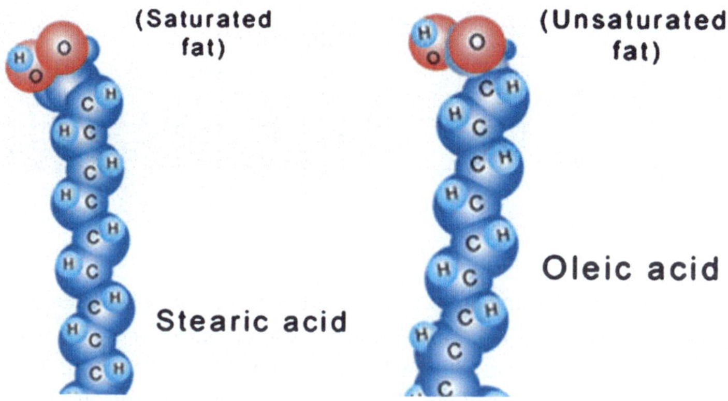


**Fig. 1.48** Saturated fat and unsaturated fat structures

within the fatty acid chain. If the fatty acid chain contains only one double bond, it is *monosaturated* and if the fatty acid contains more than one double bond, it is called *polysaturated* fatty acid. The greater the degree of unsaturation in a fatty acid (i.e., more double bonds in the fatty acid), the more vulnerable it is to *lipid* peroxidation (rancidity). Experimental observations reveal that antioxidants can protect unsaturated fat from *lipid* peroxidation [89].

Figure 1.47a shows a space-filling model of the *saturated fatty acid* and Fig. 1.47b shows its two-dimensional representation. As a matter of fact, naturally occurring *fatty acids*, *waxes*, *sterols*, and triglycerides belong to a group called *lipid*. Figure 1.48 shows the structures of *saturated* and *unsaturated fats*.

*Lipid,* on the other hand, contains the same elements, carbon (C), hydrogen (H), and oxygen (O), as carbohydrates, but these elements are present in lipid in different proportions. The building blocks of lipids are *fatty acids* and *glycerol*, and they are hydrophobic, but they are not *polymers*. Their main biological function is to store *energy* (as fats). In addition to that, they are the main structural components of *cell membranes* (as phospholipids and cholesterol), *hormones*, and other *cell-signaling molecules* (as estrogen and testosterone) [90, 91].

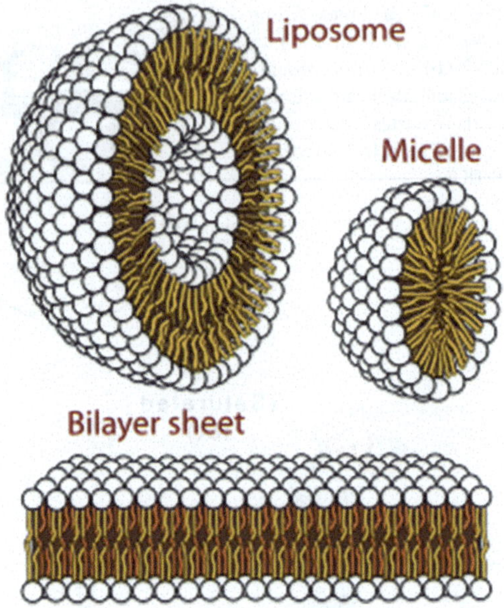


**Fig. 1.49** Formation of liposome, micelle, or bilayer sheet (Courtesy: LadyofHats, WIKI)

*Lipids* are commonly referred to as *fats* and *oils*, and are classified as *saturated* and *unsaturated*. Fats with no double bonds, termed *saturated fats*, are typically solids. As for example we can cite lard. On the other hand, fats containing double bonds are termed as *unsaturated fats*. In many cases, they are oily liquids. Olive oil is an example of an unsaturated fat (Fig. 1.48). In aqueous system, the polars heads of lipids align toward the polar aqueous environment while the hydrophobic tails minimize their contact with water and tend to cluster together forming a vesicle, depending on the concentration of the lipid. This biophysical interaction may result in the formation of liposomes, micelle, or lipid bilayers (Fig. 1.49).

*Cholesterol* (Fig. 1.50) is another example of lipid. Human body needs cholesterol for cell membranes and some fluids. However, too much of cholesterol and saturated fats in the body can cause damage to the heart and blood cells [92, 93].

## 1.14 DNA

*Nucleic acids* (NAs) are polymeric macromolecules – an important group of organic compounds (*containing carbons*) that are essential for life. Nucleic acids are complex molecules made up of chains of nucleotides that store information of living cells. A nucleotide consists of the following subunits: *a phosphate group*, *nitrogen base*, and *a simple sugar*. The elements that make up these components include carbon (C), hydrogen (H), oxygen (O), phosphorus (P), and nitrogen (N).

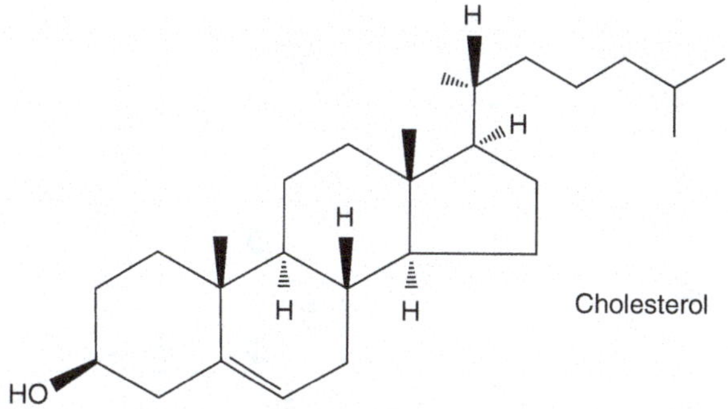


**Fig. 1.50** Structural formula of cholesterol

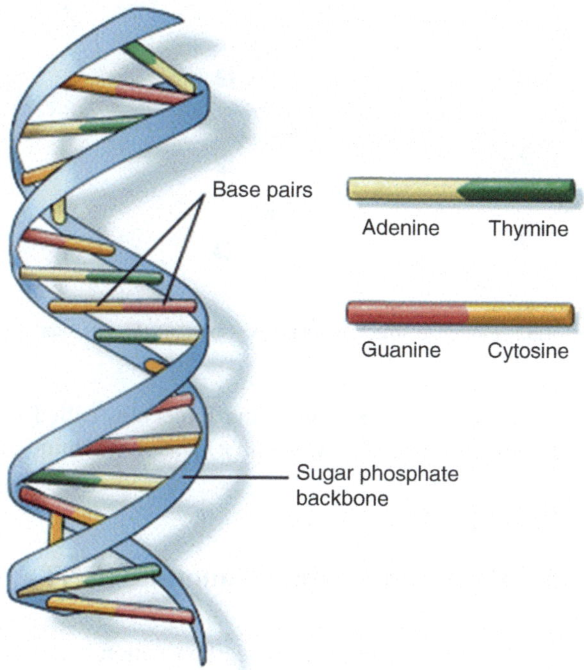


**Fig. 1.51** Tiny portion of the structure of DNA double-helix (Courtesy: National Library of Medicine, USA)

One of the most common types of nucleic acid that we heard almost every day is *deoxyribonucleic acid* or DNA (Fig. 1.51) [94].

The backbone of DNA is based on a repeated pattern of a sugar group, called deoxyribose, and a phosphate group. Figure 1.51 shows a small portion of the double-helical structure of *deoxyribonucleic acid* (DNA). The double-helical structure of DNA, DNA replication, and the fidelity of DNA transcription into *ribonucleic acid* (RNA), an another important nucleic acid involves very specific base-pairing complimentary [95].

For clarity, the flattened molecular structure of DNA is shown in Fig. 1.52. A five-carbon sugar called deoxyribose is labeled as S, a phosphate group is labeled

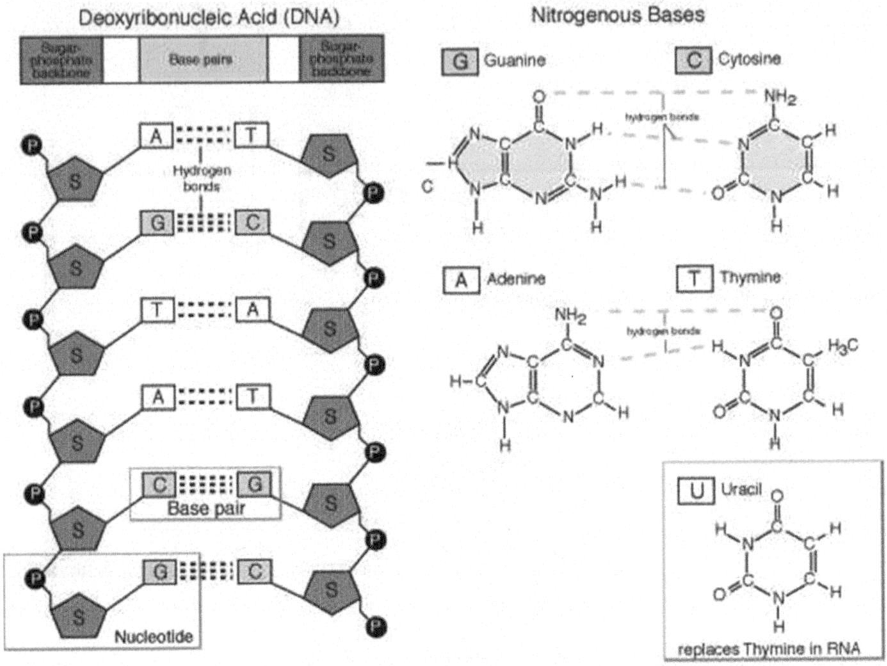


**Fig. 1.52** Flattened structure of DNA (Courtesy: U. Arizona)

as P, and four nitrogen-containing molecules called nucleotides are labeled as G (guanine), C (cytosine), A (adenine), and T (thymine).

## 1.15 DNA and Carbon Nanotube (CNT)

The ability to sort mixtures of carbon nanotubes (CNTs) based on chirality has recently been demonstrated using special short DNA sequences that recognize certain matching CNTs of specific chirality [96].

Figure 1.53 shows the picture of a *carbon nanotube* (CNT). Theory and modeling have played a central role in the development of our understanding of the properties of carbon nanostructures. Mintmire, Dunlap, and White first predicted that *fullerene nanotubes* can be either semiconducting or metallic depending on the structure [97, 98].

CNTs and their compounds exhibit extraordinary electrical properties for organic materials, and have a huge potential in electrical and electronic applications such as photovoltaics, sensors, semiconductor devices, displays, conductors, smart textiles, and energy-conversion devices (e.g., fuel cells, harvesters, and batteries). In the very near future, CNTs will be a viable replacement of indium tin oxide

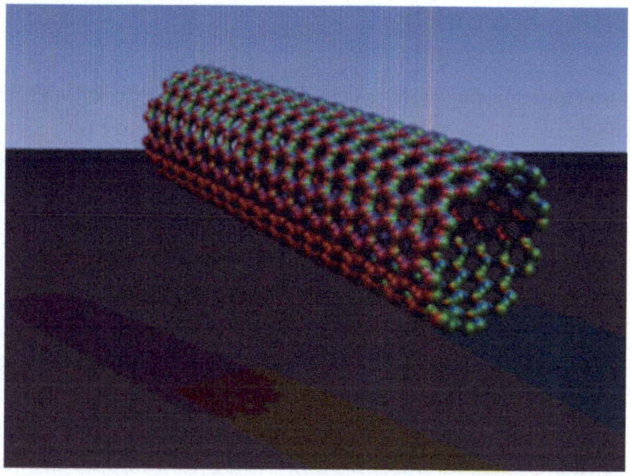

**Fig. 1.53** A carbon nanotube (CNT) (Courtesy: Amero, WIKI)

(ITO) transparent conductors in some applications. Not only that, CNTs will be a cost-effective alternative in flexible displays and touchscreens.

The large surface area, unique surface property, and needle-like shape of CNTs are of a great advantage for transporting therapeutic genetic agents including DNA and siRNAs (small/short interfering RNA/silencing (si) RNA – a class of stranded RNA molecules) to target disease sites (including difficult transfecting primary immune cells and bacteria). Because of its unique property, CNT can deliver and control the release of DNAs/siRNAs for targeted gene therapy [99].

There is strong circumstantial evidence that chemical damage to DNA can interfere with hydrogen bonding between a *pyrimidine* and a *purine* base, and can trigger the state of *uncontrolled cell division* known as *cancer*. One type of chemical damage to DNA is caused by alkylating agents, which are carcinogens (cancer-causing compounds), like methyl methane sulfonate, dimethyl sulfate, N-methyl-N-nitrosourea, and methyldiazonium ion. It has also been detected that ultraviolet (UV) radiation and other ionizing radiation (like X-ray, gamma ray, etc) [100] can be the possible sources of uncontrolled cell division in DNA.

The *biopolymers* DNA and proteins are natural biopolymers that play an essential and ubiquitous role in everyday life [101, 102]. As a matter of fact, *polymers* range from familiar synthetic plastics such as polystyrene to natural biopolymers. The polymer family can be thought of as a carbon family where a (poly) repeating unit (mers) (monomer residue also known as mers) comprises of a chain of carbon atom bound to hydrogen and oxygen atoms.

Recently, a study has been initiated to evaluate thin carbon films for use as electrode materials and pyrolyzed photorersists (*pyropolymer*) in MEMS (microelectromechanical system) [103, 104]. The successful mastering of the topological resolution of 32 nm in serial production in microelectronics is a family of polymer called *photoresists*. Photoresist-derived carbon has been successfully used in the differentiation of neuronal cells [105]. We will discuss about the polymers later.

## 1.16 Summary

Our tour in the field of carbon (C) and carbonaceous materials shows that the element carbon is everywhere, even in human body as a macronutrient, and carbon dioxide which also contains carbon (C) plays an important role to control the pH of the blood. Its many limited uses are in the form of inkjet used in ink printers. The carbon black, another form of carbon, has been used for thousands of years as a pigment in paints and coatings. Oxidative after treatment helps reduce the long dispersion times associated with the use of carbon blacks and improve their dispersion satability. Recently, carbon black filled polyimide has been used in hybrid circuits. Carbon, in the form of carbon dioxide, is used in fuzy and carbonated drinks. Dry ice, which is a solid form of carbon dioxide, is used as a cooling agent. Freon, a gas used in refrigerators and as a cooling agent, also contains carbon. Carbon is also used to manufacture many heat-resistant devices. Besides, the amorphous form of carbon is used as a bleaching agent and gas absorbent.

Of the various allotropic forms of carbon, diamond is the hardest known substance and is used for making jewelry, cutting tools, and making radiation detectors in nuclear medicine [106–108]. It is also used in high-energy physics [109]. Activated carbon clothes are used to enhance the efficacy of both advanced and active wound dressings. Activated carbon is also used in the manufacture of filters for vented ostomy bags. Carbon fibers on the other hand offer several advantages over other materials in the medical field, including the fact that it is radiolucent, transparent to X-rays, and shows as black on X-ray images. It is widely used in imaging equipment structures to support limbs being X-rayed or treated with radiation.

Carbon is also used in carbon dating, a method to determine the accurate age of fossils and minerals. The carbon nanotubes, which are sophisticated forms of carbon in nanoscale, have the potential to carry drugs in the organism as they are hollow and much smaller than the blood cells. The methods were developed for attaching DNA and protein molecules to the inside and outside of the nanotubes [110]. Carbon fiber has gone to moon on space aircraft. It is used in helicopters gliders, and in sports goods, ranging from stiffening of running shoes to ice hockey sticks, tennis racquets, and golf balls.

It is being used in military planes and missiles besides being used in protective helmets. On the other hand, carbon fiber, a cabonaceous material made out of carbon, is being used in automobile industries in internal components like instrument housing and seat frames. Last but not the least, the use of methylene blue ($C_{16}H_{18}N_3SCl$), a heterocyclic aromatic carbon compound is used in the field of biology, chemistry, and medical sciences. It acts as an antidote for cyanide poisoning. When combined with light, it is used to treat resistant plaque psoriasis, AIDS-related Kaposi's sarcoma, West Nile virus, and to inactivate *Staphylococcus aureus*, HIV-1, Duck hepatitis B, adenovirus, and hepatitis C [111–117].

## References

1. R. Daley, S. Daley, *Organic Chemistry, Part-3* (Lulu Pub, Raleigh, 2011), p. 1078
2. A.J. McMichael, *Planetary Overload* (Cambridge University Press, New York, 1993)
3. N.R. Pace, The Universal Nature of Biochemistry. Natl. Acad. Sci. I **98**(30), 805 (2001)
4. E. Smith, H. Morowitz, Universality in intermediary metabolism. Acad. Sci. USA **101**(36), 13168 (2004)
5. E. Noor, E. Eden, R. Milo, U. Allon, Central carbon metabolism as minimal biochemical walk between precursors for biomass and energy. Mol. Cell **39**(5), 809 (2010)
6. S. Baranska et al., Replicating DNA by cell factories: roles of central carbon metabolism and transcription in the control of DNA replication in microprobes, and implications for understanding this process in human cells. Microb. Cell Fact. **12**, 55 (2013)
7. G. Gottschalk, *Bacterial Metabolism* (Springer, Heidelberg, 1986)
8. K.Y. Rhee et al., Central carbon metabolism in micobacterium tuberculosis: an unexpected frontier. Trends Microbiol. **19**(7), 307 (2011)
9. N.D. Price, J.L. Reed, J.A. Papin, S.J. Wilback, B.O. Pearson, Network based analysis of metabolic regulation in the human red blood. J. Theor. Biol. **225**, 185 (2003)
10. J. Kuypers, *DNA and Carbon in Asteroids (oh my), (Periodic Table of Poetry)* (Scars Publication, Chicago, 2013)
11. J.W. Patrick, *Porosity in Carbons: Characterization and Applications* (Edward Arnold, London, 1995)
12. J.C. Serrano-Ruiz, D.J. Braden, R.M. West, J.A. Dumestic, Conversion of cellulose to hydrocarbon fuels by progressive removal of oxygen. Appl. Catal. B: Environ. **100**(1–2), 184 (2000)
13. R. Davis, M. Biddy, E. Tan, L. Tao, Biological Conversion of Sugars to Hydrocarbons Technology Pathway, Tech. Rept. NREL/TP-5100, PNNL-22318, March (2013), Golden Colorado
14. L. Pauling, The nature of chemical bond. Application of results obtained from the quantum mechanics and from the theory of paramagnetic susceptibility to structure of molecules. J. Am. Chem. Soc. **53**(4), 1376 (1931)
15. F. Weinhold, R.C. Landis, *Valency and Bonding: A Natural Bond Orbital Donor Acceptor Perspective* (Cambridge University. Press, Cambridge, London, 2005)
16. O.A. Shenderova, V.V. Zhirnov, D.W. Brenner, Carbon nanostructures. Crit. Rev. Solid State Mater. Sci. **27**(3/4), 227 (2002)
17. R.B. Heinmann, S.E. Evsyukov, Y. Koga, Carbon alloptropes: a suggested classification scheme based on valence orbital hybridization. Carbon **35**, 1654 (1997)
18. M.I. Heggie, Structural modeling, *in Properties of crystalline Silicon*, ed. by R. Hall (IEE publication, London, 1999)
19. R. Rand, S.P. Appleyard, M.F. Yardim, Design and control of structure of advanced carbon materials for enhanced performance. Proc. Nato. Advanced Study Institute on Antalya, Turkey, 10–21 May 1998
20. G. Parr, W. Yang, *Density-Functional Theory of Atoms and Molecules* (Oxford University Press, Oxford, London, 1989)
21. J. Callaway, N.H. March, Density functional methods: theory and applications. Solid State Phys **38**, 135 (1984)
22. A. Hinclliffe, *Modeling of Molecular Structure*, 2nd edn. (Wiley, Chichester, 2000)
23. R. Jones, P.R. Britton, The ab initio cluster method and the dynamics of defects in semiconductors, in *Semiconductors and Semimetals*, vol. 51, ed. by M. Stavola (Academic, San Diego, CA, 1995), pp. 287–349
24. R. Jones, Theory of hydrogen in semiconductors. Phil. Trans. Royal Soc. Lond. A **350**, 189–200 (1995)
25. G.B. Bachelet, D.R. Hamann, M. Schluter, Pseudo potentials work, – from H to Pu. Phys. Rev. B **26**, 4199–4228 (1982)

26. D. Haaland, Graphite-liquid vapor triple point pressure and the density of liquid carbon. Carbon **14**(6), 357 (1976)
27. A. Savvatimskiy, Measurements of the melting point of graphite and the properties of liquid carbon. Carbon **43**(6), 1115 (2005)
28. D.R. Lide, *Hand Book of Chemistry and Physics*, 86th edn. (CRC Press, Boca Raton, 2005)
29. US Geological Survey of Minerals Yearbook, Pub. US Department of Interor and US Geological Survey, 2009
30. H.W. Kroto, J.R. Heath, S.C. O'Brien, R.F. Curl, R.E. Smalley, $C_{60}$ Buckminsterfullerene. Nature **318**, 162 (1985)
31. G.H. Kinchin, The electrical properties of graphite. Proc. Roy. Soc. London **217**(1128), 9 (1953)
32. R. Sengupta, M. Bhattacharya, S. Bandopadhyaya, A.K. Bhowmick, A review of the mechanical and electrical properties of graphite and modified graphite reinforced polymer composite. Prog. Polym. Sci. **36**(5), 636 (2011)
33. R.R. Haering, P.R. Wallace, The electric and magnetic properties of graphite. J. Phy. Chem. Solids. **3**(3–4), 253 (1957). N.A. and also Pierre-Marie Robitalle, Blackbody radiation and carbon particle, Prog. Phys. **3**, 36 (2008)
34. Goncharuk, L. Smrcka, Electronic Properties of Graphie in Tilted Magnetic Field, vol 1598, in AIP 7th International Conference on Dimensional Structure and Devices, New York, 2014 and also Pierre-Marie Robitaille, Blackbody radiation and carbon particle, Prog. Phys. **3**, 36 (2008)
35. N. Ma et al., A new method to determine the mixing state of light absobing carbonaceous using the measured Aerosal optical properties and number size distributions. Atmos. Chem. Phys. **12**, 2381 (2012)
36. C.F. Bohern, D.R. Huffman, *Absorption and Scattering by Small Particles* (Wiley, New York, 1983)
37. T.C. Bond, R.W. Bergstrom, Light absorption by carbonaceous particles: An investigative review. Aerosol. Sci.Technol. **40**, 27–67 (2006)
38. F.J. Stevenson, M.J. Cole, *Cycles of Soil, Carbon, Nitrogen, Phosphorus, Sulfur, and Micronuitrients* (Wiley, New York, 1999)
39. M. Kohl et al., *Carbon Balance Management* (Springer, Heidelberg, 2015)
40. F.T. Mackenzie, *Our Changing Planet: An Introduction to Earth Science and Global Environmental Change* (Prentice Hall, Upper Saddle River, 2000)
41. V.A. Greulach, J.E. Adams, *Plants* (Wiley, NJ, 1962)
42. F.A. Cotton, G. Wilkinson, *Basic Inorganic Chemistry, Chapter-31* (Wiley, New York, 1976)
43. M. Hilger, C. Wairsen, G. Fuchs, C. Taylor, S. Sievert, Evidence for autotrophic $CO_2$ fixation via the reductive tricarboxylic acid cycle by members of the ε subdivision of proteobacteria. J. Bacteriol. **187**(9), 3020–3027 (2005)
44. G. Strauss, G. Fuchs, Enzymesof a noble aytotropic $CO_2$ fixation pathway in the phototropic bacterium Chloroflexus aurantiacus, the 3-hydroxipropionate cycle. Eur.J. Biochem. **215**(3), 633–643 (1993)
45. G.C. Papageorgiou, Govindjee, *Advances in Photosynthesis and Respiration*, vol 19 (Springer, Dordrecht, 2004)
46. M. Warner, I. Rose, Mechanism of *enzyme* catalyzed group transfer reactions. Annu. Rev. Biochem. **47**, 1031–1078 (1978)
47. O. Ebenhoh, R. Heinrich, Evolutionary optimization of metabolic pathways. Theoretical reconstruction of the stoichiometry of ATP and NADH producing systems. Bull. Math. Biol. **63**, 21 (2001)
48. M.C. Grisham, R.H. Garrett, *Biochemistry* (Saunders College Pub, Philadelphia, 1999)
49. H. Miziorko, G. Lorimer, Ribulose-1,5-bisphosphate carboxylase-oxygenase. Annu. Rev. Biochem. **52**, 507–535 (1983)

50. V. Cracan, R. Banerjee, Chapter 10: Cobalt and and corrinoid transport and biochemistry, in *Metallomics and the Cell: Metal Ions in Life Science 12*, ed. by L. Bana (Springer, New York, 2013)
51. A. Sigel, H. Sigel, R.K.O. Sigel, *Organometalics in Environment and Toxicolog: Metal Ions in Life Sciences* (Royal Soc. Pub, Cambridge, London, 2010)
52. L. Finney, T. O'Halloran, Transitional metal specification in the cell: insights from the chemistry of metal receptors. Science **300**(5621), 931 (2003)
53. U. Flogel, M.W. Merx, A. Godecke, U.K.M. Decking, J. Schrader, Myoglobin: a scavenger of bioactive NO. Proc. Natl. Acad. Sci. U. S. A. **98**, 735 (2001)
54. R. Phillips, J. Kondev, J. Theriot, H.G. Garcia, *Physical Biology of the Cell*, 2nd edn. (Garland Science, Taylor Fracis, New York, 1960)
55. D. van der Helm, J.R. Baker, D.L. Eng-Wilmot, M.B. Hossain, R.A. Loghry, Crystal structure of ferrichrome and a comparison with the structure of ferrichrome A. J. Am. Chem. Soc. **102**, 4224 (1980)
56. R.H. Glew, D. Rosenthal, *Clinical Studies in Medical Biochemistry*, 3rd edn. (Oxford University Press, New York, 2007)
57. W. Fredrich, *Vitamins* (Walter de Gruyter and Co., New York, 1988)
58. G.M. Loudon, *Organic Chemistry*, Chapter-5, (The Bejamin Cummings, Reading, 1995), p. 207
59. T.K. Gupta, *Hand book of Thick and Thin Film Hybrid Microelectronics* (Wiley, Hoboken, 2003), p. 107
60. L.F. Thomson, C.G. Wilson, J.M.J. Frechet, *Materials for Microlithography* (American Chemical Society, Washington DC, 1984)
61. T.K. Gupta, *Computer Aided Design and Pattern Generation Technique, in Hand Book of Thick and Thin Film Hybrid Microelectronics* (Wiley, NJ, 2002), p. 112
62. T.K. Gupta, Effect of temperature on the photosensitivity of photoresist. Euro. Poly. J. **17** (11), 1127 (1981) and also T.K. Gupta, Effect of temperature on the electrical conductivity of carbon black filled polyimide. IEEE Trans. Comp. Hybrids Manuf. **12**(4), 696 (1989)
63. S. Fakirov, *Transreactions in Condensation Polymers* (Wiley, New York, 2008)
64. P.J. Flory, Fundamental principles of condensation polymerization. Chem. Rev. **39**(10), 137 (1946)
65. H. Volger, Wettstreit um die Polyamidfasern Chemie in unserer Zeit, 47, 62 (2001)
66. J.E. Njarnason et al., Phys. Lett. **85**(4), 519 (2004)
67. Linear polyamides suitable for spinning into strong pliable fabrics, US patent, 2130,947 and 2130, 948, 20 Sept 1938
68. L. Bottenbruch, *Engineering Thermoplastics* (Hanser, Cleveland, 1996)
69. T.K. Gupta, *Copper Interconnect Technology*, Chapter-1, (Springer, New York, 2009), p. 44
70. S. Lijim, Helical Nano-tube of graphite carbon. Nat. Mater. **1**(3), 153 (2002)
71. X. Wang et al., Fabrication of ultra-long and electrically uniform single-walled carbon nanotubes on clean substrate. Nano Lett. **9**(9), 3137 (2009)
72. Z. Ren, Y. Tan, Y. Wang, *Aligned Carbon Nanotubes: Nano Science and Technology* (Springer, Heidelberg, 2013)
73. S. Coffey, *Rodd's Chemistry of Carbon Compounds*, 2nd edn. (Elsevier, New York, 2008)
74. J. Clayden et al., *Organic Chemistry* (Oxford, Oxford, 2001)
75. G.M. Loudon, *Organic Chemistry*, 3rd edn. (The Bejamin /Cummings Pub. Co, London, 1995), p. 109
76. M.J. Berg, *Biochemistry*, 6th edn. (W.H. Freeman, New York, 2007)
77. W.H. Brown, T. Poon, *Introduction to Organic Chemistry* (Wiley, Hoboken, 2005)
78. Anne-C-Eliasson Starch in Food :Structure, Function and Application, Woodhead Pub., (2004)
79. R.L. Crawford, *Lignin Biodegradation and Transformation* (Wiley, New York, 1981)

80. O.V. Stepanenko, W. Verkhusha, I.M. Kuznetsova, V.N. Uversky, K.K. Turoverov, Fluorescent proteins as biomakers and biosensors: throughing color lights on molecular and cellular processes. Curr. Protein Pept. Sci. **9**(4), 338 (2008)
81. J.A. Reynolds, C. Tranford, *Nature's Robots: A History of Proteins* (Oxford University Press, New York, 2003), p. 15
82. J. Brosnan, Interorgan amino acid transport and its regulation. J. Nutr. **133**(6 Suppl1), 2068S (2003)
83. Z. Xiang, Advances in homology protein structure modeling. Curr. Protein Pept. Sci. **7**(3), 217 (2008)
84. L. Pauling, R. Corey, H.R. Branson, The structure of Proteins: two hydrogen–bonded helical configurations of the polypeptide chain. Proc. Nat. Acad. Sci. USA **37**(5), 235 (1951)
85. J.C. Kotz, J.L. Wood, M.D. Joesten, J. Moore, Chapter 21: Chemistry of life, in *The Chemistry World*, (Saunders College Pub, Orlando, FL, 1994)
86. T. Bugg, *Introduction to Enzyme and Co-enzyme Chemistry*, 2nd edn. (Blackwell, Hoboken, 2004)
87. R.E. Reeves, D.J. South, H.J. Blytt, L.G. Warren, Pyrophosphate: D-fructose 6-phosphate 1-phosphotransferase. A new enzyme with the glycolytic function 6-phosphate 1-phosphotransferase. J. Biol. Chem. **249**(24), 7737 (1974)
88. C. Sen, L. Packer, O. Hanninen, *Hand book of Oxidants and Antioxidants* (Elsevier, Amsterdam, 2000), p. 116
89. L.E. Nachtigall, J.R. Heilman, *Estrogen*, 3rd edn. (Harper Collins, New York, 2000)
90. J.V. Craig, *Estrogen Action, Selective Estrogen Receptor Modulators, and Women Health* (World Scientific Distbr, Hackensack, 2013)
91. J.M. Lecerf, M. de Lorgeoil, Dietary cholesterol from physiology to cardiovascular risk. Br. J. Nutr. **16**(1), 6 (2011)
92. C. Dubois, M. Armand, et al., Effects of increasing amounts of dietary cholesterol on postprandial lipemia and lipoproteins in human subjects. J. Lipid Res. **35**, 1993 (1994)
93. J. Watson, A. Berry, *DNA, The Secret of Life* (Andrew Book Pub. and DNA Show LLC (P), New York, 2003)
94. G.M. Loudon, *Organic chemistry*, 3rd edn. (The Benjamin/Cummings, San Francisco, 1995)
95. D. Roxbury, T. Xiaomin, M. Zheng, A. Jagota, Recognition ability of DNA for carbon nanotubes correlates with their binding affinity. Langmuir **27**(13), 8282–8293 (2011)
96. J.W. Mintmire, B.I. Dunlap, C.T. White, Are fullerene tubeless metallic? Phys. Rev. Lett. **68**, 631 (1992)
97. S. Iijima, C. Brabec, A. Maiti, J. Bernhole, Structural flexibility of carbon nanotubes. J. Chem. Phys. **104**, 2089 (1996)
98. W. Cheung, F. Pontoriero, O. Taratula, A.M. Chen, H. He, DNA and carbon nanotubes as medicine. Adv. Drug Deliv. Rev. **62**(60), 633–649 (2010)
99. T.K. Gupta, *Radiation, Ionization, and Detection in Nuclear Medicine* (Springer, Heidelberg, 2013)
100. N.G. McCrum, N.G. Buckley, C.B. Bucknall, *Principles of Polymer Engineering* (Oxford University Press, Oxford, 1997)
101. R.D. Blake, *Informational Biopolymers of Genes and Gene Expression* (Univ. Sci Books, Sausalito, 2005)
102. J. Kim, X. Song, K. Kinoshita, M. Madou, R. White, Electrochemical Studies of Carbon Films from Pyrolyzed Photoresist. J. Electrochem. Soc. **145**(7), 2314 (1998)
103. G. Jenkins, K. Kawamura, *Polymeric Carbons-Carbon Fibre, Glass and Char* (Cambridge University Press, Cambridge, 1976), p. 15
104. H. Zhou, A. Gupta, T. Zou, J. Zhou, Photoresist derived carbon for growth and differentiation of neuronal cells. Int. J. Mol. Sci. **8**(8), 884 (2007)
105. R.J. Keddy, T.L. Nam, Diamond radiation detectors. Radiat. Phys. Chem. **41**, 767 (1993)
106. D.R. Kania, M.I. Landstrass, M.A. Plano, Diamond radiation detectors. Diamond Relat. Mater. **2**, 1012 (1993)

107. J. E. Field (ed.), *The Properties of Diamond* (Academic Press, New York, 1979)
108. R.J. Tapper, Diamond detectors in particle physics. Rep. Prog. Phys. **63**, 1273 (2000)
109. A.R. Harutyunyan, B.K. Pradhan, G.U. Sumanasekera, E.Y. Korobko, A.A. Kusnetsov, Carbon nanotubes for medical applications. Eur. Cell. Mater. **3**(2), 84 (2002)
110. M. Salah, N. Samy, N. Fadel, Methylene blue mediated photodynamics therapy for resistant plaque psoriasis. J. Drug Dermatol **8**(1), 42 (2009)
111. J.P. Tarvido, A. Del Gigilo, L.H. Paschol, M.S. Baptista, New photodynamic therapy protocol to treat AIDS-related Kaposi's sarcoma. Photomed. Laser Surg. **24**(4), 528 (2006)
112. J.F. Papin, R.A. Floyd, D.P. Dittmer, Methylene blue photoinactivation abolishes West Nile virus infectivity in vivo. Antiviral Res. **68**(20), 84 (2005)
113. P.S. Zolfagari et al., In vivo killing of Staphylococcus aureous using a light activated antimicrobial agent. BMC Microbiol. **9**, 27 (2009)
114. R.A. Floyd, J.F. Schneider, D.P. Dittmer, Methylene blue photoinactivation of RNA viruses. Antiviral Res. **61**(3), 141 (2004)
115. S.J. Wagner et al., Duck hepatitis B phtoinactivation by dimethylene blue in RBC suspensions. Transfusion **41**(9), 1154 (2001)
116. F.H.E. Schagen et al., Photodynamic treatment of adenoviral vectors with visible light: an easy and convenient method for viral inactivation. Gene Ther. **6**(5), 873 (1999)
117. E. Ben-Hur et al., The photo decontamination of cellular blood components: photosensitization in transfusion medicine. Transfus. Med. Rev. **10**, 15–22 (1996) and also A.M. Gall et al., Water borne viruses, a barrier to safe drinking water. PLoS Pathoq. **11**(6), e 1004867 (2015)

# Chapter 2
# Historical Production and Use of Carbon Materials: The Activated Carbon

## 2.1 Introduction (Activated Carbon)

The use of carbon extends far back to 3750 BC, for the reduction of copper (Cu), zinc (Zn), and tin (Sn) ores in the manufacture of bronze by the Egyptians and Sumerians. In 157 AD, carbons of both vegetable and animal origin had been used for the treatment of a wide range of diseases. In the year 1773, Car Wilhelm, a chemist from Pomerania, a Baltic coast of Europe under Swedish control, recognized the absorptive power of carbon-derived materials from different sources. However, marketing of first industrially produced *activated carbon*, *Eponit* (trade name), was first reported in the year 1911 by the Fanto Works, Austria [1]. The largest market for activated carbon is currently in the municipal water purification industry.

Activated carbon, also called *activated charcoal*, *activated coal*, or *carbo activatus*, is a microporous *inert carbon* with large internal surface area in excess of 500 $m^2/g$, which has been measured from absorption isotherms of carbon dioxide ($CO_2$) gas at room temperature (300 k) [2–3]. Activated carbons (ACs) are complex products which are difficult to classify on the basis of their behavior, surface characteristics, and other fundamental criteria. However, some broad classification is made for general purpose, based on their size, preparation methods, and industrial applications: powdered activated carbon (RI, PAC), granular AC (GAC), extruded AC (EAC), bead AC (BAC), impregnated AC, and polymer-coated AC.

Riddled with small, low-volume pores, the large internal surface area has been a fascinating quality of the magnificent elements that have revolutionized in material science and modern industry for their unique absorbing power. Since *activated carbon* (AC) has high degree of microporosity, 1 gram of AC can give a surface area between 600 and 1200 square feet.



T. Gupta, *Carbon*, DOI 10.1007/978-3-319-66405-7_2

### 2.1.1 Activated Carbon (AC)

Activated carbon when viewed under electron microscope shows high surface area structures as shown in Fig. 2.1. Individual particles are intensely convoluted and show various kinds of porosity. In some areas, one can notice graphite-like flat surfaces that run parallel to each other separated by a few nanometers or so. The micropores of activated carbon (AC) provide excellent conditions for absorption to occur, since the absorbing material can interact with many surfaces simultaneously [4].

*Activated carbon* (AC) is one of the most important microporous adsorbents due to its tremendous adsorptive capacity, an affinity for a variety of dissolved organics, and ability to be custom-tailored to suit specific applications [5]. Figure 2.2 shows some pores and their structures in AC.

It is used in gas purification, decaffeination, gold purification, metal extraction, water purification (Fig. 2.3), medicine (Fig. 2.4), sewage treatment, air filters in gas masks and respirators, filters in compressed air, and many other applications. It was James Dewar, a scientist, who first invented Dewar (vacuum flask, one of the essential commodities) and studied AC in detail, especially its absorption capacity [6].

Figure 2.2a represents the hypothetical pores of activated carbon used in water filtering system, and Fig. 2.2b represents pores of activated carbon seen under

**Fig. 2.1** shows a micrograpgh of activated carbon under an electron microscope (Photo courtesy: WIKI)

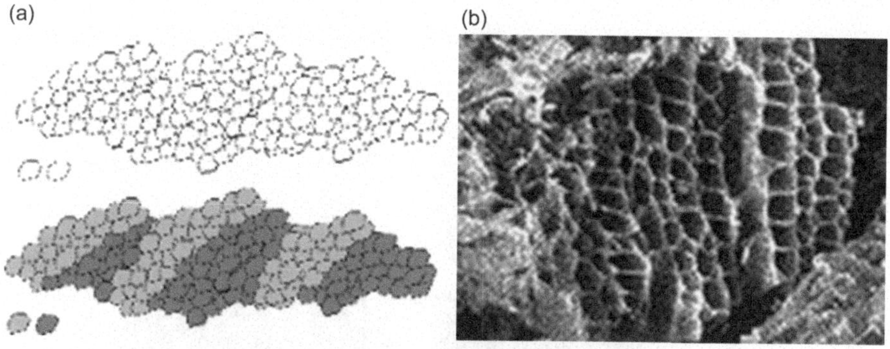


**Fig. 2.2** (**a**) Representation of the hypothetical pores of the activated carbon used in water filtering system; (**b**) Pores of activated carbon seen under electron microscope (Courtesy: WIKI)

electron microscope. Granular activated carbon (GAC) has relatively larger particle size compared to powdered activated carbon, and consequently presents a smaller external surface. These carbons are therefore used for adsorption of all gases and vapors as their rate of diffusion is faster. GAC is designated by sizes such as $8 \times 20$, $20 \times 40$, or $8 \times 30$ for liquid phase applications and $4 \times 6$, $4 \times 8$, or $4 \times 10$ for vapor phase applications. A GAC-designated size $20 \times 40$ means that the filter is made of carbon particles that would pass through a U.S. standard mesh size No. 20 sieve (0.84 mm) but be retained on a U.S. standard mesh size No. 40 sieve (0.42 mm). The structure of activated carbon consists of elementary microcrystallites of graphite, but these microcrystallites are stacked together in random orientation, and it is the spaces between the crystals which form the micropores. The pore diameter of a typical activated carbon can vary from <20 Å to >5000 Å. Porous carbons can contain several impregnated inorganic elements such as iodine ($I_2$), silver, cations such as aluminum (Al), manganese (Mn), zinc (Zn), lithium (Li), and calcium (Ca) for specific applications [7–8].

Activated carbon (AC) has been used in water filtration system using polyvinyl chloride (PVC) pipe. Figure 2.3 shows the use of AC in water filtration system which consists of pebble, palm, palm fiber, alum, activated carbon (AC), gravel, chlorine ($Cl_2$), smooth sand, and gravel (from bottom to top). Each ingredient is separated with sponge filtration and gauge to keep it at fixed place.

Activated carbon (AC) is used to treat poisonings and overdoses following oral ingestion. Figure 2.4 shows a commercial product where activated carbon is being used for medical use.

Activated carbon can be considered as a material of phenomenal surface made up of millions of pores comparable to a molecular sponge. The process in which such a surface concentrates fluid molecules by chemical or physical forces through the pores is known as *adsorption* – the adhesion of the ions, atoms, or molecules from a gas or a liquid or dissolved solid to a surface [9–10]. The word adsorption was coined in 1881 by German physicist Heinrich Kayser [10–11]. Adsorption is generally described through isotherms, that is, the amount of adsorbate on the

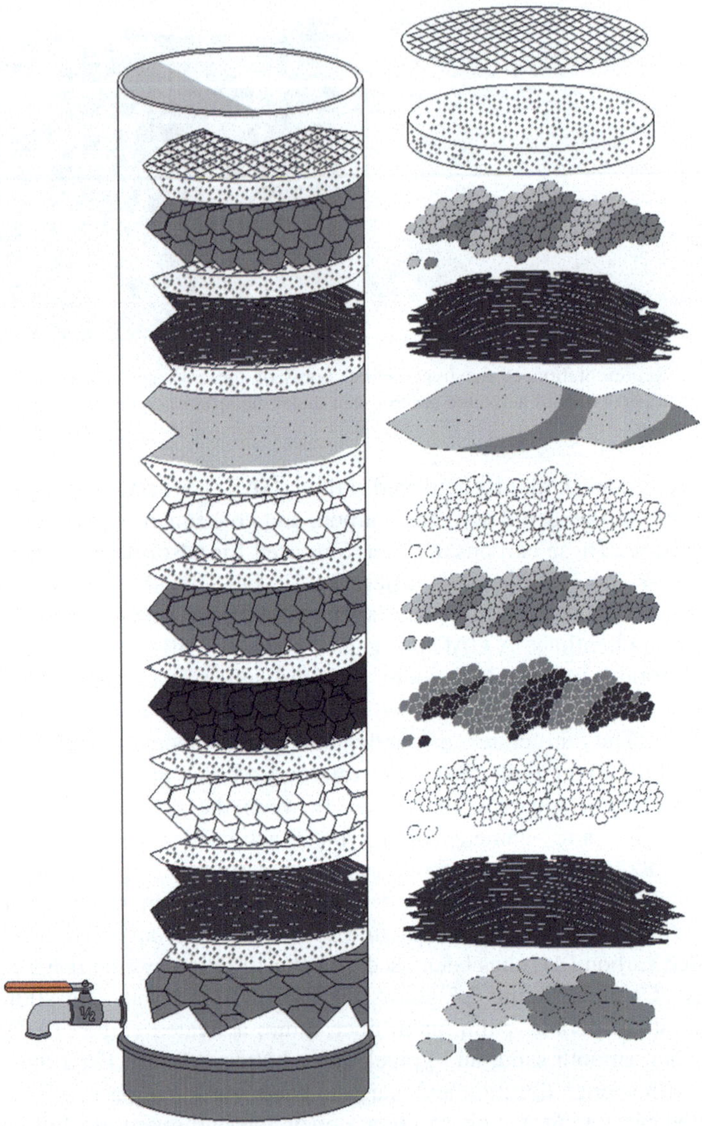


**Fig. 2.3** Water filtering system with activated carbon and other ingredients (Courtesy: Iran Akira, WIKI)

adsorbent as a function of its pressure (if gas) or concentration (if liquid) at constant temperature. The quantity adsorbed is nearly always normalized by the mass of adsorbent to allow comparison of different materials. To date, there are more than 15 models that have been put forward by different investigating scientists. We will discuss some of the models.

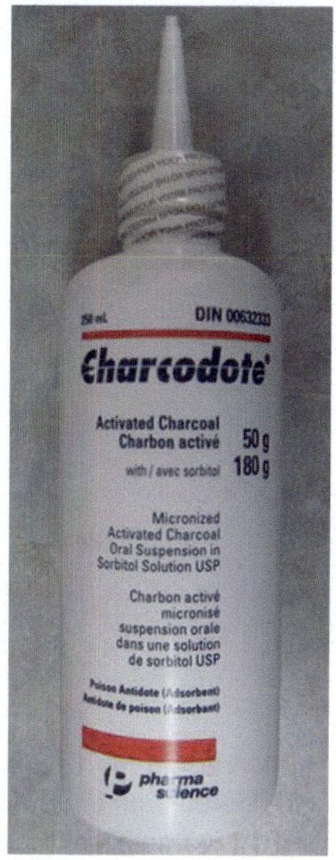


**Fig. 2.4** Activated carbon being used in medicine (Photo courtesy: S. Heilman, WIKI)

Physically, AC binds materials by van der Waals or London dispersion force. But they are unable to bind certain chemicals like alcohols, glycol, strong acids and bases, metals and most inorganics like lithium (Li), sodium (Na), iron (Fe), lead (Pb), and arsenic (As). Similar problems arise with carbon monoxide (CO).

Carbon dioxide ($CO_2$) emission from carbon-emitting fuels such as coal, petroleum, and natural gas causes lots of dreadful impacts on the ecology. Global warming is also believed to be due to the large emission of $CO_2$ gas in the atmosphere (Fig. 2.5) [12].

Global Warming Potential (GWP) was developed to allow comparisons of the global warming impacts of different gases. Specifically, it is a measure of how much energy the emissions of 1 ton of a gas will absorb over a given period of time, relative to the emissions of 1 ton of carbon dioxide. The larger the GWP, the more that a given gas warms the Earth compared to carbon dioxide over that time period. The time period usually used for GWPs is 100 years. GWPs provide a common unit of measure, which allows analysts to add up emission estimates of different gases (e.g., to compile a national greenhouse gas inventory), and allows policymakers to compare emission reduction opportunities across sectors and gases. Carbon dioxide ($CO_2$), by definition, has a GWP of 1 regardless of the time period used, because it is the gas being used as the reference. Carbon dioxide remains in the climate system for a very long time: carbon dioxide emissions cause increases in atmospheric concentrations of carbon dioxide that will last thousands of years [13].

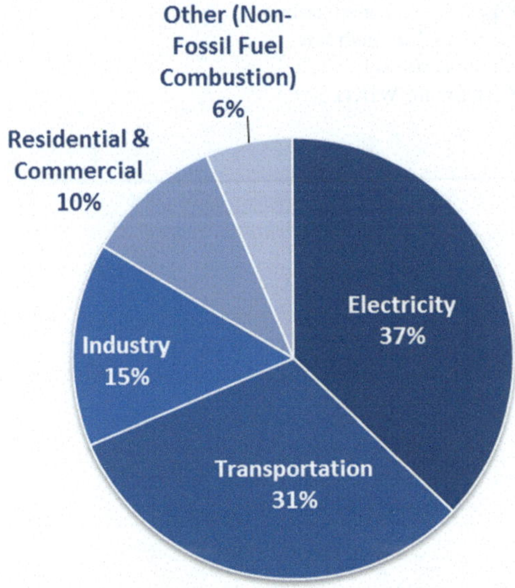


**Fig. 2.5** shows US $CO_2$ emission (inventory of US greenhouse gas emission sinks (1990–2013) (Source: EPA, USA, Ref. 13)

Mercury, a highly toxic trace element, is of particular concern because, during coal combustion and municipal waste incineration, most of the mercury present in the fuel is transferred into vapor phase, due to it high volatility. In postcombustion environment of a boiler, it is transformed to oxidized forms ($Hg^{2+}$ $X_g$). Most of the existing air pollution control technologies, such as electrostatic precipitators and baghouses, cannot remove effectively gas phase mercury emission, except for particle-bound atoms ($Hg_p$) [14].

Activated carbon materials are believed to be the most attractive absorbant for $CO_2$ gas and vapor phase mercury (Hg), due to its low cost, low sensitivity to moisture, and high stability.

Isotherm equations that deal with physical adsorption of gases and vapors give the most important characteristics of industrial sorbents that include, among others, pore volume, pore size or energy distribution, and specific surface area. The correct interpretation of experimental adsorption isotherms can be realized in terms of mathematical adsorption equations [15].

In 1914–1918, Langmuir proposed a mathematical equation for the adsorption mechanism as

$$A_g + S \leftrightarrow AS, \tag{2.1}$$

where $A$ is a gas molecule and $S$ is an absorption site. If we define surface coverage, $\theta$, as the fraction of the absorption sites occupied in the equilibrium condition, we can write

$$K = (\theta)/(1 - \theta)\, P \tag{2.2}$$

$$\text{Or } \theta = (KP)/(1 + KP), \tag{2.3}$$

where $P$ is the partial pressure of the gas or molar concentration of the solution. For very low pressure, the value of $\theta \approx KP$, and for high pressure, the value of $\theta \approx 1$. Experimentally, the parameter $\theta$ is difficult to measure. To ease the mathematical calculations, the parameter $\theta$ has been replaced by the ratios of the gas volume, as $\theta = (v/v_{mon})$, where $v_{mon}$ is the gas volume occupied at standard temperature and pressure (STP). Replacing the value of $\theta$ in Eq. (2.3), we can write Eq. (2.3) as

$$(1/v) = (1/(K\, v_{mon}))\,(1/P) + (1/v_{mon}) \tag{2.4}$$

The slope of the curve and the $y$-intercept will give the values of $v_{mon}$ and $K$, which are constant for each adsorbent/adsorbate pair at a given temperature. It is noteworthy that the Langmuir studies of gas adsorption led to the formulation of a general treatment of kinetics of reactions at the surfaces, and in 1932, he was awarded the Nobel Prize in Chemistry for his discoveries and researches in the realm of surface chemistry [16].

### 2.1.2 *Adsorption–Desorption*

Often, molecules do form multilayers because some of the molecules are adsorbed on already adsorbed molecules. Langmuir isotherm formulation was inadequate to solve the problem for multilayers. However, in 1939, Stephen Brunauer, Paul Emmett, and Edward Teller (BET) developed a model isotherm which can be extended to the finite n number of adsorbed layers [17–19].

Figure 2.6 shows the schematic of the BET model of multilayer adsorption. The distribution of the particles on the material surface is random. The BET theory, despite many restrictions, was the first attempt to create a universal theory of physical adsorption [20]. The theory goes by the three names and hence called BET theory, the modified Langmuir's mechanism, and it follows as

$$A_g + S \leftrightarrow \; A_1S \ldots \text{(for single layer following Langmuir)} \tag{2.5}$$

$$A_g + S \leftrightarrow \; A_2S \ldots \text{(for double layer)} \tag{2.6}$$

$$A_g + S \leftrightarrow \; A_3S \ldots \text{(for triple layer)} \tag{2.7}$$

$$A_g + S \leftrightarrow \; A_nS \ldots \text{(for } n\text{-layers following Langmuir)} \tag{2.8}$$

**Fig. 2.6** Schematic of the BET model of multilayer adsorption (Courtesy: Riely, Creative Comm., WIKI)

The derivation of the formula is much more complicated than Langmuir's single layer model, and for multiple layers we can write the equation as

$$\{p/(v(1-p))\} = 1/v_{\text{mon}}c + \{p\,(c-1)\}/\{v_{\text{mon}}c\} \tag{2.9}$$

where $p = (P/P_0)$ is pressure divided by the vapor pressure for the absorbate at that temperature, $v$ is the STP volume of the adsorbed adsorbate, $v_{\text{mon}}$ is the STP volume of the amount of adsorbate required to form a monolayer, and $c$ is the equilibrium constant ($K \times$ the vapor pressure of the absorbate). The Langmuir isotherm is usually better for chemisorption, and BET isotherm works better for physisorption.

The BET theory has been further modified by Dublin and Radushkievich (DB) [21], and later by Polanyi [22]. According to their concept, multilayer adsorbed layer is not a two-dimensional problem as followed by Langmuir theory, but it possesses a definite volume and applies Van der Waals equation. Polanyi assumed that the adsorption potential is independent of temperature over a wide range. Such a statement follows from the fact that the van der Waals forces are also independent of temperature, which does not hold always for polar adsorbates.

Considering different two-dimensional (2D) adsorptional isotherms, different theories have been put forward from time to time. Some of these theories are worth mentioning, e.g., the Henry law for ideal non-localized monolayer, the Langmuir equation for ideal localized monolayer [23], the Volmer and the Hill-de-Boer [24] equations for the nonideal and nonlocalized monolayers, and the Fowler-Guggenheim [25] equations for the nonideal. However, among all of these theories the localized monolayers are worth mentioning. The 2D equations of state are very useful for investigating the phase transitions in the adsorbed layer [26–27].

The graph in Fig. 2.7 depicts monolayer adsorption (or type I adsorption isotherm), where $x$ is the mass of the gas adsorbed on mass $m$ of the adsorbent at pressure $P$ [28–29]. The ratio of $(x/m)$ is plotted against pressure ($P$). $P$ denotes the saturation pressure. Type I adsorption isotherm is the adsorption of nitrogen ($N_2$) and hydrogen ($H_2$) on charcoal at temperature near to 1800 °C. In BET Eq. (2.9), if

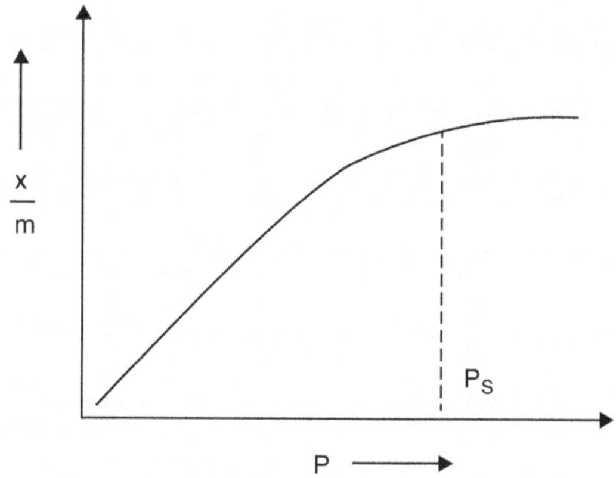


**Fig. 2.7** Type I adsorption isotherm

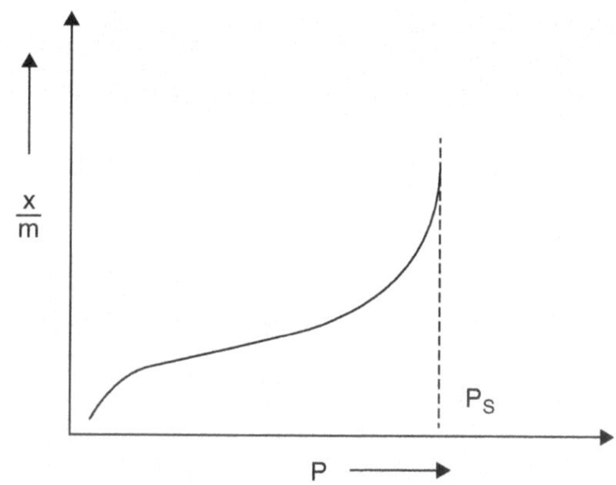


**Fig. 2.8** Type II adsorption isotherm

the values of $(\underline{p}) < < 1$ and $c > > 1$, then it leads to monolayer formation of type I (Fig. 2.7).

In type II adsorption isotherm (Fig. 2.8), we can see a large deviation from the Langmuir model of adsorption. The intermediate flat region is the isotherm that corresponds to monolayer formation. Examples of type II adsorption isotherms are $N_2$ gas adsorption at 1950 °C on iron (Fe) catalyst and $N_2$ adsorbed at 1950 °C on silica gel. Here, the value of $c$ in Eq. (2.9) is very large.

Type III adsorption isotherm (Fig. 2.9) also deviates largely from the Langmuir model. If $c <<< 1$ in BET equation, we can expect an adsorption isotherm similar to type III. In this curve, there is no flat portion which indicates that there is no formation of monolayer. Examples of type III adsorption isotherms are bromine ($Br_2$) or iodine ($I_2$) adsorbed at 790 °C on silica gel.

At low-pressure regions, type IV adsorption isotherm (Fig. 2.10) shows similarities with type II adsorption isotherm, indicating that at low pressure it forms monolayer and multilayers of adsorption. The saturation level is reached at a pressure below the saturation vapor pressure. This can be explained on the basis of a possibility of gases getting condensed in the tiny capillary pores of adsorbent at

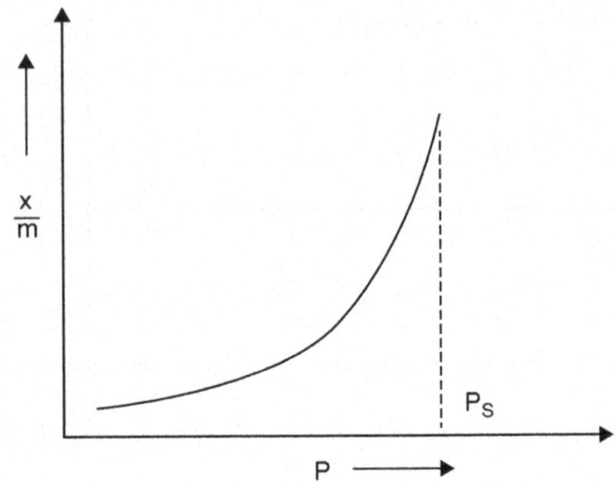


**Fig. 2.9** Type III adsorption isotherm

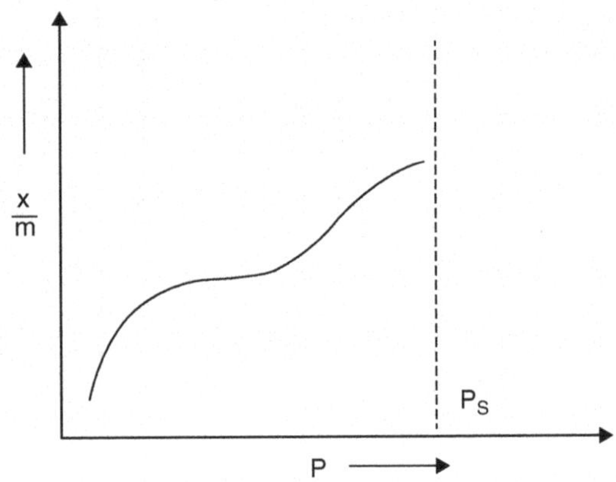


**Fig. 2.10** Type IV adsorption isotherm

a pressure below the saturation pressure of the gas. Examples of type IV adsorption isotherm are of adsorption of benzene on iron oxide ($Fe_2O_3$) and on silica gel at 590 °C.

The graph for type V adsorption isotherm (Fig. 2.11) is similar to type IV. Example of type V adsorption isotherm is adsorption of water on charcoal at 1000 °C. The understanding of the adsorption isotherms is important in the solution of a large class of industrial problems on efficient and safe storage, and handling of gases during transportation. It has been proposed that flammable gases could be adsorbed on to a variety of adsorbents and transported and then released safely after their arrival [30].

In the case of adsorption where capillary condensation occurs, hysteresis loops appear [31–32]. Very recently, these hysteresis loops have been classified in a way similar to the International Union of Pure and Applied Chemistry (IUPAC) adsorption isotherm scheme (Fig. 2.12) [33].

The major progress made in the understanding of the adsorption, pore condensation, and hysteresis behavior of the fluids in the nanoporous materials has led to

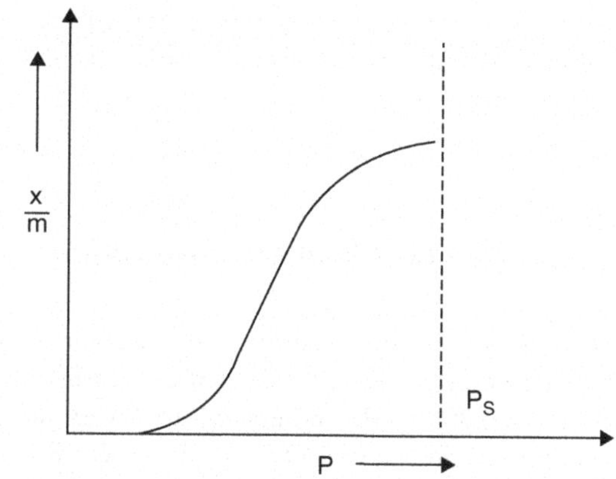


**Fig. 2.11** Type V adsorption isotherm

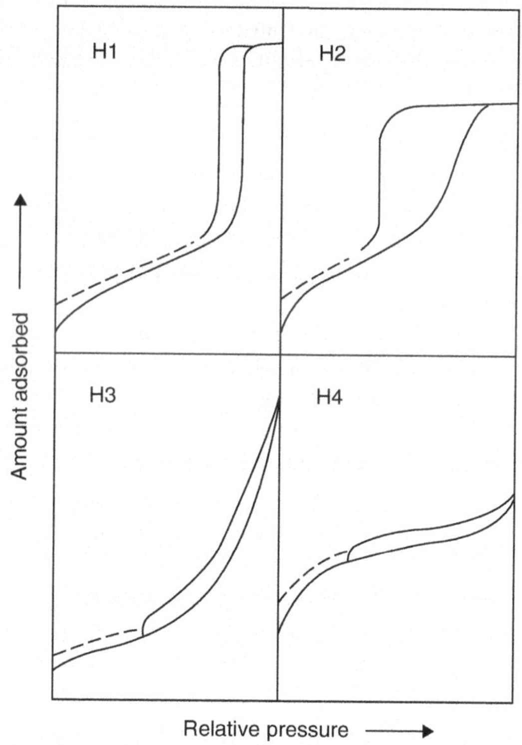


**Fig. 2.12** Four IUPAC classifications of adsorption isotherms: Hysteresis loops

the development in the structural characterization by physical adsorption. The development and availability of advanced theoretical models on statistical mechanics like nonlocal density functional theory (NLDFT) and molecular simulation also had helped in this endeavor. In spite of the progress made on the theoretical models and computer simulation works, there still exist major problems in the characterization of disordered porous materials, especially mesoporous with inhomogeneous surface chemistry.

### 2.1.3 Kinetic Theory of Adsorption

According to some experimental observers the kinetic theory of adsorption in activated carbon, Fick's interparticle diffusion equation has been accepted as a more accurate model compared to linear driving force model for solving the desorption kinetics of $CO_2$ and Hg gases. Rates of adsorption and desorption in porous adsorbents are generally controlled by transport within the pore network, rather than by the intrinsic kinetics of sorption at the surface. It is convenient to consider interparticle transport as a diffusive process and to correlate kinetic data in terms of diffusivity defined by Fick's first law as [34–35]:

$$J = -D_{\mathrm{eff}}(\mathrm{d}C/\mathrm{d}r) \tag{2.10}$$

where $J$ is the diffusion flux, $D_{\mathrm{eff}}$ is the diffusivity {(length)$^2$/time}, $C$ is the concentration of the substance (amount of substance per unit volume (mol/m$^3$)), and r is the radius of the porous catalyst (assuming to be spherical). We will focus our attention on the spherical shell of porous material contained within the region r and $(r + \delta r)$.

Now, {rate of input by diffusion at $(r + \delta r)$} = rate of efflux by diffusion at r + lost amount due to chemical reaction, assuming that the stoichiometric coefficient of reactant is $(-1)$

and that there is no change in the number of moles on reaction. The balanced equations in mathematical form can be presented by

$$\left[4\pi r^2 (D_{\mathrm{eff}}(\mathrm{d}C/\mathrm{d}r)\right]_{(\mathrm{r}+\delta\mathrm{r})} = \left[4\pi r^2 D_{\mathrm{eff}}(\mathrm{d}C/\mathrm{d}r)\right]_{\mathrm{r}} + K\,\rho\,S_{\mathrm{g}}\left(4\pi r^2 \delta r\right) C \tag{2.11}$$

where $K$ is the absorption rate constant, $\rho$ is the apparent density of the catalyst particle (mg/cm$^3$), and $S_{\mathrm{g}}$ is surface area per mg of the absorbent (cm$^2$/mg). The differential equation governing simultaneous chemical reaction and diffusion becomes

$$(\mathrm{d}/\mathrm{d}r)\left(r^2 D_{\mathrm{eff}}\mathrm{d}C/\mathrm{d}r\right) - K\,\rho\,S_{\mathrm{g}} r^2 C = 0 \tag{2.12}$$

If the effective diffusivity is assumed to be constant, Eq. (2.12) will transform to

$$\left(\mathrm{d}^2C/\mathrm{d}r^2\right) + \{(2/r)\,(\mathrm{d}C/\mathrm{d}r)\} - \{(K\,\rho\,S_{\mathrm{g}}C)/D_{\mathrm{eff}}\} = 0 \tag{2.13}$$

The above equation can be represented by Thiele modulus $h_{\mathrm{T}}$ (which is equal to {(diffusion time)/(reaction time)} [36–37] as

$$\left(\mathrm{d}^2C/\mathrm{d}r^2\right) + \{(2/r)\,(\mathrm{d}C/\mathrm{d}r)\} = (h_\mathrm{T}/R)^2C \tag{2.14}$$

where $h_\mathrm{T} = R\{(K\ \rho\ S_\mathrm{g}C)/(D_\mathrm{eff})\}^{1/2}$. Now, we can solve this linear differential equation by introducing a new variable $k_1$, where $k_1 = Cr$. Substituting this to Eq. (2.14) gives us

$$\left(\mathrm{d}^2k_1/\mathrm{d}r^2\right) = (h_\mathrm{T}/R)^2k_1 \tag{2.15}$$

The solution of Eq. (2.15) can be expressed in exponential form as

$$k_1 = \mathrm{Cr} = C_1\mathrm{e}^{\left(h_\mathrm{T}\ r/R\right)} + C_2\mathrm{e}^{-\left(h_\mathrm{T}\ r/R\right)} \tag{2.16}$$

On the basis of pseudo-first order equation of Lagergren, kinetics of sorption on solid surfaces for liquid–solid phase sorption has been attempted by many investigators [38–41]. The linearized Lagergren pseudo-first order is generally expressed mathematically as

$$\log\,(q_\mathrm{e} - q_\mathrm{t}) = \log\,(q_\mathrm{e}) - (k_1/2.303)\,t \tag{2.17}$$

where $q_\mathrm{t}$ and $q_\mathrm{e}$ are the sorption capacity at time $t$ and at equilibrium, respectively (mg/g), and $k_1$ is the rate constant. Due to the presence of several variables, like agitation speed, sorbent diameter, solute concentration, sorbent mass, and solute temperature, the data with activated carbon following the pseudo-first order rate mechanisms show accuracy between 73% and 98%. In spite of many improvements in Langmuir, BET, and Lagergren theories, distinct deviation exists between the theory and the experimental observations. According to some critics, this might be due to heterogeneity of the most real solid sorbents [42].

Going back to 3759 BC, active carbon was first widely used as absorbent in the form of carbonized wood (charcoal). Porous solids are rapidly growing group of inorganic and organic compounds, and the most representative porous solids are activated carbon [43–47]. Discovery of molecular carbon structures as fullerenes, heterofullerenes, and carbon nanotubes (CNTs) has boosted the search for new selective adsorption applications of such materials.

### 2.1.4 Carbon Molecular Sieves (CMS)

Astonishing developments took place in the field of carbon-based materials such as carbon molecular sieve (CMS) and microporous and mesoporous carbons: MAXSORB (superhigh surface area microporous carbon, Kasai Coke and Chemicals Company, Japan) type material that has reached the theoretically possible limit of specific surface area, approaching 4000 $\mathrm{m}^2$/g. The *carbon molecular*

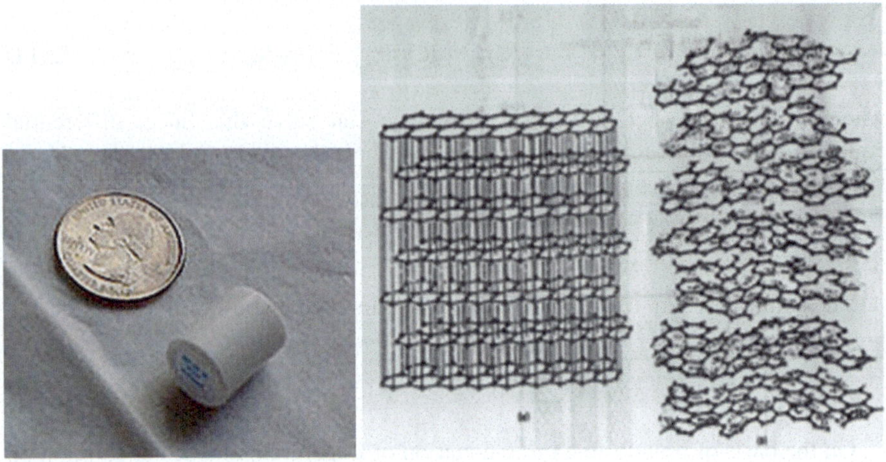

**Fig. 2.13** A typical carbon molecular sieve, with pore size (3 Å–5 Å) (Courtesy: WIKI)

*sieves* (CMS) (Fig. 2.13), which are special kind of activated carbon (AC) can discriminate moles on the basis of their shape and size. The main difference between these two carbonaceous structures may refer to their *adsorption kinetics*. The essential feature of AC is separation of molecules on the basis of differences in the adsorption equilibrium constant, whereas CMS separate molecules through differences in the adsorption rate of molecules [48]. As a matter of fact, CMS possess smaller pore size than AC; as a result, they are suitable for separation and purification of gas mixtures, e.g., nitrogen and oxygen from air; hydrogen and nitrogen from ammonia [49–50].

The production of CMS from lignocellulosic biomass involves three steps: pyrolysis of the starting material, activation of the prepared char, and modification of the pores of the activated carbon. Chemical vapor deposition (CVD) is typically applied to deposit carbon at the pore mouth to narrow down (modify) the size of the pores. Benzene, methyl pentane, and methane have been used as carbon sources to deposit carbon at the pore mouth [51–52].

The preparation for activated carbons (ACs) generally is performed by following two steps: (a) carbonization and (b) carbon activation. There are two methods that are followed to produce activated carbon. The first method is called physical reactivation, where the carbonaceous material is thermally decomposed in the range 600–900 °C in the absence of oxygen (inert atmosphere, like nitrogen or argon), and the noncarbon species are eliminated during the operation. The carbon mass thus obtained consists of closed pores or micropores that are activated in an oxidizing atmosphere, e.g., in the presence of oxygen, carbon dioxide, or steam, to enlarge the diameter of the small pores [53].

In chemical activation process, the carbonaceous material is treated with either phosphoric acid ($H_3PO_4$), sulfuric acid ($H_2SO_4$), nitric acid ($HNO_3$), or with strong

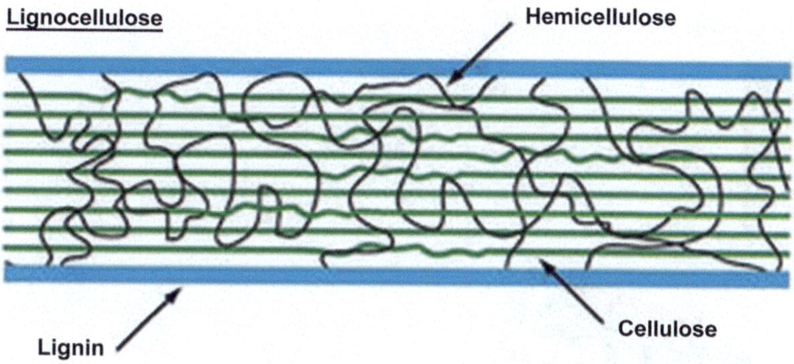


**Fig. 2.14** Structure of lignocellulose (Courtesy: Shahrzad.khayat,Microbe WIKI)

bases like sodium hydroxide (NaOH) or potassium hydroxide (KOH), and then carbonization process is performed at lower temperatures (450–700 °C) [54–55].

The concentration of acid in chemical activation process is approximately 50% (w/w), and the activation temperature varies between 350 and 600 °C. Phosphoric acid with the maximum impregnation ratio has been found to produce the largest $S_{BET}$ (BET stands for Brunauer-Emmett-Teller theory, which aims to explain the physical adsorption of gas molecules on a solid surface and serves as the basis for an important analysis technique for the measurement of surface area, $S_{BET}$, ~1785–1802 $m^2\ g^{-1}$). However, with KOH and NaOH, one can obtain higher specific surface areas (>100 $m^2\ g^{-1}$), but as these two chemicals (KOH and NaOH) are very corrosive, utmost care should be taken during operations [56].

The carbonaceous materials are of lignocellulosic origin. Lignocellulosic materials consist of cellulose, hemicellulose, and lignin (Fig. 2.14), and are the family of wood and wood products (Fig. 2.14). These materials are natural, abundant, and renewable resources essential to the functioning of the industrial societies and critical to the development of a sustainable global economy [57]. All types of cellulose microfibrils are composed of linearly linked ß-(1-→4) D glucopyranose units and differ only by the degree of polymerization. The remaining polysaccharides are known collectively as hemicellulose.

Lignocellulosic biomass (LB) is a potential inexpensive source of fuel which can replace the more expensive biohazard petroleum. All the chemicals derived from it (LB) will have a lower environmental impact than petrochemicals. Additionally, it is considered carbon dioxide-neutral, because burning it with coal in power plant does not add carbon to the environment beyond what is required for the process of growth. The future of extracting chemicals from lignocellulosic biomass appears to be bright, and it is hoped that use of this biopotential material will reduce US dependence on fossil fuels and ultimately develop a renewable resource based economy [58].

Figure 2.15 shows the molecular structure of lignin, cellulose, and hemicellulose. Lignin is a complex polymer of aromatic alcohols, having high molecular weight (HMW) and formulated by enzyme-initiated free radical polymerization of

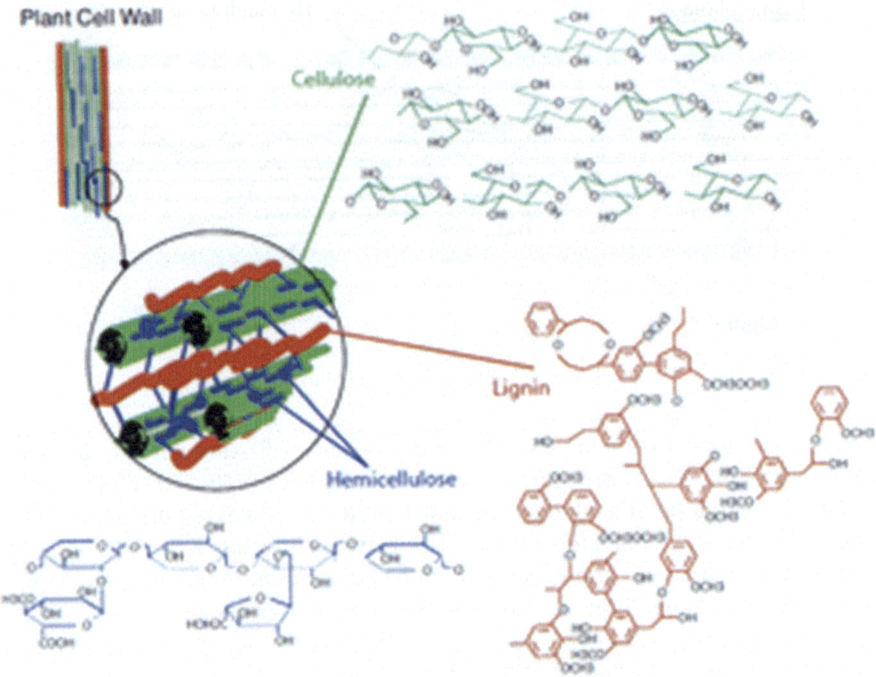


**Fig. 2.15** Molecular structure of lignin, cellulose, and hemicellulose (Courtesy; WIKI)

coniferyl alcohol plus sinapyl alcohols. Lignin acts as a binding agent for the cellulose and hemicellulose fibers through a variety of linkages involving either carbon–carbon bonds of aromatic rings or propyl side chains [59–60].

### *2.1.5 Applications of Activated Carbon*

Activated carbons and their products are very complex and hard to differentiate on the basis of their behavior, surface characterization, and preparation methods. However, some classifications have been made on the basis of their physical characteristics.

The first one chosen here is powdered activated carbon (PAC), having fine granules of diameter less than 1 mm in size, with an average diameter, surface to volume ratio, and a small diffusion distance. These ACs are used in raw water intakes, rapid mix basins, clarifiers, and gravity filters. The next one is called granular activated carbon (GAC). It has larger particles than the powdered AC. Because of larger diameters, they are preferably used for absorption of all gases, deodorization, and in rapid mix basins.

The third one is a powered AC mixed with a binder and shaped into a cylinder to be used for gas phase applications, and they are named as extruded AC or EAC. The diameter of the cylinder varies from 0.8 to 130 mm. The cylinders have high mechanical strength and contain very less dust particles. Similar to EAC, there is another AC family member, named as bead ACs or BAC. They are very similar to EAC in mechanical strength and dust content. For its spherical shape, it is preferred for fluidized bed applications.

Several inorganic chemicals such as iodine (I), silver (Ag), aluminum (Al), manganese (Mn), zinc (Zn), iron (Fe), lithium (Li), and calcium (Ca) are impregnated inside the pores of the ACs and used for specific applications. As for example, silver, known for its antimicrobial and antiseptic properties, is impregnated in ACs and used as an absorbent in the purification of water.

Porous carbon or the activated carbon is coated with polymer. They can be used in hybrid circuits as thick film resistors, and nitrogen-rich porous carbon anode with high performance has been used in sodium ion batteries [61]. The porous carbon coated with polymer has also been used for hemoperfusion. Hemoperfusion is a treatment technique in which large volume of patient's blood is passed over an absorbent substance to remove toxic substances from blood. It has also been very popular in making special cloths and fibers used in personnel protection for the military. Charcoal hemoperfusion has been used to treat liver (hepatic) failure, various types of poisoning, and certain autoimmune diseases when coated with antigens or antibodies.

The degree of exhaustion of activated carbon bed is measured by the amount of milligrams of iodine solution absorbed by a gram of activated carbon. However, the use of iodine number as a measure of the degree of exhaustion of carbon bed can only be recommended if it has shown to be free of chemical interactions with the adsorbates. The iodine number of the activated carbon varies between 500 and 1200 mg/g. Typically, water treatment activated carbons have iodine numbers ranging from 600 to 1100 mg/g.

To evaluate a series of active carbons for their rates of adsorption, a method called decolorization of a standard molasses solution has been taken as a standard method, from which is derived the molasses number. Due to the size of color bodies, molasses number represents the potential pore volume available for larger absorbing species. A high molasses number indicates high adsorption of big molecules (within the range 100–600). Thus, given two active carbons with similar pore volumes for adsorption, the one having the higher molasses number will usually have higher feeder pores, resulting in a more efficient transfer of adsorbate into the adsorption space. It is interesting to note that the European molasses number is inversely related to the North American molasses number.

The iodine numbers and the molasses numbers are the indicators of the pore sizes. According to IUPAC (International Union of Pure and Applied Chemistry) definitions, three groups of pores are distinguished: (a) macropores (>50 nm), (b) mesopores (2–50 nm), and (c) micropores (<2 nm) [62]. Macropores and mesopores can be generally regarded as the highways into the carbon particle, and are crucial for kinetics, whereas micropores usually contribute to the major part of the internal surface area. The pore size distribution in activated carbon is crucial for practical applications, such as adsorption of tannin, methylene blue, and dechlorination.

The desired pore structure, distribution of pores, specific surface area, texture, and specific adsorption are related to the surface functional groups [63–64]. It has been reported that the formation of a large number of basic and acidic groups on the surface of oxidized carbon to sorption and other properties can differ significantly from the unoxidized carbon [65–66].

The adsorption and desorption nature of the activated carbon enables it to work as a compressor and a heat pump in an adsorption refrigeration system. It consists of a solar collector, a condenser, and an evaporator placed in a refrigerator box. An activator soaked with methanol is placed inside the solar collector. During daytime, the solar collector is heated due to irradiation and the methanol is desorbed from the activated carbon. The methanol vapor thus desorbed from the activated carbon is condensed in the evaporator. At night, the collector temperature decreases to ambient temperature, and the activated carbon adsorbs the methanol from the evaporator. The liquid methanol in the evaporator vaporizes and absorbs heat from the water contained in the trays.

Figure 2.16 shows the schematic of desorption process and adsorption working principle. During desorption, the storage is charged, while it is discharged in the adsorption mode. Q in Fig. 2.16 represents energy (Des = desorption, Cond = condensation, Ads = adsorption, and Evap = evaporation). In an open sorption storage system, air transports water vapor and heat out of the packed bed of solid

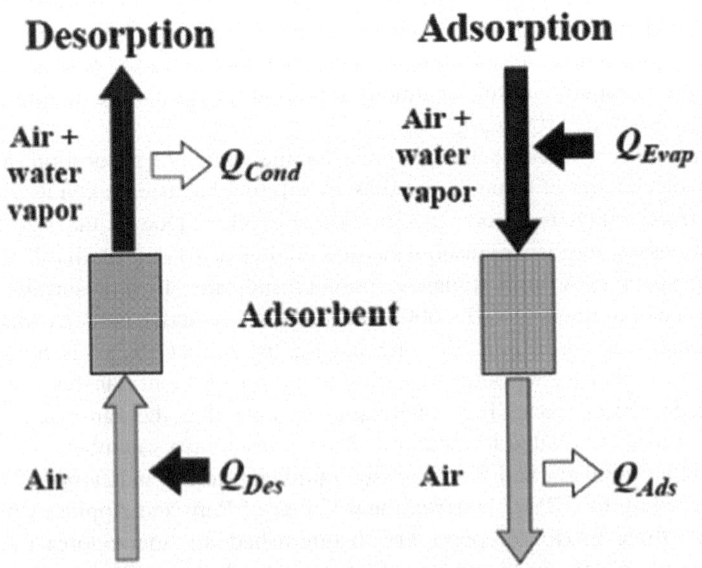


**Fig. 2.16** Schematic of desorption process and the adsorption working principle. Q denotes energy (Adapted from F. Sadar, Theoretical heat pump)

absorbents. In desorption mode, a hot air stream enters the packed bed and desorbs the water from the adsorbent. The coefficient of performance (COP) can be defined as

$$\text{COP} = (\text{latent heat of assortion})/(\text{heat of desorption}) \tag{2.18}$$

Equation (2.18) defines the first system criteria in terms of adsorption materials. To fulfill the requirement, several criteria may exist, e.g., maximum cooling that can be achieved, the amount of heat that can be transferred, and the overall operating conditions of the heat pump. A large surface area is preferable for providing large adsorption capacity. Carbonaceous absorbants are nonpolar, and these absorbents have more affinity for oil and hydrocarbon than water.

Activated carbon, due to its adsorption nature, is used in the spill cleanup, drinking water filtration, air purification, and medical science. While it is useful in acute poisoning, it has been shown not to be effective in long-term accumulation of toxins, such as with the use of toxic herbicides. Ingestion of activated carbon has been a long-debated forum among the scientists. However, in the beginning of nineteenth century, it has been used in England as an antidote to flatulence and stomach trouble [67]. There is some evidence of its effectiveness to prevent diarrhea in cancer patients who have received irinotecan [68–69]. The Department of Energy (DOE) is exploring different possibilities of the use of nanoporous carbon, and the study of nanoporous carbon has been a fascinating field in modern materials science. Figure. 2.17 shows the world's largest reactivation plant located at Feluy, Belgium.

Figure 2.18 shows a typical mobile adsorber on lorry from Chemviron, Belgium, and a typical adsorber unit from Donau carbon, Frankfurt, used in printing works. Both the Chemviron carbon, Feluy, Belgium and Donau carbon, Frankfurt, Germany, are the manufacturers of activated carbon (AC) and related materials.

### 2.1.6 Summary

Adsorption, catalysis, membrane separation, ion exchange, or process scale chromatography has played an important role in reducing the ecological load. There is no doubt that the modern adsorption technologies are rapidly improved and adopted to contemporary tasks of mankind. It is now widely accepted that the modern adsorption science is playing a dominant role in modern industry under environmental, economical, and energy-saving aspects. Adsorption-related domains have a major impact in many areas central to the question of our future, and in this context, it may be esteemed as the technology of the twenty-first century.

Application of solid adsorbents requires elaborate characterization of the material, which comprises determination of chemical composition, crystallographical and geometrical structure, surface and mechanical properties, as well as the shape and size distribution of the pores in porous materials. BET equation has been

**Fig. 2.17** World's largest reactivation plant located at Feluy, Belgium (Photo Courtesy: Franck Honore, Creative Comm.,WIKI)

**Fig. 2.18** A typical mobile adsorber on lorry and a typical adsorber in printing works (Photo Courtesy: Chemviron carbon and Donau carbon, & photo credit: D. Bathen)

accepted by IUPAC as the conventional method for determining the adsorbent surface area [70]. On the other hand, from the adsorption/desorption isotherms of nitrogen at 77 K, the suitable pore distribution can be calculated by means of Kelvin's equation. The specific surface area evaluation requires only a monolayer part of the adsorption isotherm, but the pore size distribution is measured by means of either the multilayer part of isotherm or in terms of mercury porosimetry.

On the basis of transformed desorption curves, it is possible to obtain the total pore volume of porous materials and pore size distribution curve (PSD) [71, 72]. Pore size distributions, mean pore radii, and the total pore volumes derived from thermogravimetric (TG) data are very similar to those calculated from adsorption isotherms measured at static conditions, especially from low-temperature nitrogen adsorption. Besides, methods based on density functional theory (DFT) and molecular simulations allow obtaining an accurate pore size analysis over the complete micropore/mesopore size range with a single method. The fundamental practical applications of adsorption and related domains are: separation and purification of liquid and gas mixtures, drying of gases and liquids before loading them into the industrial system, as well as removal of impurities from the liquid and gas media.

## References


1. F. Derbyshire, M. Jagtoyen, M. Thwaites, Activated Carbons-Production and Applications, in *Porosity in Carbons*, ed. by Patrick, J.W., Edward Arnolds, London, (1955) and also E.P. Leimkuehler, Production, Characterization and Applications of activated carbon, Ph.D. Thesis, University of Missouri, (2010)
2. R.H. Bradley, I. Sutherland, E. Sheng, Carbon surface: area porosity, chemistry and energy. J. Colloid Interface Sci. **179**(2), 561 (1996)
3. K.Y. Foo, B.H. Hameed, A short review of activated carbon assisted electrosorption process: an overview, current stage and future prospects. J. Hazard. Mater. **170**, 352 (2009)
4. F. Cecen, O. Aktas, *Activated Carbon for Water and Waste Water Treatment, Integration Adsoptin and Biological Treatment* (Wiley-VCH, Weinheim, 2011)
5. S. Ismadji et al., Activated carbon from char obtained from vacuum pyrolysis of teak dust: por structure development and characterization. Bioresource Tech. **96**, 1364 (2005.) and also S. Gao et al., Coal based magnetic activated carbon as a high performance adsorbent for methylene blue. J. Porous Mat., **23**(4), 877 (2016)
6. G.D. Living, J. Dewar, On the separation of the least volatile gases of atmospheric air, and their spectra. Proc. R. Soc. London **68**(442–450), 389 (1901)
7. Y.H. Kang, A. Shiue, S.C. Hu, C.-Y. Haung, H.T. Chen, Using phosphoric acid impregnated activated carbon to improve the efficiency of chemical filters for the removal of Air-borne Molecular Contaminants (AMCS) in the Make Up Air Unit (MAU) of a clean room. Building and Environ **45**(4), 929 (2010)
8. W. Liu, R.D. Vidic, T.D. Brown, Optimization of Sulfur impregnated protocol for fixed bed application of activated carbon based Sorbentsfor gas phase mercury removal. Environ. Sci. Technol. **32**, 531 (1998)
9. D.M. Rutheven, *Principles of Adsorption and Adsorption Process* (Wiley, New York, 1994)
10. M. Suzuki (ed.), *Fundamentals of Adsorption* (Elsevier, Amsterdam, 1993)
11. H. Kayser, *Annalen der Physik und Chemie* (cerlag von Jonann Ambrosius Barth, Leipzig, 1835)
12. N.A. Rashidi, S. Yusup, L.H. Loong, Kinetic studies on carbon dioxide capture using activated carbon. Chem. Eng. Trans. **35**, 381 (2013)
13. Overview of Green House Gasses. Annual report of Environmental Protection Agencies, (EPA) U.S. April 14, 2013
14. G. Skodras, I. Diamantopoulou, G. Pantoleontos, G.P. Sakellaropoulos, Kinetic studies of elemental mercury adsorption in activated carbon fixed bed reactor. J. Hazard. Mater **158**, 1 (2008)

15. A. Dabrowski, Adsorption – from theory to practice. Adv. Colloid Interf. Sci. **93**, 135 (2001)
16. K.J. Laider, *The World of Physical Chemistry* (Oxford University Press, Oxford, London, 1993), p. 309
17. S. Brunauer, L.S. Deming, E. Teller, On the theory of van der Waals adsorption of gases. J. Am. Chem. Soc. **62**(7), 1723 (1940)
18. S. Brunauer, *The Adsorption of Gases and Vapors*, vol 1 (Oxford University Press, Oxford, London, 1945), p. 154
19. S. Brunauer, P.H. Emmet, E. Teller, Adsorption of gases in mutuimolecular layers. J. Am. Chem. Soc. **60**(2), 309 (1938)
20. K.S.W. Sing, in *Physical Adsorption : Experiment, Theory, and Applications, NATO ASI Series*, ed. by J. M. Fraissard, C. W. Conner (Kluwer, Dordrecht, 1996)
21. M.M. Dublin, L.W. Radushkevich, The equation of the characteristic curve of the activated charcoal. Proc. Acad. Sci., USSR Phys. Chem. Sect. **55**, 331 (1947)
22. M. Polanyi, Adsorption of gases (vapors) by a solid non-volatile adsorbant. verh. Deutsch. Phys. Ges **16**, 55 (1916)
23. D.M. Young, A.D. Crowell, *Physical Adsorption of Gases* (Butterworth & Co., Ltd., London, 1962.) and also J. Oscik, Adsorption, Elis Horwood, Chichester, PWN, Warsaw (1982)
24. J.H. de Boer, *Advances in Catalysis*, vol 8 (Academic Press, New York, 1956)
25. R.H. Fowler, E.A. Guggenheim, *Statistical Thermodynamics* (Cambridge University Press, Cambridge, London, 1943)
26. A.W. Anderson, *Physical Chemistry of Surfaces* (John Wiley, New York, 1990)
27. D.M. Young, A.D. Crowell, *Physical Adsorption of Gases* (Butterworth, London, 1962)
28. H. Marsh, F. Rodriguez-Reinoso, *Activated Carbon* (Elsevier, Amsterdam, 2006), p. 155
29. R.C. Bansel, M. Goyal, *Activated Carbon* (Taylor and Francis Pub, New York, 2005)
30. S. Biloe, V. Goetz, S. Mauran, Characterization of adsorbent composite blocks for methane storage. Carbon **39**, 1653 (2001)
31. M. Thommes et al., Adsortion hysteresis of nitrogen and argon in pore network and characterization of novel micro-and mesoporous silicates. Langmuir **22**, 756 (2006)
32. M. Thommes, Physical adsorption characterization of nanoporous materials. Chemie Ingenieeur Technik **82**(7) (2010)
33. F. Rouquerol, J. Rouquerol, K. Sing, *Adsorption by Powders and Porous Solids* (Academic Press, San Diego, CA, 1999)
34. H.G. Charles, *Chemical Engineering Kinetics and Reactor Design* (John Wiley, NY, 2009), pp. 447–451
35. D.M. Ruthevan, *Principles of Adsorption and Adsorption Process* (John Wiley, New York, 1994.) Chapt-5
36. E.W. Thiele, Relation between catalytic activity and size of particle. Indust. Eng. Chem. **31**, 916 (1939)
37. C. Hill, *An Introduction to Chemical Engineering and Reactor Design* (John Wiley, New York, 1977), pp. 440–446
38. S. Lagergren, About the Theory of So-Called Adsorption of Soluble Substances, Kungliga Sevenska Vetenskapsakademiens. Handlinger **24**(4), 1 (1898)
39. Y.S. Ho, G. McKay, A comparison of Chemiabsorption kinetic models applied to pollutant removal on various sorbents. Trans. IChemE **76**(B), 332 (1998)
40. Y.C. Sharma, G.S. Gupta, G. Prasad, D.C. Rupainwar, Use of Wollastoneite in the removal of Ni(II) from aqueous solutions. Water Air Soil Pollut. **49**, 69 (1990)
41. N.A. Rashidi, S. Yusup, L.H. Loong, Kinetic studies on carbon dioxide capture using activated carbon. Chem. Eng. Trans. **35**, 361 (2013)
42. A. Dabrowski, Adsorption – from theory to practice. Adv. Colloid Interf. Sci. **93**, 135 (2001)
43. M. Inagaki, *New Carbons: Control of Structure and Functions* (Elsevier, Amsterdam, 2000)
44. T. D. Burchell (ed.), *Carbon Materials for Advanced Technology* (Oxford, London, 1999)
45. J. W. Patrick (ed.), *Porosity in Carbon* (Edward Arnold, London, 1995)

46. P.A. Thrower (eds.), *Chemistry and Physics of Carbon, Marcel Dekker, NY (1989) and zeolites Zeolites: Facts, Figures, and Future: Proceedings of 8th International Conference Amsterdam*, The Netherlands, (July 1989),
47. H. Chon, S. I. Woo, S.-E. Park (eds.), *Recent Advances and New Horizons in Zeolite Science and Technology* (Elsevier, Amsterdam, 1996)
48. E. Vilaplana-Ortego, J. Alica-iz Monge, D. Cazoria-Amors, A. Linares-Solano, Effect of stabilization time of pitch fibres on the molecular sieve properties of carbon fibres. Microporous Mesoporous Mater. **109**, 21 (2008)
49. M.A. Ahmed, W.M.A. Wan Daud, M.K. Aroua, Adsorption Kinetics of Various Gases in Carbon Molecular Sieves (CMS) Produced from Palm Shell. Colloids Surf., A, Physicochem. Eng. Asp **312**, 131 (2008)
50. S.J. Son et al., Development of carbon dioxide adsorbent using carbon materials prepared from coconut shell. Korean J. Chem. Eng. **22**, 291 (2005)
51. E. Davida, A. Talaie, V. Stanciu, A.C. Nicolae, Synthesis of carbon molecular sieves by benzene pyrolysis over microporous carbon materials. J. Matter. Process. Technol. **157**, 290 (2004)
52. T. Zhang, W.P. Walawender, L.T. Fan, Preparation of carbon molecular sieves by carbon deposition from methanol. Bioresour. Technol. **96**, 1929 (2005)
53. M. Kiyonb, Carbon Molecular Sieve Membranes for Natural Gas Separation, Ph.D. Thesis, Georgia Tech, Georgia, Atlanta, (2010)
54. A. Aworn et al., Preparation nad characterization of agricultural waste activated carbon by physical activation having micro and mesopores. J. Anal. and Appl. Pyrolysis **82**(2), 279 (2008)
55. D.C.S. Azevedo et al., Microporous activated carbon prepared from coconut shells using chemical activation with $ZnCl_2$, Microporous and Mesoporous Materials, 100 (1–3), 361 (2007)
56. K.S.W. Sing, Adsorption methods for the characterization of porous materials. Adv. Colld. Interf. Sci. **76**, 3 (1998)
57. T.O. Hu, *Characterization of Lignocellulosic Materials* (John Wiley, NJ, 2008)
58. L.A. Lucia, Lignocellulosic biomass: a potential feedstock to replace petroleum. BioResources **3**(4), 981 (2008)
59. A.Tolbert et al., Characterization and analysis of the molecular weight of lignin for biorefining studies, Biofpr, Soc. Chem Ind., John Wiley, NJ., (2014)
60. N.D. Bonawitz et al., Disruption of mediator rescues the stunted of a lignin deficient Arabidopsismutant. Nature **509**, 376 (2014)
61. T.K. Gupta, Effect of temperature on electrical conduction of carbon –black filled polyimide. IEEE Trans on Comp. Pack. And Manuf **12**(4), 696 (1989.) and also J. Ou, L. Yang and X. Xi, Nitrogen Rich Porous Carbon Anode with High Performance for Sodium Ion Batteries. **10**, 1007 (2016)
62. K. Schaeffer, R. Potwora, *Advantage of Coconut Shell Versus Bituminus Coal Activated Carbon, Water Conditioning and Purification*, June 2008
63. S. Philippe, J.L. Figueiredo, *Carbon Materials for Catalysis* (Wiley, New Jersey, 2009.) and also L. Zhang, Y. Liu, S. Wang, B. Liu and J. Peng, The removal of Sodium Dodecyl Benzene Sulfonate by Activated Carbon Modified with Quaternary Amonium from Aqueous Solution. **10**, 1007 (2016)
64. S. Hu, Y.L. Hsieh, Ultrafine microporus and mesoporous activated carbon fibers from Alkai lignin. J. Matter. Chem. A **1**, 11279 (2013)
65. M. Molina-Sabio, M.T. Gonzalez, F. Rodriguez-Reinoso, A. Sepulveda-Escribeno, Effect of steal and carbon dioxide activation in the macropore distribution of activated carbon. Carbon **34**(4), 505 (1996)
66. R.H. Bradley, I. Sutherland, E. Sheng, Carbon surface: area, porosity, chemistry, and energy. J. Colloid and Int. Sci **179**(2), 561 (1996)

67. L.J. Rolland, C. Sherman, *The Food Encyclopedia: Over 8000 Ingredients, Tools, Techniques and People* (Robert Rose Inc, Toronto, Canada, 2006), p. 148
68. M. Steam, *Warts and All, Straight Talking Advice on Life's Embarrassing Problems* (Murdoch Books, London, 2007)
69. M. Michael et al., Phase II study of activated charcoal to prevent irinotecan –induced diarrhea. J. Clin.Oncol. **22**(21), 4410 (2004)
70. M. Thommes, Physical adsorption characterization of nanoporous materials. Chemie Ingenieur Technik **82**(7), 1059 (2010)
71. A. Debrowski (ed.), *Adsorption and Its Applications in Industry and Environmental Protection*, vol I (Elsevier, Amsterdam, 1999)
72. S.S. Shrotri, C.C. Harris, L. Haung, P. Somasundaran, A graphical technique for calculating adsorption/desorption isotherms for different solid/liquid ratios. Colloids Surf. **A-141**, 189 (1998)

# Chapter 3
# Carbon Composites and Related Metal Matrix

## 3.1 Introduction: History of Carbon Composites

The history of carbon composites dates back to the late 1800s. Thomas Edison used carbon composite filament in light bulbs. A carbon composite is composed of two or more materials to create a superior and unique material. Edison's carbon composite filaments were made out of cellulose-based materials, such as cotton or bamboo, unlike the petroleum-based precursors used today. It was not until late 1950s that high tensile strength carbon fibers were discovered [1–3]. However, the first truly modulus commercial grade carbon fibers were invented in 1964. The benefits of these high-strength carbon-based composites are that they weighed a fraction of the weight of steel, yet contained much greater tensile strength than steel [4].

Another important benefit of carbon composite materials is its high modulus or resistance to stretching. This inelasticity plays an important role in reinforcing rigid structures such as the nose cones in hypersonic aircraft (Fig. 3.1a) [5]. Picture in Fig. 3.1b shows a honeycomb sandwich construction which is considered to be one of the most valued structural engineering innovations developed by the carbon composite industry.

One of the important materials belonging to carbon family, which was developed in the early 1970, was the composite form of carbon, which is also called *reinforced carbon–carbon* (RCC). The US Air Force and NASA did not wait long to capitalize on RCC technology. These composites allowed aircraft to become faster and more efficient. It is hard to imagine how many modern innovations have been possible with this technology. As a result, the study of the *composite carbon* (C–C) and the related products is of particular interest in modern engineering as the material is considered to represent a microcosm of science and technology of carbon. The National Academy of Engineering considered the technology as the greatest engineering achievement of the twentieth century [6]. Figure 3.2 shows



T. Gupta, *Carbon*, DOI 10.1007/978-3-319-66405-7_3

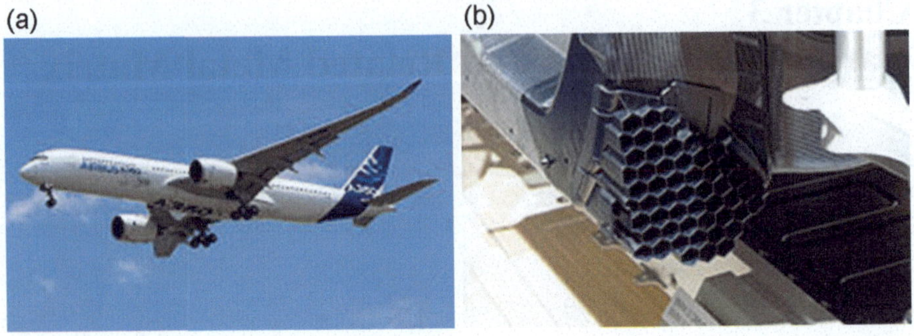


**Fig. 3.1** (**a**) An airbus A-350 and (**b**) honeycomb structure, both made out of composite materials (Courtesy: WIKI)

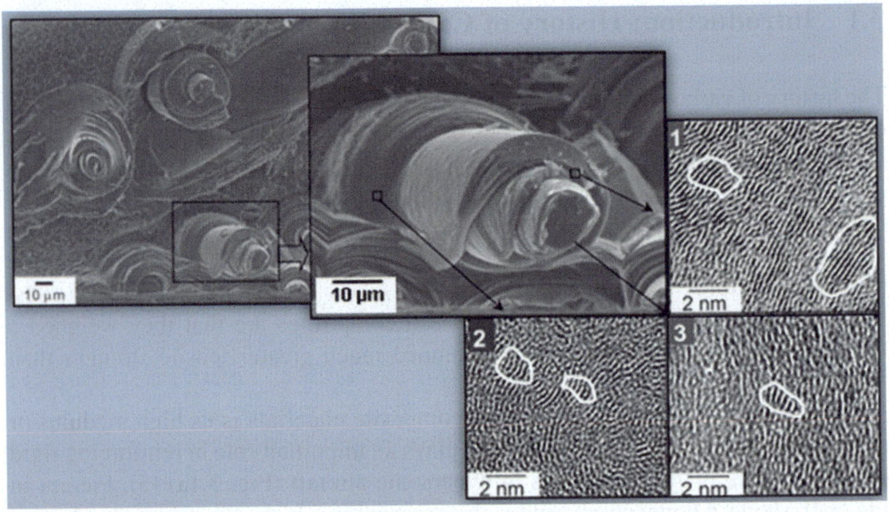


**Fig. 3.2** Microstructure of carbon–carbon composite on various length scales (Reproduced with permission from Sukrov Igor, Mech. Eng., UNH, New Hampshire)

microstructure of carbon–carbon composite on various length scales. The transmission electron microscopy shows coherent domains of the stacks of parallel well-aligned carbon layers.

## 3.2 Fabrication of Carbon Composites

The fabrication of C–C composite and its technology can be described as elements of polycrystalline carbon processing, pyrolytic carbon manufacture, textile processing, chemical and thermal treatments. As a matter of fact, the C–C processing can proceed along two avenues as shown in Fig. 3.3 [7, 8].

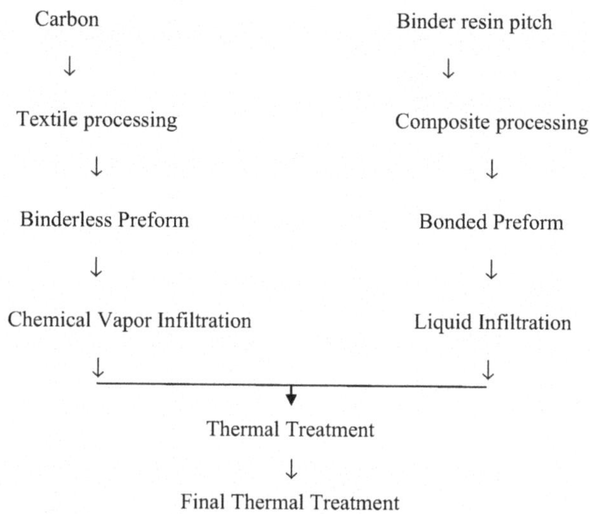


**Fig. 3.3** Typical processing for carbon–carbon (CC) composite

Figure 3.3 shows the typical processing procedure for manufacturing C–C composite. The two-pronged process was developed for the production of carbon fiber reinforced carbon matrix composites in composites [9]. The process described above consists of densification, namely, the repeated carbonization and re-impregnation of the composite with resin and pitches, till the composite material forms the basis for C–C composite fabrication [10].

Conventional bulk polycrystalline graphite has a relatively high value of specific heat capacity with structural stability. It also possesses high values of thermal conductivity and low values of thermal expansion yielding high thermal shock resistance [11]. The principal motive of the development of C–C composite is to combine the excellent bulk properties of graphite together with its high values of specific strength and Young's modulus. These C–C composites may be nongraphitic or graphitic, nongraphitizing or graphitizing, and the high-temperature (from 100 to 3000 °C) treatment to these composites can give rise to different degrees of graphitization.

A matrix formed by the assemblage of two or more nonmiscible material may show distinct physical and mechanical properties. Depending on the type of matrix, the composite carbon can be grouped into four main classes as carbon/polymer/plastic composites, carbon/metal composites, carbon/carbon composites, and carbon/ceramic composites. As a matter of fact, composite carbon is the generic name of carbon fiber based composites.

Carbon fiber reinforced polymers have been developed in the late 1950s for the replacement of glass fiber plastic composites. In fact, aircraft industry was looking for a long time to replace glass fiber composites with a higher stiffness and lower specific weight composite [12, 13]. Figure 3.4 shows the tail of a helicopter made of carbon fiber reinforced polymer.

**Fig. 3.4** The tail of a helicopter made of carbon fiber reinforced polymer (Courtesy: Larsen 25, GNU, WIKI)

## 3.3 Polyacrylonitrile (PAN)

Carbon fibers are generally prepared from different precursors, such as polyacrylonitrile (PAN), mesophase pitch, and rayon. Due to the low productivity of rayon-based carbon fibers, PAN and mesophase pitch have captured a major portion of the market. However, mesophase pitch-based carbon fibers are inferior to PAN-based carbon fibers in tensile strength [14, 15].

PAN is a synthetic, semicrystalline organic polymer resin, with linear chemical formula $(C_3H_3N)_n$. It is a thermoplastic and can be melted at 300°C if the heating rate is controlled [16]. Commercial production is made by free radical polymerization of acrylonitrile (AN) [17]. In the production of carbon fibers containing 600 tex (6 k) PAN tow, the specific requirements of the filaments are diameter of 11.6 μm and linear density of 0.12 tex [18]. In the year 2013, researchers adopted a new technique called electrospinning and fabricated thin nanofibers from PAN [19].

The processing of PAN-based carbon fibers can be divided into five steps: (a) spinning of the copolymer to form fiber, (b) stretching, (c) oxidation and stabilization at temperatures between 220 and 280 °C under tension, (d) carbonization in an inert atmosphere at 1600°C, and finally (d) graphitization at 3000 °C.

The spinning of the copolymer to form fiber can be wet or dry. The wet spun fiber is only used as a precursor. Unfortunately, during the dry spinning process, the

**Fig. 3.5** Picture of an instrument that performs carbonization (Photo courtesy: A. Oagle, Clemson Univ., SC)

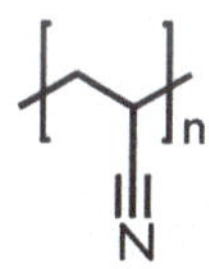

**Fig. 3.6** Structure of polyacrylonitrile (PAN) (Courtesy: Di Carmine, Creative Comm., WIKI)

spun material collapses. So, the wet spinning process is the only alternative to PAN-based carbon fiber manufacturer. The wet spun PAN is used as a precursor, which contains a copolymer that apparently catalyzes during the cyclization in air and helps the carbonization process [20]. A picture of a typical machine that performs carbonization is shown in Fig. 3.5.

The spinning process of PAN is followed by stretching, which results in an elongation of 500–1300%. This is an important part of the process because it results in a high-strength fiber. Figure 3.6 shows the structure of polyacrylonitrile (PAN).

The stabilization process of PAN consists of heating the stretched fiber to 200–280°C in an oxygen atmosphere under tension, leading to cyclization, dehydrogenation, and oxidation. During this period, PAN molecules are cyclized to transform into a nonmeltable ladder structure, which is critical to avoid melting during carbonization [21]. To increase the stabilization rate, oxygen is premixed with ammonia [22]. Stabilization is considered to be one of the most important

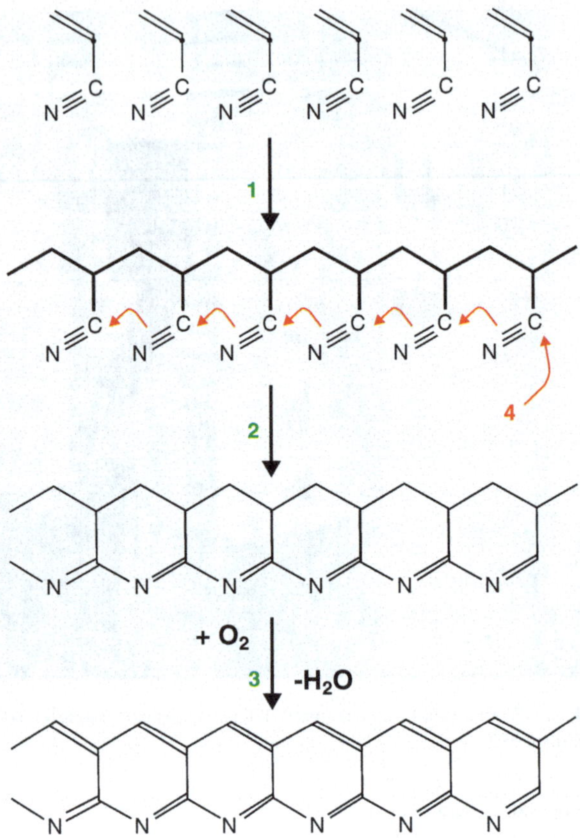


**Fig. 3.7** Synthesis of carbon from polyacrylonitrile (PAN) (Courtesy: Timor Lenk, Creative Comm., WIKI)

factors that determine the mechanical strength of the PAN-based carbon fibers [23]. The molecular changes occurring during stabilization is shown in Fig. 3.7. Experimental studies of FTIR show that intramolecular and intermolecular reactions occurred simultaneously [24].

Figure 3.7 shows synthesis of carbon composite from polyacrylonitrile (PAN). The number (1) in the figure represents polymerization of acrylonitrile to PAN; the second one is the cyclization during low-temperature process (after stabilization); and the third one is the high-temperature oxidative carbonization (hydrogen is removed). After this, the process of graphitization starts, where nitrogen is removed and the chains are joined into graphite planes. Graphitization is done at 1000–1500 °C, and the temperature is raised slowly at the rate of 15–20 degrees/minute. During this process, several volatile by-products are released. The circular morphology of the fiber is maintained, and the final diameter of the circular fiber varies between 5 and 10 micrometers. Experimental evidence from Raman spectroscopy reveals the existence of $sp^3$ hybrid bonding with a bond energy of 370 kJ/mol and a bond length of 0.15 micrometer in PAN-based fibers [25].

**Fig. 3.8** Picture of a graphitization furnace (Photo courtesy: A. ogale, Clemson Univ. NC)

Figure 3.8 shows the picture of a graphitization furnace. The furnace can go up to a temperature of 2700 °C. Sometimes, high-temperature heat treatment (>2000 °C) is necessary to induce microcracks in the matrix of the crystal structure (graphitization) [26]. These cracks are filled by chemical vapor infiltration (CVI) process.

A sketch of a conventional chemical vapor infiltration (CVI) system is shown in Fig. 3.9. The gases and the matrix material enter the reactor from the feed system at the bottom of the reactor. The fibrous preform undergoes a chemical reaction at high temperature with the matrix material, and thus the latter infiltrates in the fiber [27, 28].

The CVI growth mechanism is shown in Fig. 3.10. As the fiber surface and the matrix material react, a coating of the matrix is formed on the surface of the fiber. The unreacted reactants along with the gases exit the reactor via the outlet system and are transferred to an effluent treatment plant.

## 3.4 Composite Carbon Fibers

The composite carbon fibers are also being manufactured from pitch which is a by-product of the destructive distillation of coal, crude oil, or asphalt. The composition of pitch includes aliphatic compounds, naphthalene aromatic, the polar aromatic, and asphaltenes. The C–C fibers produced from these isotropic pitch are having low strength and modulus (~ 40–70 GPa) and suitable for filler and

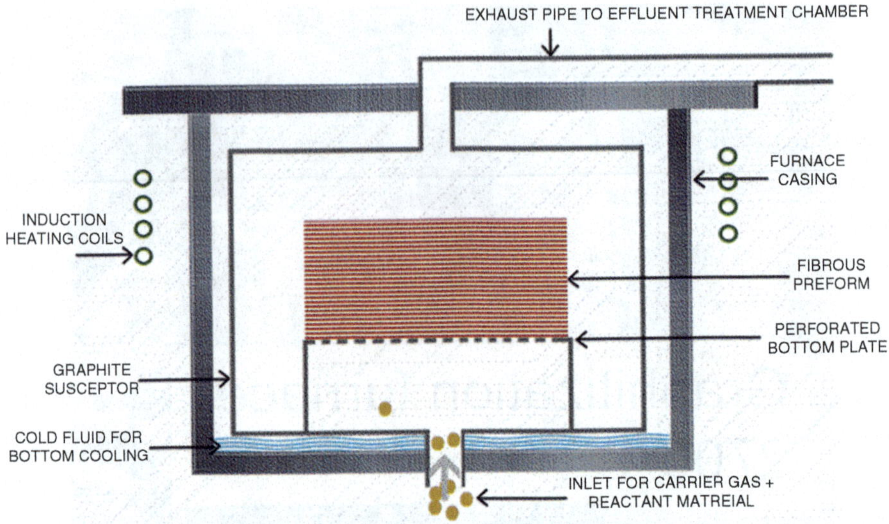


**Fig. 3.9** A typical chemical vapor infiltration process (Courtesy: V4711, Creative Comm., WIKI)

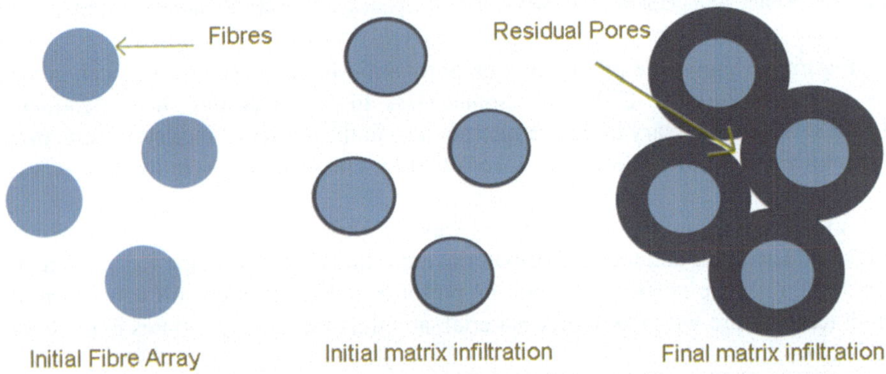


**Fig. 3.10** Chemical vapor infiltration growth mechanism (Courtesy, Sjlandge, Creative Comm., WIKI)

insulation applications. However, when the pitch contains high amounts of asphalt, it is called mesophase pitch [29, 30]. It has high modulus and medium strength. The fibers made out of the mesophase pitch are more costly than the PAN-based fiber.

The processing of mesophase pitch can be summarized as follows: (1) Initial polymerization of the isotropic pitch to produce mesophase pitch. (2) Second, spinning of the mesophase pitch is performed to obtain green fiber. (3) Next step is thermosetting of the green fiber. (4) The final step is the carbonization and graphitization of the thermoset fibers to obtain a high-modulus carbon fiber. After heat treatment, the large crystallite mesophase fibers resemble the graphite crystal well-aligned along the fiber axis. The heat-treated mesophase fibers provide low

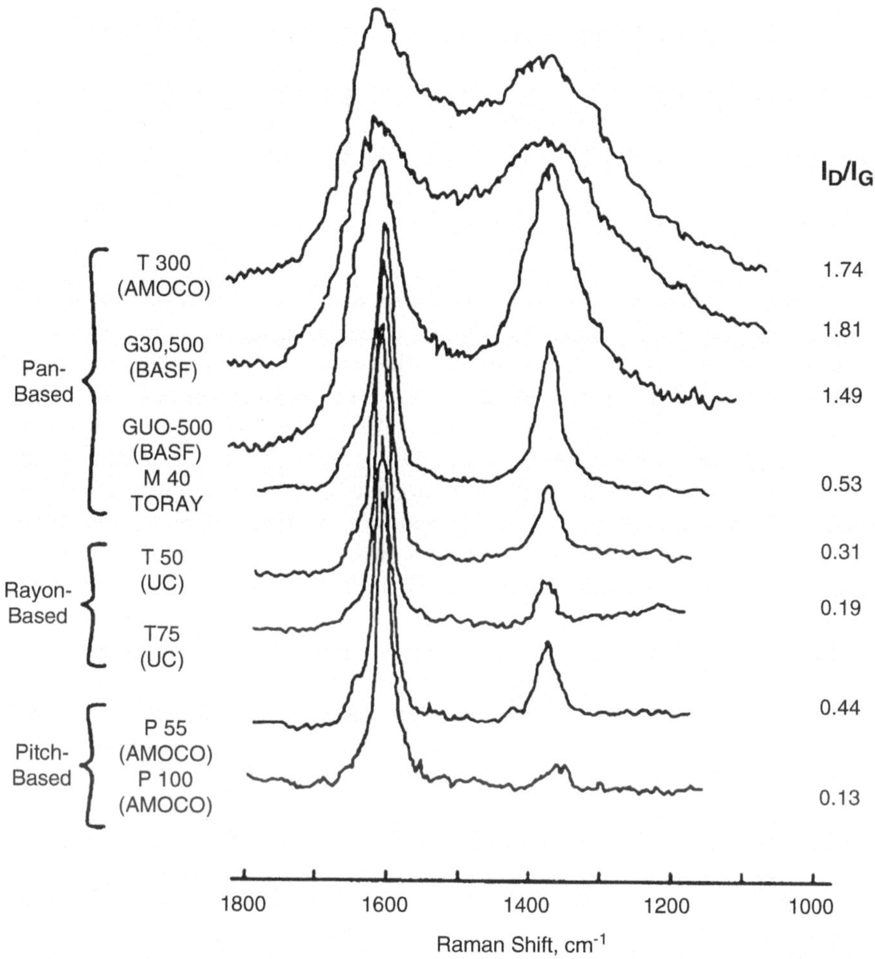


**Fig. 3.11** Raman laser spectroscopy for PAN-based, rayon-based, and pitch-based carbon fibers (Ref. [32], Reproduced with permission from Elsevier)

scattering sites for phonons and ultimately possess high thermal conductivity. In addition to high thermal conductivity, the high degree of crystallinity also results in low shear and compressive strengths. Unfortunately, these fibers also have pits, scratches, striations, and flutes that are detrimental to tensile properties, ultimately affecting the modulus and thermal conductivity of the fibers. Figure 3.11 represents Raman laser spectra for PAN-based, rayon-based, and pitch-based carbon fibers [31, 32].

The cellulose-based carbon fibers present a network structure with high disorder, which gives the fibers a very low thermal conduction coefficient and the ability to be used in the manufacture of rayon, a synthetic carbon fiber. Figure 3.12 shows the repetitive units of cellulose in rayon [33].

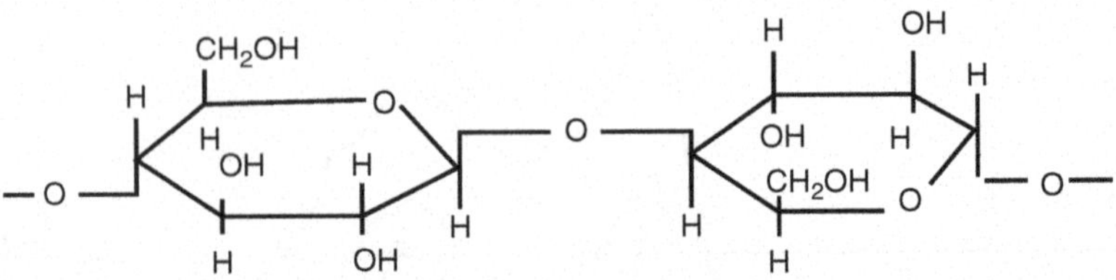


**Fig. 3.12** Repetitive units of cellulose in rayon

The reinforced carbon fiber is well suited to structural applications at high temperatures, or where thermal shock and/or a low coefficient of thermal expansion are needed. The heat treatment for high tensile strength (HT) fiber does not exceed 1200 °C, and the tensile strength of these fibers is in the range between 3 and 7 GPa. The elastic modulus of these fibers is between 200 and 320 GPa. However, the use of carbon-reinforced polymers is limited at a very high temperature due to the presence of polymers. The replacement of organic resins by a carbon material leads to the development of carbon–carbon composites, and the composites have shown exceptional refractory properties in an inert atmosphere [34]. Unfortunately, it lacks impact resistance. A vivid example of this is the Space Shuttle Columbia's disaster during its atmospheric re-entry. A suitcase-sized biopod foam broke off from the external tank, striking Columbia's left wing with reinforced carbon–carbon (RCC) panels, leaving a 6–10 inches diameter hole (Fig. 3.13).

An ample selection of carbon fiber architecture is now available, and it is categorized into four different groups: (a) discrete, (b) linear, (c) laminar, and (d) integrated. The characteristic of each category is shown in Table 3.1 [35, 36].

Discrete fibers are short-length fibers (a few centimeters) and are randomly oriented. They are in the form of a mat having low strength. The continuous carbon fibers are approximately 10 μm in diameter. The laminar 2D pattern is a weaved biaxial structure woven 00 and 900 (wrap and fill) in three basic patterns, plain, satin, and twill. The highest frequency of yarn interlacing is found in plain weave. The integrated 3D weave consists of an orthogonal nonwoven fabric produced by placing fibers in three or more orthogonal directions.

## 3.5 Metal Matrix Composites

Another composite of C–C in metal matrix (typically aluminum or copper) is very attractive because of their high thermal and electrical conductivities compared to carbon fiber/polymer matrix composites. Moreover, the coefficient of thermal expansion of the metal matrix can be tailored by varying the composition of the metal. Undoubtedly, the most efficient composites of heat sinks are those based on copper or aluminum matrices with a different number of reinforcements. Copper–carbon and aluminum–carbon composites are in high demand for effective heat-dissipating materials. These composites possess high thermal conductivity and

**Fig. 3.13** Brittle fracture and the removed portion of the wing of Space Shuttle *Columbia* (Photo courtesy: Rick Stiles, NASA, WIKI)

**Table 3.1** Carbon fiber architecture

| Type of reinforcement | Textile construction | Fiber length | Fiber weave | Fiber orientation |
|---|---|---|---|---|
| Discrete | Chopped fiber | Short | None | Random |
| Linear | Filament<br>Yarn | Continuous | None | Linear |
| Laminar | Simple fabric | Continuous | 2D | Planar |
| Integrated | Advanced fabric | Continuous | 3D | 3D |

thermal expansion coefficient comparable with chip materials ensuring the reliability of power module [37–39].

Figure 3.14 shows a power device attached to a heat sink made out of a composite of aluminum (Al)–copper (Cu)–carbon fiber (CF). Al–CF composite is fabricated by conventional powder metallurgy method, where 50 vol % of CF powder is hot-pressed with 5 vol % of Al–Si (silicon) alloy between the melting points of Al and Al–Si alloy. With 50 vol % of CF, thermal conductivity and CTE of Al–CF composite are 258 W/m.K and 7.0 × 10–6/K to in-plane direction of carbon fiber [40–41]. Very recently, Cu–Al–CF, Cu–diamond, and Al–diamond composite films by powder metallurgy and hot-pressing process are found to be very promising materials as next generation thermal management materials in power electronic packages.

Unfortunately, there are some drawbacks of the metal matrices. As for example, aluminum (Al) and other metals in the molten stage can form intermetallics, and the diffusion of the metal degrades the strength of the fiber and considerably reduces its

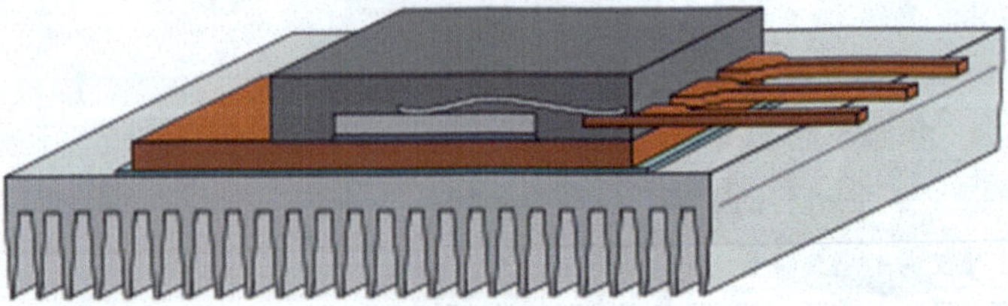

**Fig. 3.14** A power device attached to a heat sink made out of a composite of aluminum–copper–carbon (Courtesy: WIKI)

ability to reinforce the composite. The interface chemistry of the metal matrix can be complicated due to the formation of preferential grain boundary, and the formation of microcracks can further complicate the matrix. With Al–matrix, the galvanic couple can cause galvanic corrosion. In the year 1980, the maintenance staff of the Statue of Liberty noticed the corrosion that had taken place between the copper skin and the wrought iron support structure (Fig. 3.15) [42, 43].

The diffusion of the metals during the formation of intermetallics can partially be minimized by applying a barrier layer of titanium, boron, nickel, or niobium carbide prior to processing the composite. The niobium carbide has shown excellent barrier properties. Fabrication of metal matrix includes diffusion coating, electroplating (copper) followed by hot-pressing, and liquid metal infiltration. Finally, additional copper is deposited by plasma spraying, and the final composite is hot-pressed [44, 45].

Oxide of aluminum ($Al_2O_3$) is not suitable as a matrix material in carbon–fiber composite, but the carbides of silicon (SiC) and boron ($B_4C$) form good matrix materials. Though the carbon fiber–SiC composite shows sufficiently higher tensile strength (Carbon fiber–SiC matrix tensile strength is 515 MPa (ksi), whereas SiC fiber–SiC matrix tensile strength is 310 MPa (ksi)), they degrade at high temperatures as they show fracture modes. Figure 3.17 shows some of the fracture modes of the carbon fiber–SiC composites. In Japan, these carbon fiber–ceramic matrices are being used in building walls, floor panels, footbridges, and other architectural components [46].

Figure 3.16 shows a frame of a car wheel. The metal parts are made out of silicon carbide and carbon composites. The carbon fiber (CF) reinforced silicon carbide (SiC) composites have been found lately as the potential candidates for some of the parts of the automobiles (brakes, wheels), high efficiency car engine, chemical reactors, aircrafts, and re-entry space vehicles due to their high specific strength with suitable fracture toughness at elevated temperatures [47–54]. Figure 3.17 shows some fractures that have developed in a carbon fiber–SiC matrix.

Though the properties of carbon composites are strongly controlled by the nature of the fiber–matrix interfaces, matrix texture, and porosity, it can be argued that the fiber–matrix interface is the most important factor regarding tensile and shear

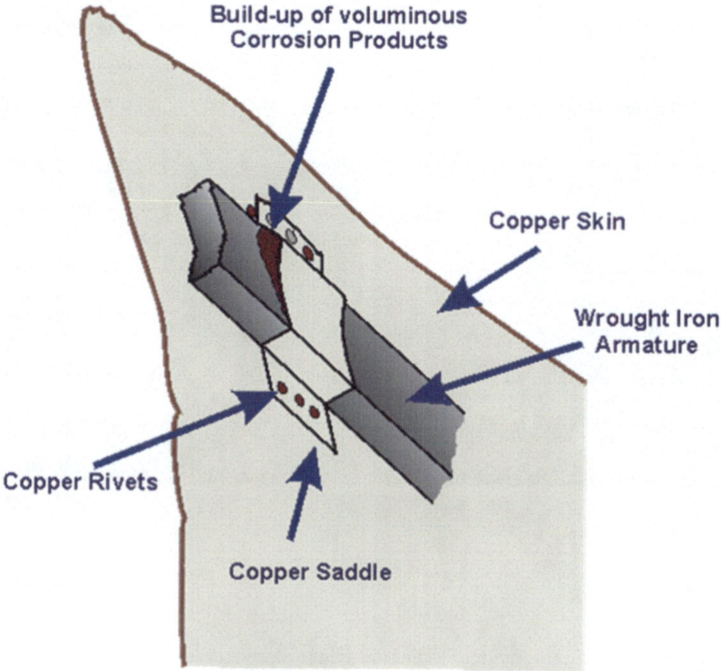


**Fig. 3.15** Galvanic corrosion (the Statue of Liberty) between the wrought iron support structure and the copper skin (Photo courtesy: Peterlewis, WIKI)

**Fig. 3.16** Frame of a car wheel. The metal parts are made out of silicon carbide and carbon composites (Courtesy: Nrblex, GNU, WIKI)

properties. A number of techniques have been developed to characterize interfacial behavior in ceramics, polymer, and metal–matrix composites through the measurements of the mechanical properties of the fiber matrix.

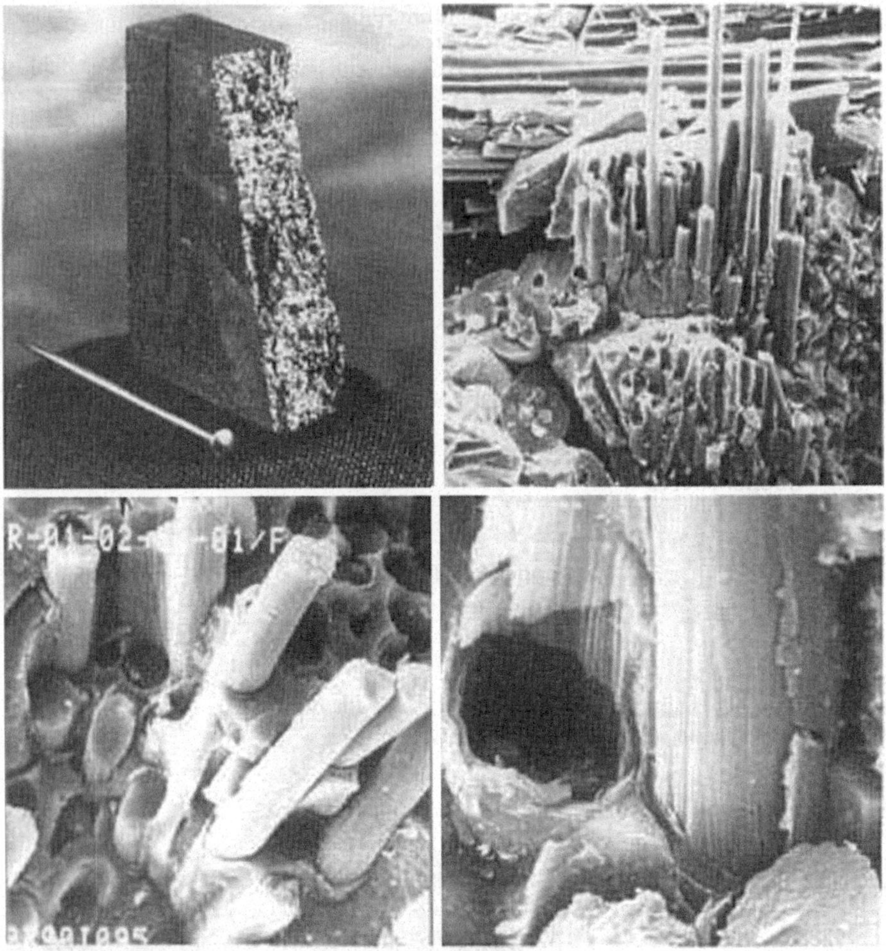

**Fig. 3.17** Some fractures that have developed in a carbon fiber–SiC matrix (Courtesy: Rockwell Int., Canoga Park, CA)

Carbon fibers are being used as electromagnetic interference (EMI) shields. The woven carbon fibers are used as high-temperature insulation materials, and they are also used in batteries.

## 3.6 Summary

A carbon composite (C–C) is like other composite materials composed of two or more materials to create a superior and unique material. The important benefit of carbon composite materials is its high modulus or resistance to stretching which plays an important role in reinforcing rigid structures such as the nose cones in hypersonic aircraft. The carbon composite (C–C) materials belong to carbon family and are called *reinforced carbon–carbo*n (RCC).

The C–C composite is the generic name of carbon fiber, and the aircraft industry was looking for a long time to replace glass fiber composites with higher stiffness and lower specific weight composites. Carbon fibers are generally prepared from different precursors, polyacrylonitrile (PAN), and the processing of PAN-based carbon fibers is performed by spinning, stretching, oxidation, and stabilization, followed by carbonization in an inert atmosphere and graphitization at 3000°C.

Another composite of C–C in metal matrix (typically aluminum or copper) is very attractive because of their high thermal and electrical conductivities compared to carbon fiber/polymer matrix composites. These materials are used in packaging IC chips for power modules because of their reliability and stability at high temperatures.

Unfortunately, there are some drawbacks of the metal matrices such as intermetallics formation and the diffusion of the metal. In addition, the interface chemistry of the metal matrix can be complicated due to the formation of preferential grain boundaries and the formation of microcracks. Moreover, with Al–matrix, the galvanic couple can cause galvanic corrosion.

## References

1. High performance carbon fibers, National Historic Chemical Landmarks, Am. Chem. Soc., (2014)
2. A.K. Kaw, *Mechanics of Composite Materials*, 2nd edn. (CRC Press, Boka Raton, 2005)
3. J.M. Corum, R.L. Battiste, K.C. Liu, M.B. Ruggles, Basic properties of reference crossply carbon fiber composite, Oak Ridge Nat. Lab. ORNL/TM-2000/29, DE-ACo5-96OR-22464
4. N. Baral, H. Guezenoc, P. Davies, C. Baley, High modulus carbon fibre composites: correlation between transverse tensile and module I interlaminar fracture properties. Mater. Lett. **62** (6–7), 1096 (2008)
5. Z.Y. Liu, B.L. Xiao, W.G. Wang, Z.Y. Ma, Tensile strength and electrical conductivity of carbon nanotube reinforced aluminum matrix composites fabricated by powder metallurgy combined with friction stir processing. J. Mater. Sci. Technol. **30**(7), 649 (2014)
6. H. Hatta, T. Aoi, I. Kawaha, Y. Kogo, I. Shiota, Tensile strength of carbon-carbon composites: 1-effect of C-C density. J. Compos. Mater. **38**(19), 1667 (2004)
7. H.M. Stoller, E.R. Frye, 73rd Syp. On Am Chem. Soc., Adv. Mater. Composite and Carbon, Annual Meeting, Am. Ceram Soc. (1973), p. 163
8. L.M. Manocha, High performance carbon carbon composite. Sadhana **28**, 349 (2003)

9. The Int. Handbook Committee, *Vol. 1, of Engineered Materials Handbook*. (ASM Int. Pub, Ohio, 1971) and also P. Morgan, *Carbon Fibers and Their Composites* (Taylor and Francis, FL, 2005)
10. H.M. Stoller, E.R. Eyre, Symposium of the 73rd Annual Meeting of the American Ceramic Society, Advanced Materials Composite and Carbon, Am. Ceram. Soc., (1973), p. 163
11. D.F. Pedraza, P.G. Klemens, Effective conductivity of polycrystalline graphite. Carbon **31**(6), 951 (1993)
12. R. Jain, L. Lee, *Fiber Reinforced Polymer (FRP) Composites for Infrastructure Applications. Focussing on Inovation, Technology Implementation and Sustainability* (Springer, New York, 2012)
13. M.J. Hinton, P.D. Soden, A.S. Kaddour, The world wide failure exercise, in *Failure Criteria in Fiber-Reinforced-Polymer Composite*, (Elsevier, Amsterdam, 2004)
14. J.D. Buckley, D.D. Edie, *Carbon-Carbon Materials and Composites* (Noyes, Park Ridge, 1993)
15. M.K. Seo, S.H. Park, S.H. Kang, S.J. Park, Carbon fibers (III), recent technical and patent trends. Carbon Lett. **10**(1), 43 (2009)
16. A.K. Gupta, D.K. Paliwal, P. Bajaj, Melting behavior of acrylonitrile. J. Appl. Polym. Sci. **70**, 2703 (1998)
17. P. Bajaj, T. Sreekumar, K. Sen, Effect of reaction medium on radical polymerization of acrylonitrile with venyl acids. J. Appl. Polym. Sci. **79**, 1640 (2001)
18. A. Serkov, M. Radishevskii, Status and prospects for production of carbon fibers based on Polyacrylonitrile. Fiber Chem. **40**, 24 (2008)
19. D. Papkov et al., Simultaneously strong and tough ultra fine continuous nanofibers. ACS Nano **7**(4), 3324 (2013)
20. E. Fitzer, PAN based carbon fibers –present state and trend of the technology from the view point of possibilities and limits to influence and to control the fiber properties by the process parameters. Carbon **27**(5), 621 (1989)
21. E. Fitzer, W. frobs, M. Heine, Optimization of stabilization and carbonization treatment of PAN fibers and structural characterization of the resulting carbon fibers. Carbon **24**(4), 387 (1986)
22. G.S. Bhat et al., New aspects in stabilization of acrylic fibers for carbon fibers. Carbon **28**(2–3), 377 (1990)
23. S. Lee, J. Kim, B.C. Ku, J. Kim, Y. Chung, Effect of process conditionon tensile properties of carbon fiber. Carbon Lett. **12**(1), 26 (2011)
24. S. Lee, J. Kim, B.C. Ku, J. Kim, Y. Chung, Structural evolution of polyacrylonitrile fibers in stabilization and carbonization. Adv. Chem. Eng. Sci. **2**, 275 (2012)
25. H.O. Pierson, *Hand Book of Carbon, Graphite, Diamond and Fullerenes* (Noyace, NJ, 1984)
26. R.J. Diefendrof, Continuous Carbon Fiber Reinforced Carbon Matrix Composites in Composites, Ed. ASM Int. Handbook Committee, *Vol.1 of Engineered Materials Handbook* (ASM International, OH, 1987), p. 911
27. M. Balasubramanium, *Composite Materials and Processing* (CRC Press, Boka Raton, 2013)
28. K.H. Bang, G.-Y. Chung, H.H. Koo, Preparation of C/C composites by chemical vapor infiltration (CVI) of propane pyrolysis. Korean J. Chem. Eng. **28**(1), 272 (2011)
29. V. Liedike, K.J. Hunttinger, Pitches as matrix precursor of carbon fiber reinforced carbon: II Stabilization of mesophase pitch matrix by oxygen treatment. Carbon **35**(9), 106 (1996)
30. M. Hoffstetter, E. Wintermantel, Meso fiber –a novel approach to reinforce micro parts. Biomaterials **41**(12), 1011 (2010)
31. X. Haung, Fabrication properties of carbon fibers. Materials **2**, 2369 (2009)
32. T. Roberts, The Carbon Fiber Industries: Global Strategic Market Evaluations, 2006–2010, Mater. Technol. Pub. Watford, UK., pp. 93–177, (2006) and E. Fitzer, Pan-Based Carbon Fibers –Present State and Trend of the Technology from the View Point of Possibilities and Limits to Influence and to Control the Fiber Properties by the Process Parameters, Carbon **27** (5), 621 (1989)

33. G.B. Kauffman, Rayon, first synthetic fiber product. J. Chem. Educ. **70**(11), 887 (1993)
34. J.D.E. White, A.H. Simpson, A.S. Shteinberg, A.S. Mukasyan, Cobustion joining of refractory materials: carbon-carbon composite. J. Mater. Res. **23**(1), 160 (2007)
35. F.K. Fo, Preform fiber architecture for ceramic matrix composite. Ceram. Bull. **68**(2), 401 (1989)
36. R. Warren, *Ceramic Matrix Composites* (Blackie and Son, London, 1992)
37. D.D.L. Chung, *Carbon-Carbon Composites* (Butterworth Heinemann, Newton, 1994), p. 131
38. J.F. Silvain, A. Veillere, Y. Lu, Copper-carbon and aluminum-carbon fabricated by powder metallurgy processes, J. Physics, Eurotherm Seminar, 102, Conf. Sr. No. 525 (2014)
39. K. Kuniya, H. Arakawa, T. Kanai, T. Yasuda, Development of copper-carbon fiber composite for electrodes of power semicinductor devices. IEEE Trans. Comp. Hybrid. Manuf. Technol. **6** (4), 467 (1983)
40. J.F. Silvain, A. Veillere, Y. Lu, Copper-carbon and aluminum-carbon fabricated by powder metallurgy processes, J. Physics, Eurotherm Seminar, 102, Conf. Sr. No. 525 (2014)
41. P.M. Geffroy, J.F. Silvan, Structural and thermal properties of hot-pressed Cu/C matrix composites materials used for the thermal management of high power electronic devices. Mater. Sci. Forum **534-536**, 1505–1508 (2007)
42. A. Burkert, *Materials and Corrosion* (Wiley, New York, 2001)
43. H.P. Hack, *Galvanic Corrosion* (ASTM Intl, Philadelphia, 1988)
44. K. Upadhya, Ceramics and composites for rocket engines and space structures. J. Metals **44**, 15–18 (1992)
45. M. Singh, T.P. Shpargel, G.N. Morscher, R. Asthana, Active metal brazing and characterization of brazed joints in titanium to carbon-carbon composites. Mater. Sci. Eng. **412**(1–2), 123 (2005)
46. R. Naslain, Materials design and Procrssing of high temperature ceramic matrix composites: state of the art and future trends. Adv. Compos. Mater. **8**(1), 3 (1999)
47. Q. Zhang, G. Li, A review of the application of C/SiC composites in thermal protecting systems. Multidiscip. Model. Mater. Struct. **5**(2), 199 (2009)
48. R.B. Deo, J.H. Starnes, R.C. Holzwarth, Low cost composite materials and structures for aircraft applications, Scitific and Technical Aerospace Rept. (STAR), 41 (22): 1-1-1-11, NASA, USA (2003)
49. R. Naslain, Preparation and properties of non-oxide CMCs for application in engines and nuclear reactors, an overview. Compos. Sci. Technol. **64**, 155 (2004)
50. Z. Rak, CF/SiC/C composites for tribological application, in *High Temperature Ceramic Matrix Composites*, ed. by W. Krenkel, R. Naslain, H. Schneider (Wiley, Weinheim, 2001), p. 820
51. W. Krenkel, Carbon fibre reinforced silicon carbide composites, in *Handbook of Ceramic Composites,* ed. by N. P. Bansal (Springer, Heidelberg, 2005), p. 117
52. C.P. Ju, C.K. Wang, H.Y. Cheng, J.H.C. Lin, Process and wear behavior of monolithic SiC and short carbon fiber-SiC matrix composite. J. Mater. Sci. **35**, 4477 (2000)
53. W. Kucler, *Chemische Technik: Prozesse*, vol 8, ed. by R. Dittmeyer et al. (Wiley-VCH, Weinham, 2005), p. 11166
54. C.A. Nannetti, A. Ortona, D.A. de Pinto, B. Riccardi, Manufacturing SiC-fibre reinforced SiC matrix composites by improved CVI/slurry infiltration/polymer impregnation and pyrolysis. J. Am. Ceram. Soc. **87**, 1205 (2004)

# Chapter 4
# Polymer Families and Their Extended Activities

## 4.1 Introduction

The term polymer (many monomers) is derived from the ancient Greek word πολύζ (polus, meaning many, much) and μέροζ (*meros*, meaning *parts*), and refers to a molecule whose structure is composed of multiple repeating units of *carbon*, mostly hydrogen and/or nitrogen and oxygen (Figs. 4.1, 4.2 and 4.8). Indeed, the term *polymer* was first introduced in 1833 by the Swedish chemist, Jons Jakob Berzelius. He also introduced the term *isomer* (from the Greek *isos* meaning *equal*, and *meros* meaning *part*) to describe substances having identical compositions but differing properties. In 1922, the German chemist, Herman Staudinger, felt it necessary to coin the word *macromolecule* to describe large covalently bonded organic chain molecule containing more than $10^3$ atoms. A macromolecule is a very large molecule commonly created by polymerization of smaller subunits, monomers. The most common macromolecule in biochemistry is biopolymers (nucleic acids, proteins, etc.). According to IUPAC (International Union of Pure and Applied Chemistry) definition, the term macromolecule as used in polymer science refers only to a single molecule.

Polymer is still used in its original sense in the literature dealing with glasses, ceramics, and inorganic polymers, many of which are noncovalently bonded and may contain not only infinitely polymerized chains but also infinitely polymerized layer and framework structures. As a matter of fact, *carbon* (C) is the backbone of a polymer material [1–4].

Figure 4.1a shows the repeating units of a polymer and b shows a large carbon-based molecule made of repeating units representing a polymer. Figure 4.2 shows the repeating units of the polymer polypropylene. The important structural feature of a polymer is its architecture and shape, which relates to the way branch points lead to a deviation from a simple linear chain [5].

As a matter of fact, the discovery of nitrocellulose from the joint work of Henri Braconnot (in the year 1777), a French chemist and pharmacist, and Christian Schonbein, a German-



T. Gupta, *Carbon*, DOI 10.1007/978-3-319-66405-7_4

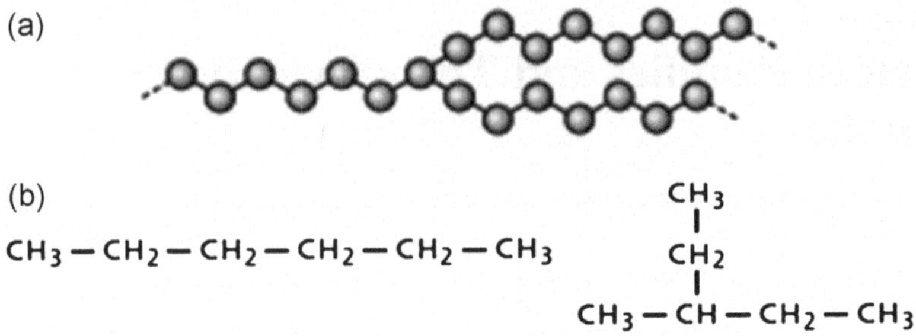


**Fig. 4.1** (**a**) Repeating units of a polymer (Courtesy: Ilmari karonen, WIKI) and (**b**) Large carbon-based molecules made of repeating units representing a polymer

**Fig. 4.2** Repeating units of polymer, polypropylene

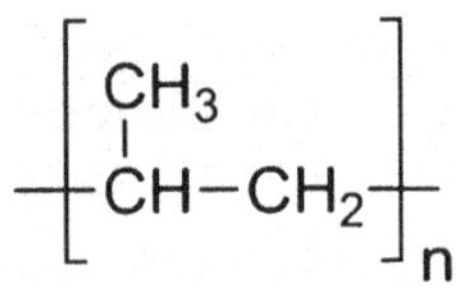


Swiss chemist (in the year 1846) was the first carbonaceous material known at that time. The material mixed with camphor and acetone was used as a dressing material since the U.S. Civil War. However, in 1922, Hermann Staudinger (of Worms, Germany) was the first to propose that polymers consisted of long chains of atoms held together by covalent bonds. In 1974, Paul Flory, an American chemist, was awarded the Nobel Prize in Chemistry for his work in polymer random coil configuration (Fig. 4.3) [6].

Figure 4.3 shows the random walk model of polymer together with the random coil configuration, where, $\alpha$ is the expansion factor. The fully extended length of the chain is $L$, which is equivalent to $N \times l$, where $N$ is the number of subunits of a freely joined chain, each of length $l$, which occupies zero volume, so that no part of the chain excludes another from any location. The average (root mean square) end-to-end distance for the chain is $\{\sqrt{} < r_1{}^2 >\}$, and it turns out to be $l$ times $\sqrt{N}$. In Fig. 4.3a, vector $R$ is the total end-to-end vector of an ideal chain of $r_i . . . . . . r_n$ (the vectors) corresponding to individual monomers.

Polymers in therapeutics and nanomedicine improved solubility, decreased toxicity, controlled release, and can help to enable site-specific delivery. Polymers are used to generate numerous biomolecule polymer conjugates, small molecule drug–polymer conjugates, supramolecular drug delivery systems, and can be used as a platform for controlled release [7–12]. Recently, functionalized multivalent polymer micelles, with a well-defined microstructure, has been a versatile nanomedicine platform with greatly improved drug pharamacokinetics and efficacious response in cancer treatment. Figure 4.4a shows a schematic illustration of the core–shell structure of a polymer micelle and b shows the functionalized polymer micelles with active targeting to tumor.

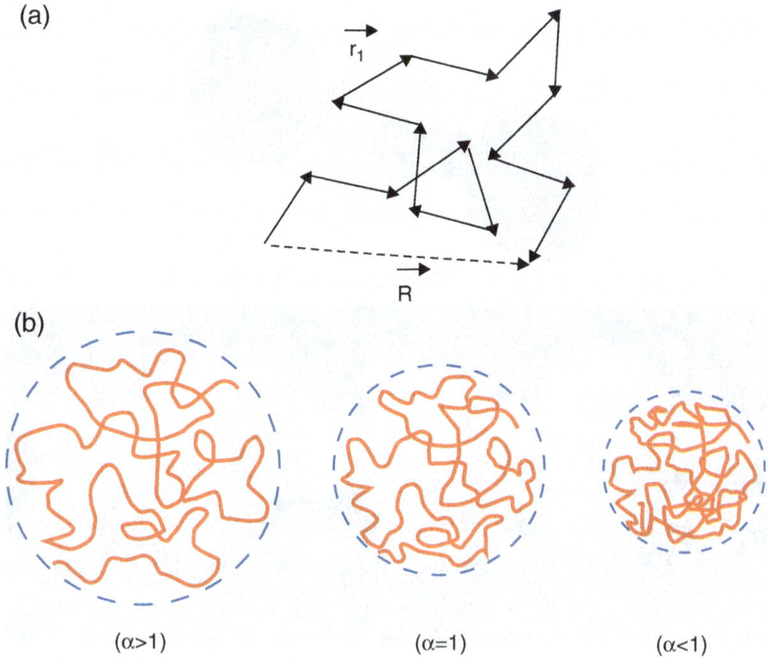


**Fig. 4.3** (**a**) Random walk model of polymer and (**b**) random coil configuration, where $\alpha$ denotes the expansion factor (Courtesy: WIKI)

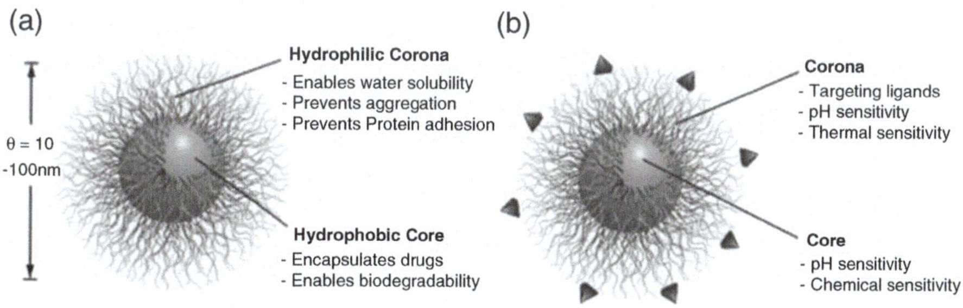


**Fig. 4.4** (**a**) Schematic illustration of the core–shell structure of a polymer micelle and (**b**) the functionalized polymer micelles with active targeting to tumors (Reproduced with permission, Springer, Ref. [7])

## 4.2 Different Varieties of Polymers and Their Applications

As a matter of fact, polymers (plastics) shape our lives because polymers can be shaped. It is the major driving force behind most of the research that tries to design plastics fit for electronic applications [13, 14]. Each polymer type has already found widespread applications in many optical and electronic devices such as batteries,

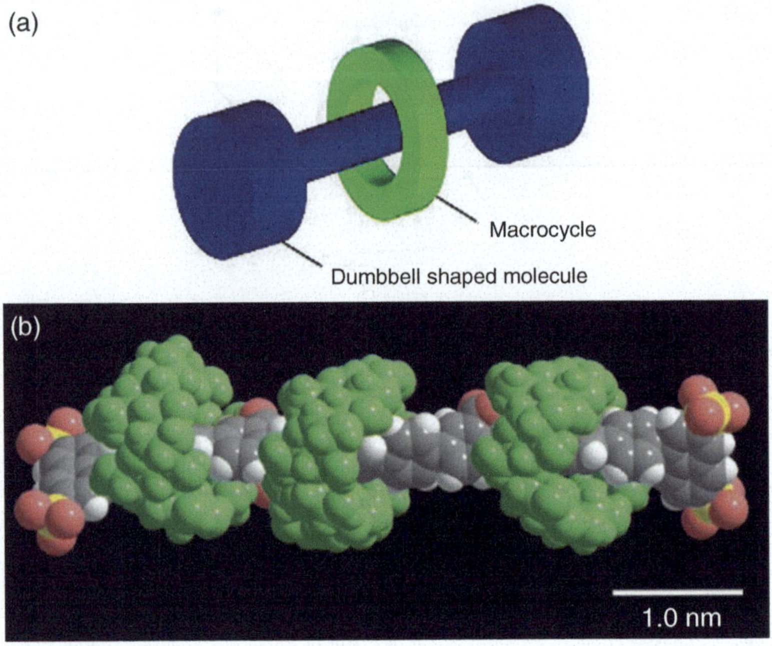


**Fig. 4.5** (**a**) Graphical representation of rotaxane and (**b**) dubbbell-shaped molecule of conjugated polyrotaxane (Courtesy: WIKI)

displays, plastic wires, optical signal processing, information storage, solar energy conversion, and in molecular biochemicals, which shape the life of the living organisms (DNA, RNA, protein, peptides, glycogen, etc.), and in molecular electronics. A classical example is Rotaxane-based molecular machines that have been used as logic molecular switching elements and as molecular shuttles [15]. Rotaxane has been used in biological systems including: cystine, knot peptides, cyclotides, or lasso peptides such as microcin J25 protein, and a variety of peptides with rotaxane substructure [16]. Figure 4.5a shows the graphical representation of a Rotaxane, and b shows the dumbbell-shaped molecule of conjugated polyrotaxanes.

In 2012, unique pseudo-macrocycles consisting of double-lasso molecular machines (also called rotamacrocycles) have been reported in *Chemical Science*. These structures can be tightened or loosened depending on the pH [17].

Figure 4.6 artistically illustrates a variety of drug delivery systems, including functionalized multivalent polymer micelles, well-defined nanostructures, and polyplexes flowing through the blood vessel. In the course of our discussions and presentations in this chapter, we will be able to find the facts about polymer, a complex chain of repeating carbon atoms. Figure 4.7 shows the repeating units of *carbon* in a *polymer*.

Another example of a *polymer* is *protein*, which performs a vast array of functions within living organisms, including catalyzing metabolic reactions,

Fig. 4.6 Artistic illustration of a variety of drug delivery systems (Courtesy: Aldrich-Sigma)

Fig. 4.7 The repeating unit of carbon in a polymer

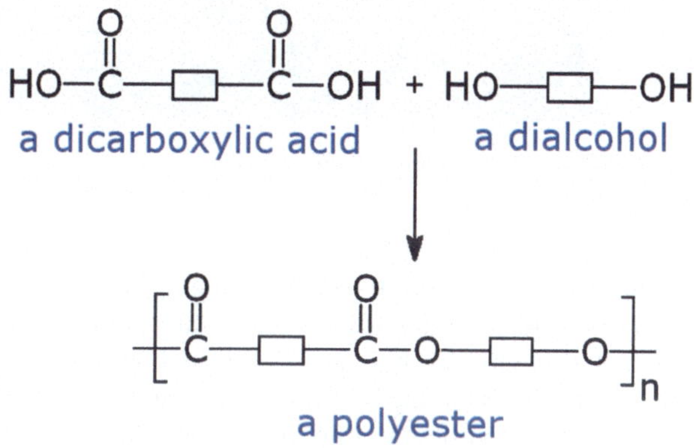


Fig. 4.8 Structure of an amino acid with one carbon atom

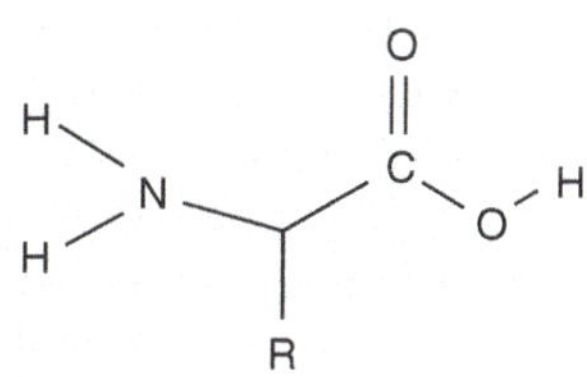

replicating DNA, responding to stimuli, and transporting molecules from one location to another [18, 19]. In proteins, *amino acids* are the monomers (*mono* means *one*, and *mer* means *part*) and are arranged in a linear chain joined together by peptide bonds. A monomer is a molecule that may bind chemically to other molecules to form a polymer. Many proteins are enzymes that catalyze the chemical reactions in metabolism. The structure of an *amino acid* can be represented by Fig. 4.8.

Figure 4.8 shows the structure of an amino-acid with one carbon atom. The symbol R is used when the exact identity of the group is not important. The carbon between amino group and the acid group always has one hydrogen atom on it. A *ball and stick model* of *amino acid* is shown in Fig. 4.9. The blue ball represents

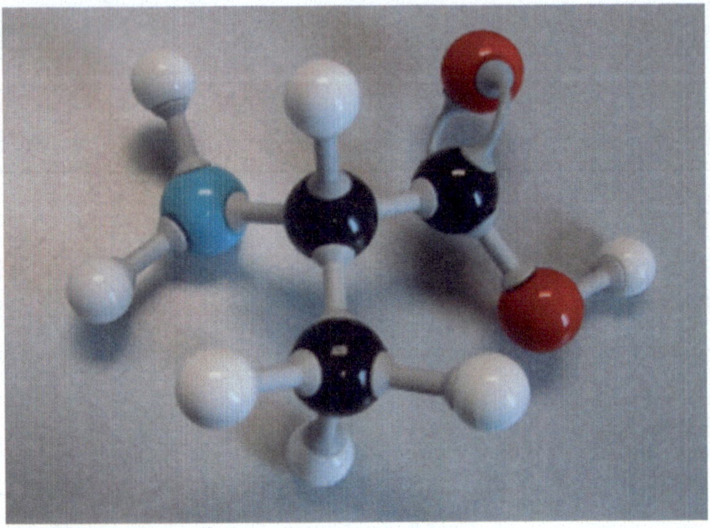

**Fig. 4.9** Ball and stick model of amino acids

nitrogen atom, the black balls represent carbon, the white balls are hydrogen atoms, and the red balls represent oxygen atoms.

Amino acids in living organisms are all α-amino acids (because amino group is attached to the first carbon atom after the one in the –COOH group). In biochemistry, amino acids having both amine and carboxylic acid group attached to the first $\alpha$-carbon have particular importance. Because of their biological significance, amino acids are important in nutrition, and are commonly used in nutritional supplements, fertilizers, and food technology.

Polyamino acids (PAA) are typically synthesized by the ring-opening polymerization of amino acid N-carboxyanhydrides (NCAs) initiated by an amino-terminated reagent shown in (Fig. 4.10) [20].

PAA is used in drug delivery and biomedical applications. PAA are similar to proteins (and peptides) in that they consist of amino acid building blocks linked by means of amide bonds. Amino acids have diverse chemical properties and can be subdivided according to the nature of their side chains, including those that are aliphatic, aromatic, sulfur-containing, and those that contain ionizable groups. When designing a PAA for use in drug delivery, amino acids are selected based on the chemical properties that meet the needs of both the drug and the solvent environment. Typically, amphiphilic block copolymers are formed from a hydrophilic (ionizable) and a hydrophobic amino acid. There are 21 naturally occurring amino acids (Fig. 4.11).

Asparagine (-$CH_2CONH_2$) and aspartic acid (-$CH_2COOH$), both are found in the vegetable asparagus. Cysteine (-$CH_2SH$, S = sulfur) has come from Greek *kustis*, and is found in urine salts. Methionine (-$C_2H_4SCH_3$, S = sulfur), an amino acid named from Greek *theion*, means sulfur. Serine (-$CH_2OH$) is a Latin word, and it

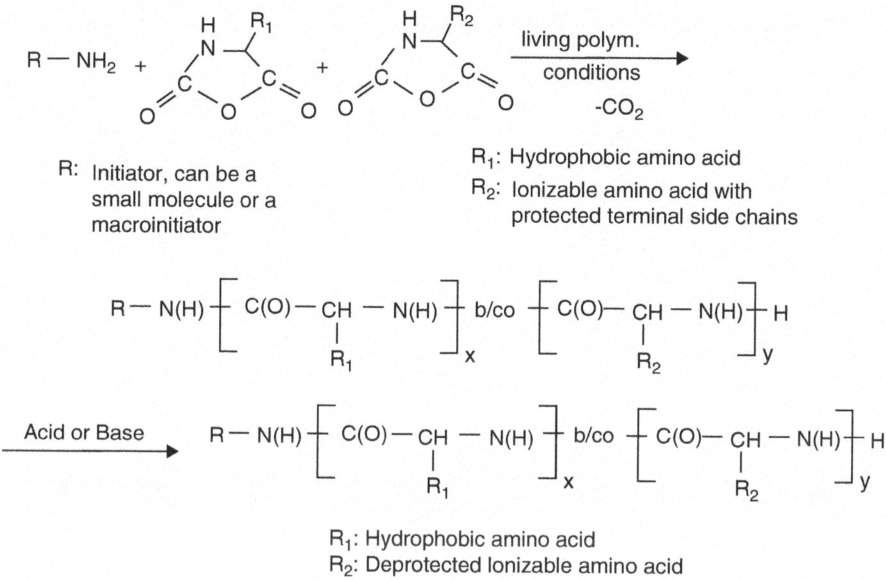


**Fig. 4.10** NCA polymerization leading to the synthesis of PAA block copolymers and PAA random copolymers (Courtesy: Aldrich Sigma, Ref. [21])

means silk. The chemical serine is found in silk. The chemical threonine (-CH ($CH_3$)OH)) is the name of a sugar called *threose*. Tryptophan (-$CH_2C_8H_5N$), a form of amino acid, is secreted by the pancreas and acts in the small intestine. Tyrosine (-$CH_2$PhOH, Ph = phenyl group) comes from a Greek word, and it means cheese. The compound is found in cheese. Valine (-CH($CH_3$)$_2$), one form of amino acid, is naturally found in liquid fatty acids, in some plant oils, and also found in a particular plant called valerian which is used as a drug.

## 4.3 Structure of Special Type of Polymers

There are several families of natural polymers available in the market. One of the best examples of natural polymers is chitosan, which is a nontoxic, semicrystalline, biodegradable, and biocompatible polysaccharide, and has been used as a drug carrier for controlled drug delivery [21–23]. The cationic properties of chitosan have enabled its extensive use in gene delivery. While nucleic acid loaded chitosan nanoparticles fabricated from native chitosan have shown low buffering capacity and limited stability, chemically modified chitosan, on the other hand, can help to improve in vivo transfection efficacy [23]. Figure 4.12 shows the chemical structure of chitosan.

The development of both natural and synthetic polymers possesses benefits that make them valuable components in medicine (therapeutics) and in the industry. As

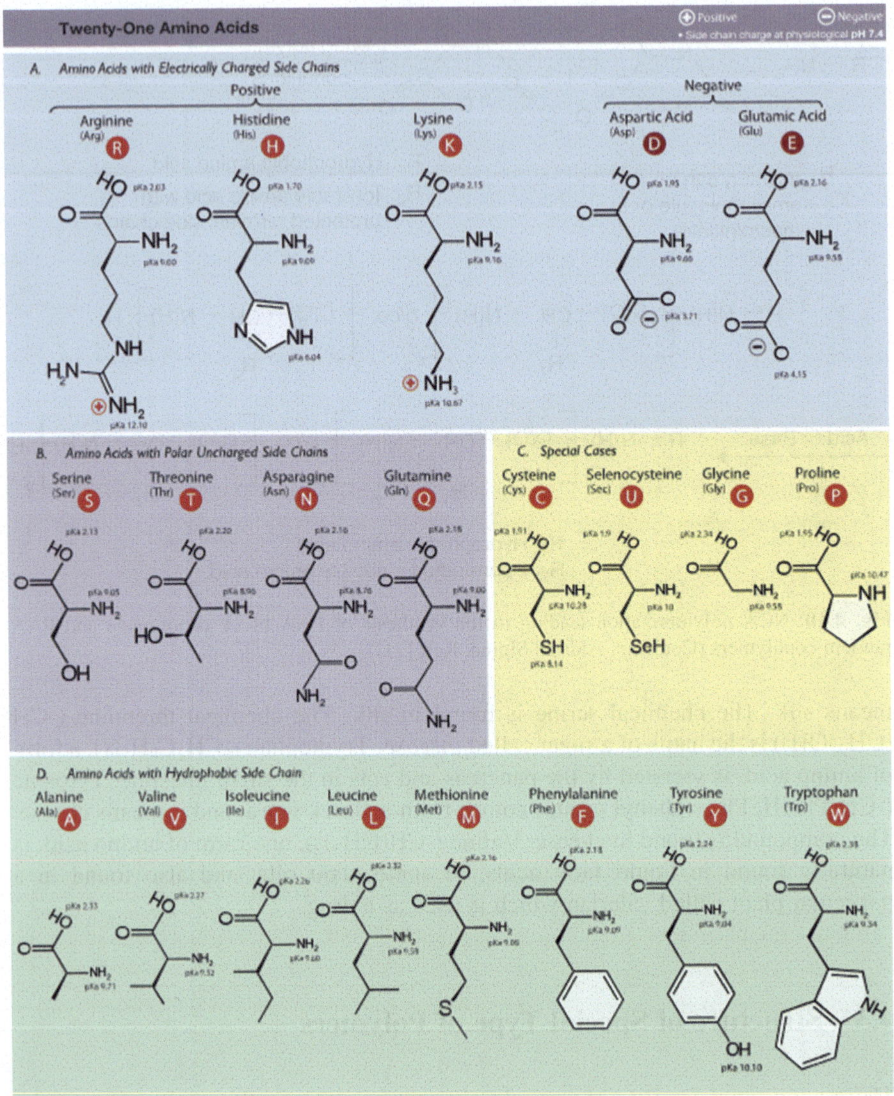


**Fig. 4.11** Twenty-one naturally occurring amino acids (Courtesy: Dancojocari, Creative Comm. WIKI)

a matter of fact, the role played by organic chemistry, a chemistry of carbon atoms, in the pharmaceutical industry continues to be one of the main drivers in drug discovery process. However, the precise nature of that role, particularly in drug metabolism and chemical toxicology, depends on the metabolic activities of enzymes, the macromolecular biological catalysts. They are responsible for thousands of metabolic processes that sustain life [24, 25]. A lethal illness can be caused

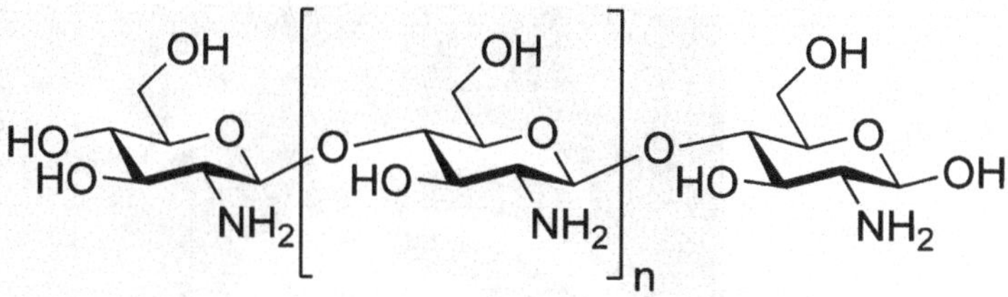


**Fig. 4.12** Chemical structure of chitosan (Courtesy: Warraich Sahib, WIKI)

by the malfunction of just one type of enzyme out of thousands of types present in our bodies.

RNA polymerase (RNAP or RNApol), also known as DNA-dependent RNA polymerase, is an enzyme that produces primary transcript RNA. In cells, RNAP is necessary for constructing RNA chains using DNA genes as templates, a process called transcription. The process is used to detect a deadly virus inside the blood of an affected person. Recently (in the year 2014), Ebola has been an epidemic in the southern parts of Africa. The transcription process has been successfully applied to detect the Ebola virus infection on a US doctor when he was treating the patients in South Africa. The 2006 Nobel Prize in chemistry was awarded to Roger D. Kornberg for creating detailed molecular images of RNA polymerase during the various stages of transcription [26, 27].

A mutation of a single amino acid in the enzyme phenylalanine hydroxylase, which catalyzes the first step in the degradation of phenylalanine, can lead to intellectual disability if the disease is untreated [28]. Carbonic anhydrase II (gene name CA2) is one of the 14 forms of human $\alpha$-carbonic anhydrases. Carbonic anhydrase catalyzes the reversible hydration of carbon dioxide. Defects in this enzyme are associated with osteopetrosis and renal tubular acidosis (Fig. 4.13) [29, 30].

Another organic compound (where carbon is the backbone) is folic acid or folate (molecular formula is $C_{19}H_{19}N_7O_6$). It is also referred to as vitamin $\varphi$, vitamin M, vitamin $B_9$, vitamin $B_c$ or (folacin), pteroyl-L-glutamic acid, and pteroyl-L-glutamate. Vitamin $B_9$ is essential for numerous bodily functions, such as to synthesize DNA, repair DNA, and methylate DNA as well as to act as a cofactor in certain biological reactions [31]. Human body cannot synthesize folate de novo; therefore, folate has to be supplied through the diet.

Figure 4.14a shows the structure of folic acid and b shows the ball and stick model of folic acid (left) and anticancer drug methotrexate (right). As both have similar structures, methotrexate is a competitive inhibitor of many enzymes that use folate. In chemical and drug industries, protein engineering is an active area of research to create new enzymes with novel properties, either through rational design or in vitro evolution [32, 33].

Most of the industrial polymers that are vinyls are called carbon chain polymers. As a matter of fact, *carbon* can be regarded as the *backbone of a polymer*, and the

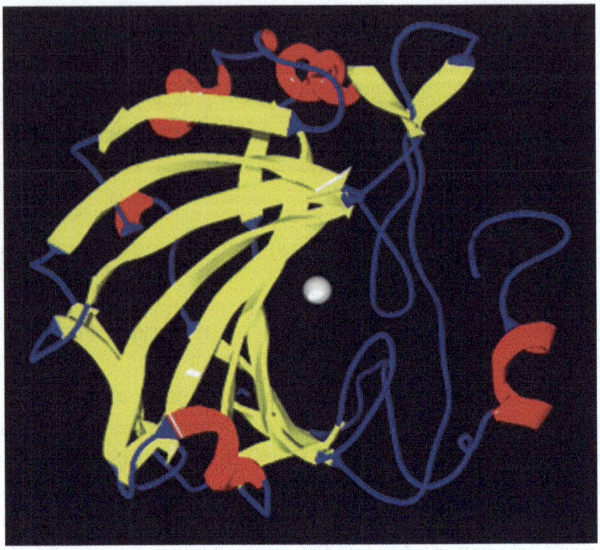

**Fig. 4.13** Ribbon diagram of human carbonic anhydrase II (Courtesy: GFDL, GNU, WIKI)

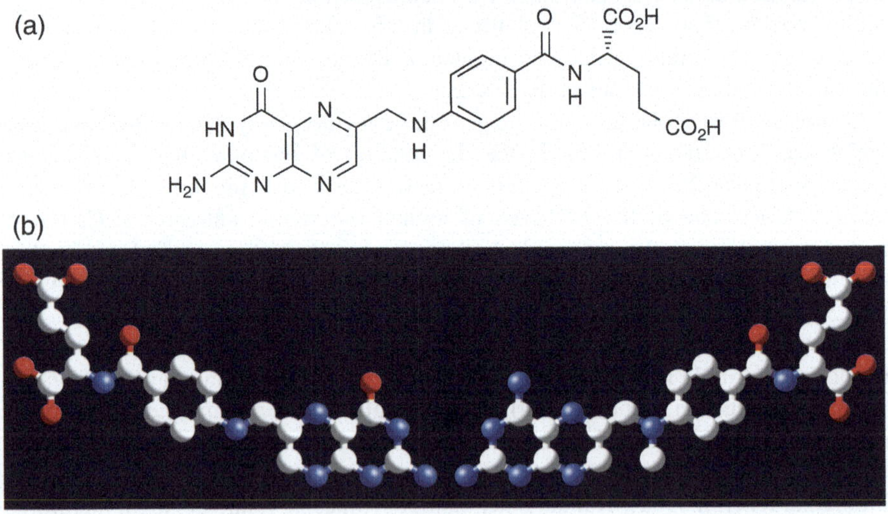


**Fig. 4.14** (**a**) Structure of folic acid and (**b**) ball and stick model of folic acid (*left*) and anticancer drug methotrexate (*right*) (Courtesy: WIKI)

different types of polymer structures are classified as *linear*, *branched*, and *network* polymers (Fig. 4.15).

Polyethylene is called linear or straight polymer, because it consists of a long string of carbon–carbon bonds. The structural formula of polyethylene (Fig. 4.16a) is shown here together with a space fill model (Fig. 4.16b).

About more than 21 billion pounds of polyethylene is manufactured in the United States, of which more than half is made by free radical polymerization.

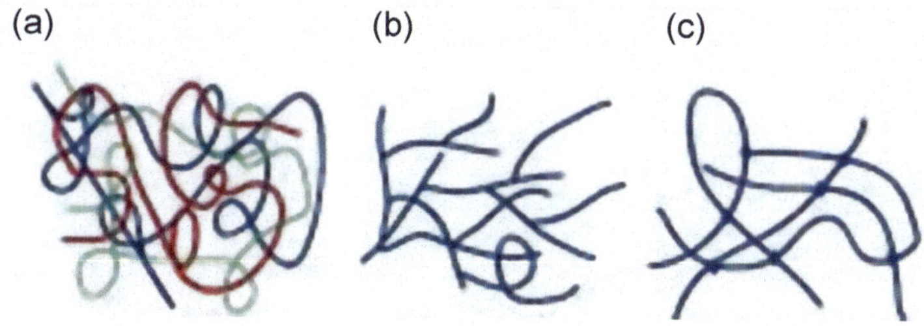


**Fig. 4.15** Different types of carbon backbone structures: (**a**) linear (**b**) branched, and (**c**) network

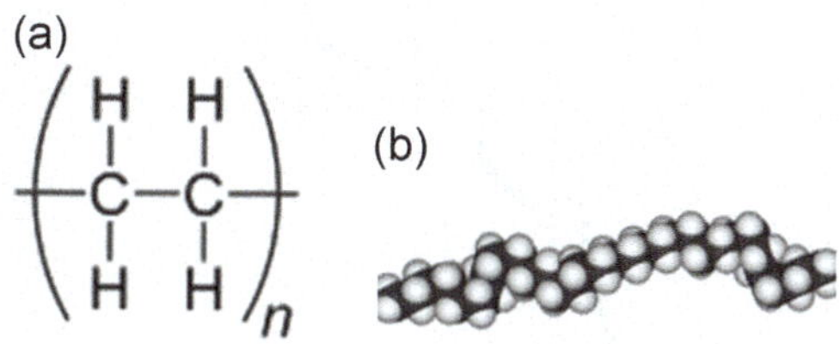


**Fig. 4.16** Structural formula of polyethylene (**a**) together with a space fill model (**b**)

The free radical process yields a very transparent polymer called low-density polyethylene, used in films and packaging.

Many of the commercially important polymers, such as polyvinyl chloride (PVC) ($CH_2$=CH-Cl), polychlorotrifluoroethylene (Kel-F) ($CF_2$=CF-Cl), polyacrylonitrile (Orlon, Acrilan) ($CH_2$=CH-CN), and poly(methyl methacrylate) (Plexiglass, Lucite) ($CH_2$=C (CH)-$CO_2CH_3$), are produced from alkene monomers by free radical polymerization.Many of the articles we use in our daily life, such as telephones, computers, automobiles, sports equipment, stereo systems, food packages, etc. are the products of alkene polymers.

## 4.4 Branched Polymers

*Branched* polymers on the other hand are those polymers that branch at irregular intervals. Branching is seen to occur naturally during enzymatically catalyzed polymerization of glucose to form polysaccharides such as glycogen. Because of their irregular packaging, the polymers have high densities, high tensile strength, and high melting points. Figure 4.17 shows glycogen, a branched polysaccharide [34, 35].

The last variety of the polymer structures is the network polymer that has trifunctional monomer units, having three active covalent bonds that allow these polymers to form three-dimensional networks instead of linear chain framework. The network polymer is also referred to as *cross-linked* or *thermoset polymer*. A single large network is seen to form during polymerization, and initially accomplished by heating the raw materials and bringing them together. These polymers

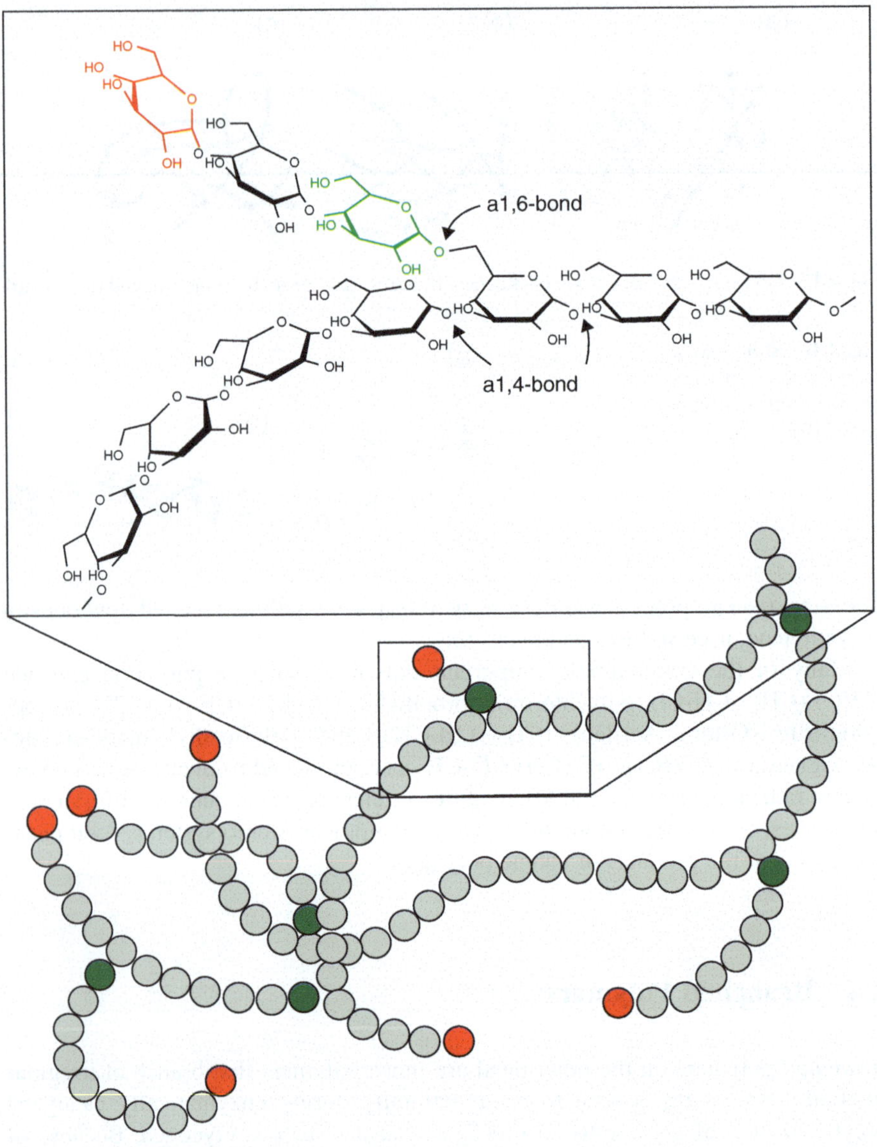


**Fig. 4.17** Glycogen, a branched polysaccharide (Courtesy: Maskin, WIKI)

are hard, rigid, and brittle because of network structure, e.g., Bakelite, melamine, formaldehyde resin, etc. Figure 4.18 shows the picture of a network polymer.

The modern concept of polymers is as covalently bonded macromolecular structures. Historically, products arising from the linkage of repeating units by covalent bonds have been the primary focus of polymer science. Polyisoprene of latex rubber and polystyrene of Styrofoam are examples of polymeric natural/

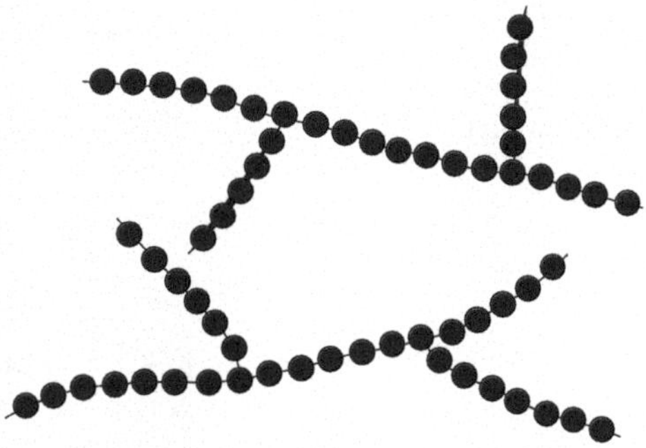

**Fig. 4.18** A network polymer

biological and synthetic polymers, respectively. In biology, all biological macromolecules like proteins (polyamides), nucleic acids (polynucleotides), and polysaccharides are purely polymeric, or composed of mainly polymer components. In isoprenylated/lipid-modified glycoproteins, small lipidic moleculeS and oligosaccharide modifications occur on the backbone of the polyamide (protein) [36].

## 4.5 Different Types of Polymers

Polymers are of two types: *natural* and *synthetic*. Naturally occurring polymers are cotton, starch, wool, silk, and natural rubber [37, 38]. *Natural rubber* is a mixture of hydrocarbons, and the empirical composition is $C_5H_8$. *Natural rubber*, or the *latex rubber*, that comes from the latex of *Hevea brasiliensis*, is mainly poly-cis-isoprene [39–42]. Although natural rubber exhibits many excellent properties in terms of mechanical performance, in certain respects it is inferior to the synthetic rubber made by the polymerization of a variety of petroleum-based precursors called monomers. Figure 4.19 shows how latex is being collected from a trapped rubber tree.

Natural rubber is highly unsaturated, and careful study of the ozonolysis of rubber leaves little doubt that it is a clear linear polymer of isoprene (polyisoprene, $CH_2 = C\ (CH_3)\ CH = CH_3$) in which the residual bonds are located exclusively between C2 and C3 of the isoprene units. Although isoprene is not the biological precursor of rubber, the latter may be thought of as the product of the 1, 4-addition polymer of isoprene. The average molecular weight of rubber is estimated to be in the range of 60,000 to 350,000, which corresponds to a degree of polymerization (number of monomer units per molecule) of 1000 to 5000.

Among the natural polymers are polymeric hydrocarbons that include *cellulose*, *starch*, and *glycogen* (the carbohydrates stored in animal livers). All the known polysaccharides are glycosides in which the acetal carbon atom of a monosaccharide unit is linked by way of an oxygen to one of the nonacetal carbon atoms of

**Fig. 4.19** Collection of latex from a rubber tree (Courtesy: PRA, Creative Comm. WIKI)

another monosaccharide. Typical linkages are found in simple disaccharides such as $\alpha$- and $\beta$-maltose and cellobiose, which are the dimers of glucose.

The *synthesized polymers* on the other hand are *Bakelite* ($C_{16}H_{26}O_2$), a phenol formaldehyde resin; *neoprene,* a polymerized chloroprene($C_4H_5Cl$); *nylon* ($C_{12}H_{22}N_2O_2$); *polyvinylchloride* (PVC) ($CH_2$-CHCl)n; and *polystyrene* ($CH_2$-CH ($C_6H_5$))n, a polymerized free radical of *styrene* (ph-CH=$CH_2$, ph is phenyl group). Most commonly, the continuously linked backbone of a polymer used for the preparation of plastics consists mainly of carbon atoms. Both of the polymers, natural and synthetic, are created via *polymerization* of many small molecules, known as monomers. The synthetic polymers are prepared by two methods: step growth polymerization and chain growth polymerization [43]. In 1940, Waldo Semon, a scientist from B.F. Goodrich, developed a cheaper version of synthetic rubber known as Ameripol. During World War II, production of synthetic rubber (polymerization of 1, 3 butadiene gives polybutadiene, the synthetic rubber $CH_2$-CH=CH-$CH_2$) expanded greatly since the Axis powers controlled nearly all the limited supplies of natural rubber by mid-1942 once Japan conquered Asia [44]. By mid-1950s, Caltech researchers began to use synthetic rubber mixed with ammonium perchlorate and high proportions of aluminum powder in the large. Nowadays, titanium dioxide is used with copolymerization and volatile matter in producing synthetic rubber, which is used a great deal in printing textiles.

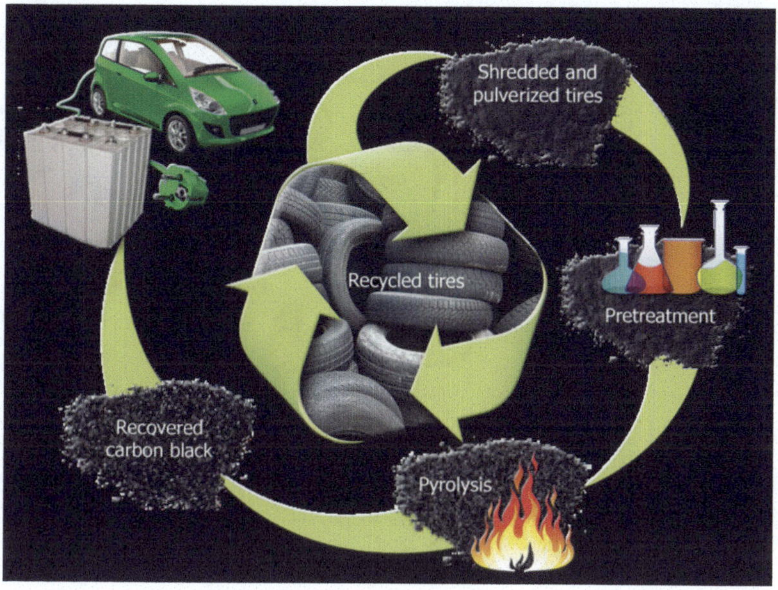


**Fig. 4.20** Recovery of synthetic rubber to carbon black and fabrication of anodes for lithium-ion batteries (Courtesy: Oak Ridge National Lab, TN, USA)

Recently, scientists from Oak Ridge National Laboratory (ORNL), TN recycled tires to recover carbon black and subsequent use of the pyrolytic carbon black to develop anodes for lithium-ion batteries (Fig. 4.20). They claimed that after 100 cycles the capacity measures nearly 390 milliampere hours per gram of carbon anode, which exceeds the best properties of commercial graphite [45]. This kind of performance is highly encouraging in the light of the fact that the global battery market for vehicles and military applications is approaching $78 billion and the materials market is expected to hit $11 billion in 2018.

Polymers are studied in the fields of biophysics and macromolecular science (which includes polymer chemistry and polymer physics). In the context of the polymeric natural/biological and synthetic polymers, we can refer to polyisoprene of *latex rubber* and the *polystyrene* of *Styrofoam*. In biological contexts, we can refer to *proteins* (polyamides), *nucleic acids* (polynucleotides), and polysaccharides – which are purely polymeric, or composed of the large part of polymeric components – e.g., isoprenylated/lipid-modified glycoproteins.

## 4.6 Biopolymers

*Proteins* are large biological molecules (peptides), or macromolecules, consisting of one or more long chains of amino acid residues. There are 23 naturally occurring amino acids. The peptide backbone is the repeating sequence of nitrogen (N), $\alpha$-

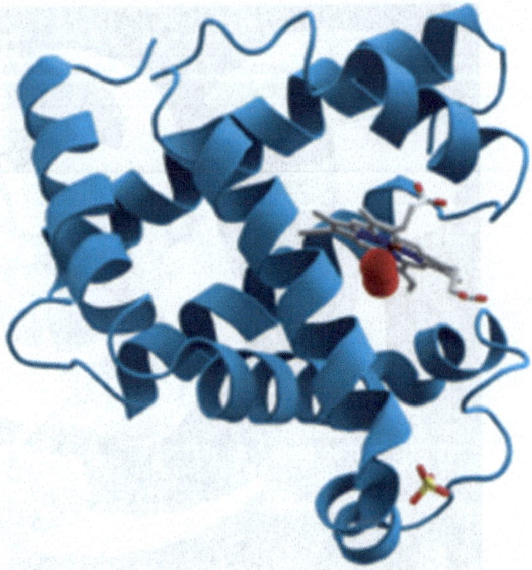

**Fig. 4.21** 3D structure of protein showing turquoise alpha helices (Courtesy: Aza Toth, WIKI)

carbon, and carbonyl group. Proteins perform a vast array of functions within living organisms, including catalyzing metabolic reactions, replicating *DNA*, responding to stimuli, and transporting molecules from one location to another (Fig. 4.21). Many structure prediction methods have served to inform the emerging field of protein engineering, in which novel protein folds have already been designed [46]. Constituent amino acids can be analyzed to predict secondary, tertiary, and quaternary protein structure, in the case of hemoglobin containing heme units (Fig. 4.22).

As of 2014, the protein Data Bank has over 90, 000 atomic resolution structures of proteins [47]. On the other hand, the biopolymers in living cells may be synthesized by enzyme-mediated processes, such as the formation of DNA catalyzed by DNA polymerase. DNA is a polymer, and carbon is the backbone of a polymer. The monomer units of DNA are nucleotides. There are four different types of nucleotides, such as adenine (A), guanine (G), cytosine (C), and thymine (T) (Figs. 4.23 and 4.24).

The deoxyribose sugar, the backbone of the DNA has five carbon atoms and three oxygen atoms. The carbon atoms are numbered as C1', C2', C3', C4', and C5' to distinguish from the numbering of the atoms of the purine and pyrimidine rings. The hydroxyl groups on the C5' and C3' ends are linked to the phosphate groups to form the backbone of DNA [48]. Deoxyribose lacks hydroxyl group at the C2' position compared to ribose, the sugar component of RNA [49] (Fig. 4.25).

The deoxyribose sugars are joined at both the 3′ hydroxyl and 5′ hydroxyl groups to phosphate groups in ester links also known as *phosphodiester* bonds (Fig. 4.26). In the year 1902, Archibald Garrod made an important proposition attributing a defective enzyme to a defective gene, and it was the first ever proposition to link

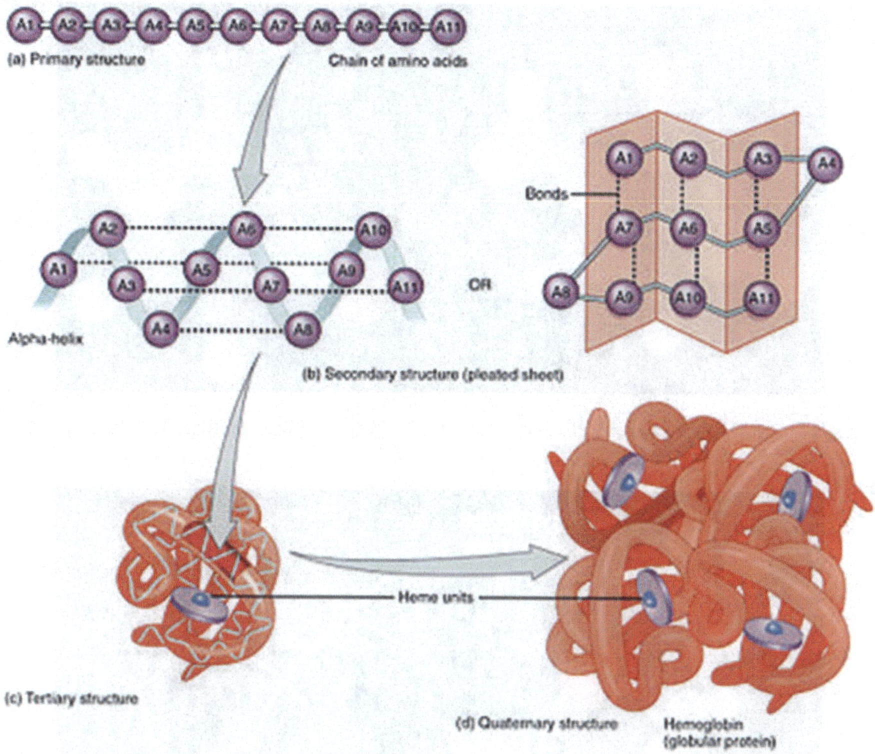


**Fig. 4.22** Analysis of the amino acids, which predicts three varieties of protein structures (Courtesy: Open Stax College, Creative Comm. WIKI)

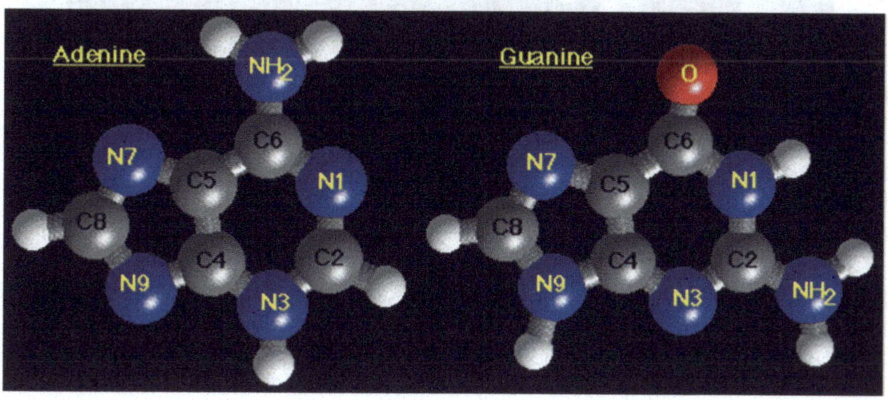


**Fig. 4.23** Structure of adenine (A) and guanine (G) (Courtesy: R.B. Hallick, U. Arizona)

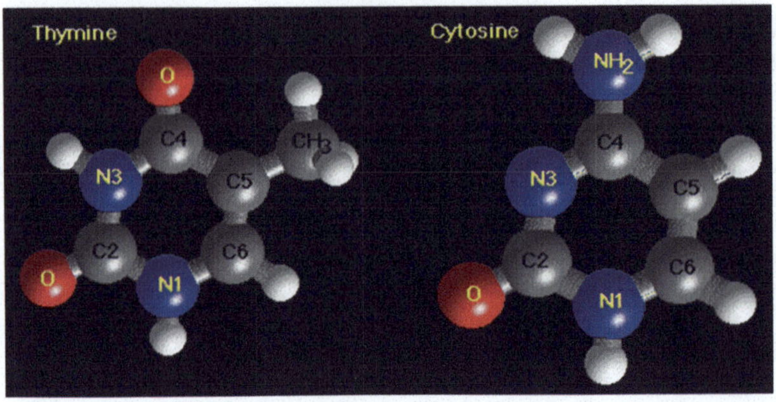


**Fig. 4.24** Structure of thymine (T) and cytosine (C) (Courtesy: R.B. Hallick, U. Arizona)

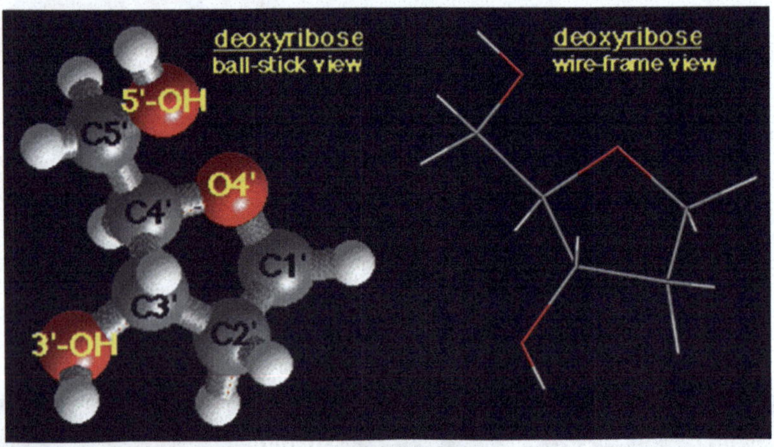


**Fig. 4.25** Structure of deoxyribose (Courtesy: R.B. Hallick, U. Arizona)

gene and proteins [50]. The identification of RNA and the crucial missing link between DNA and protein was subsequently confirmed by cell fabrication experiments [51].

The synthesis of proteins involves multiple enzyme processes to transcribe genetic information from DNA to RNA and subsequently translate that formation to synthesize the specified proteins from amino acids. Gene encoded in DNA are first transcribed into pre-messenger RNA (mRNA) by proteins as RNA polymerase. Most organisms then process the pre-mRNA using various forms of post-transcriptional modification to form the mature mRNA, which is then used as a template for protein synthesis by the ribosome (Fig. 4.27).

Lipids are a group of naturally occurring molecules that include fats, waxes, sterols, fat-soluble vitamins (A, D, E, and K), monoglycerides, diglycerides, triglycerides, and others. Their main biological functions are storage of energy,

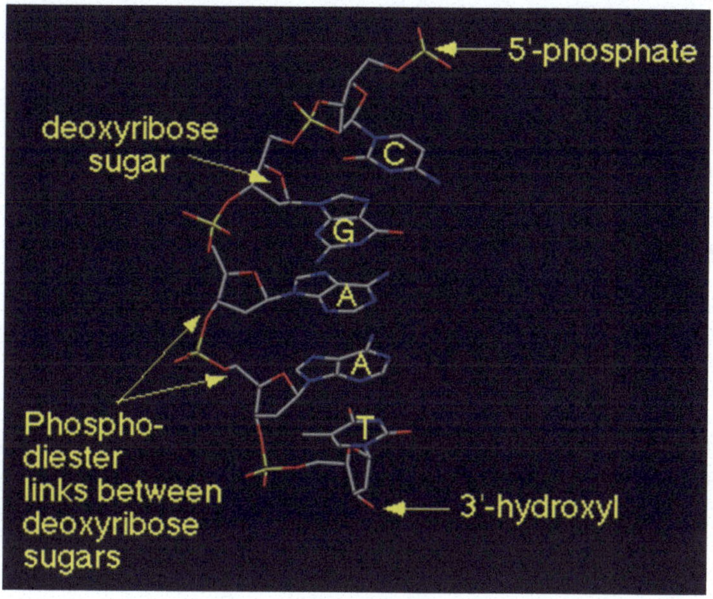


**Fig. 4.26** Deoxyribose sugar joined to phosphate and hydroxyl groups (Courtesy: R.B. Hallick, U. Arizona)

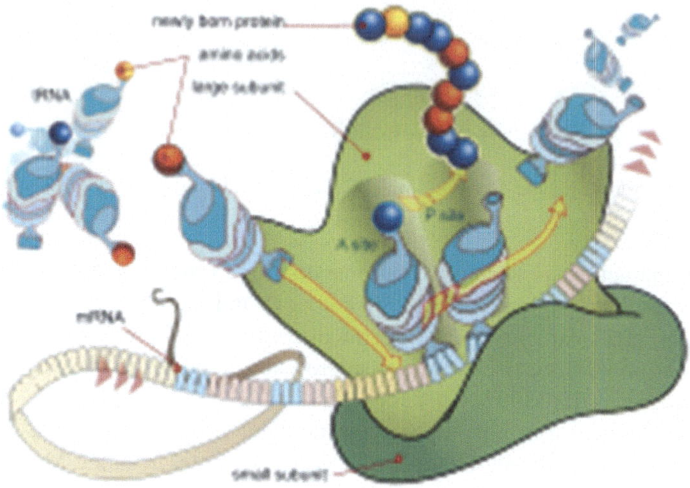

**Fig. 4.27** How a ribosome produces a protein using mRNA as template (Courtesy: Lady of Hats, WIKI)

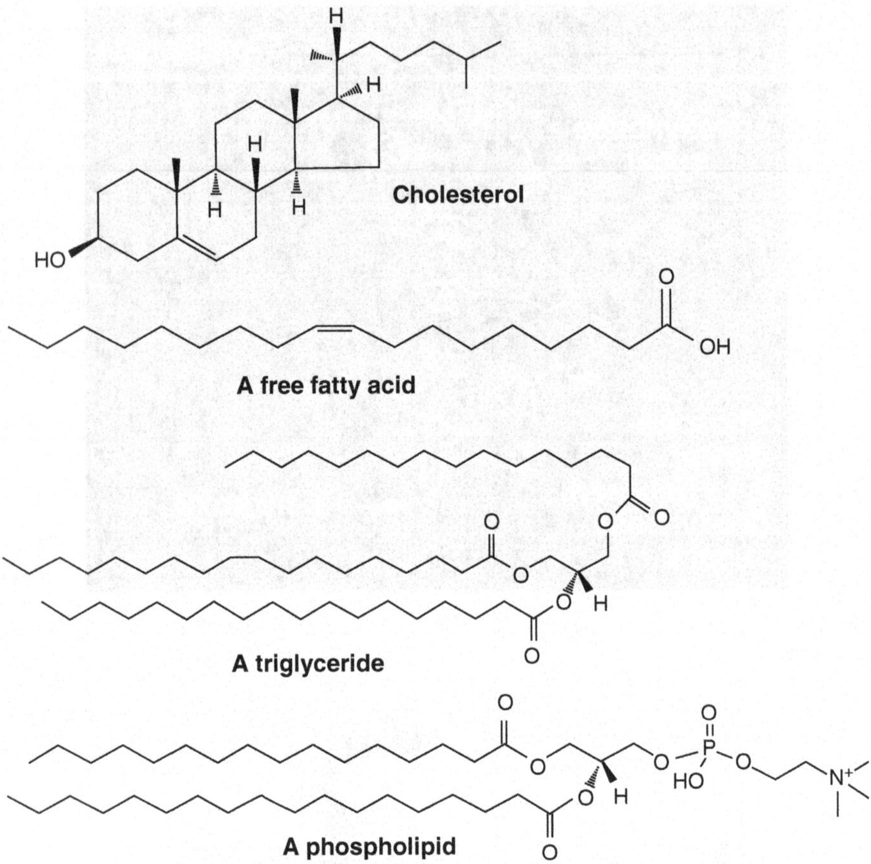


**Fig. 4.28** Structures of some common lipids (Courtesy: Eoin Fahy, Creative Comm. WIKI)

signaling and acting as structural components of cell membranes [52]. Biological lipids are the biochemical subunits of ketoacyl and isoprene groups and can be divided into eight categories. The chemical formation of prebiological lipids and their formation into protocells are considered key steps in the models of abiogenesis, the origin of life [53, 54].

The biochemical assembly lipoprotein contains protein and lipids, whose purpose is to transport hydrophobic lipid (a.k.a fat) molecules in water. Many enzymes, transporters, structural proteins, antigens, adhesions, and toxins are lipoproteins. Examples include the plasma lipoprotein particles classified under high-density (HDL) and low-density (LDL) lipoproteins, which enable fats to be carried in blood stream; the transmembrane proteins of the mitochondrion and chloroplast; and bacterial lipoproteins (Fig. 4.29) [55]. LDL carries 3000–6000 fat molecules around the body, and is referred to as bad lipoprotein. HDL on the other hand collect the fat molecules and take them back to the liver, and is referred to as good lipoprotein.

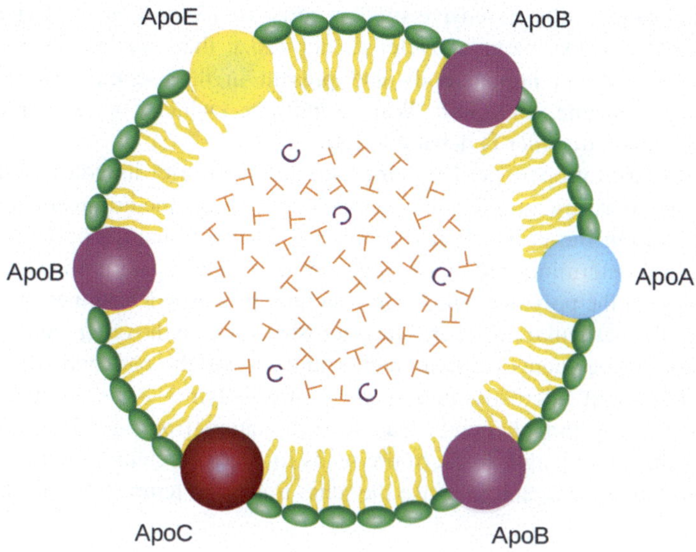


**Fig. 4.29** Structure of a lipoprotein (Courtesy: Xvazquez, GNU, WIKI)

The densities of LDL and HDL for a young healthy human weighing ~70 kg (~150 lb) are between 1.006 and 1.019 and >1.063 g/mL [56, 57].

Biosynthesized by animal cells, lipid is a sterol, or modified steroid. It is an essential structural component of animal cells. The word *cholesterol* derived from Greek *chole* (bile) and *stereos* (solid), followed by the chemical suffix – *ol* for an alcohol, is an organic compound which has 27 carbon atoms.

The elucidation of its chemical structure is one of the foremost achievements in organic chemistry during 1910s and 1920s. In 1928, German chemists Windaus and Wieland received Nobel Prizes in chemistry for their work on the structure of cholesterol and the closely related bile acids. The four-ring carbon skeleton characteristic of cholesterol was later found not only in the sterols of plants and animals, but also in the precursors of vitamin D, in the male and female sex hormones, and in the hormones from the adrenal cortex. These discoveries opened up a new field of research that has engaged a number of scientists in different disciplines.

Konrad Bloch and collaborators were able to show in a series of brilliant investigations how the two carbon atoms of acetic acid ($C_2H_4O_2$) are used to synthesize a long hydrocarbon with 30 carbon atoms, *squalene* ($C_{30}H_{50}$), which in turn is cyclized in a novel type of reaction to a steroid with 30 carbon atoms, lanosterol. The lanosterol is then transformed in a complicated series of reactions into cholesterol ($C_{27}H_{46}O$), with 27 carbon atoms [58, 59]. The 30-carbon compound squalene is a hydrocarbon and a triterpene, and is a natural and vital part of synthesis of all plants and animal sterols. It is an important part of the Mediterranean diet and thought to be a chemopreventive substance that protects people from cancer [60]. Extensive study reports from World Health Organization (WHO) and US Department of Defense emphasize that squalene is a chemical naturally occurring in human body and present in flu vaccines. However, the linkage of squalene to Gulf War Syndrome which is a chronic and multisymptomatic disorder is debatable [60].

The most basic properties of polymer are the identity of its constituent monomers and the microstructure of the polymer which means the arrangement of these monomers within the polymer at the scale of a single chain. The length of the polymer chain determines the physical properties (melting and boiling points, and the viscosity) of the polymer. The basic structural property determines the physical behavior of the particular polymer. The microstructure, on the other hand, relates to the physical arrangements of monomer residues along the backbone of the chain [61]. An important microstructural feature of a polymer is its architecture and relates to the way branch points lead to a deviation from simple linear chain. Increasing the chain length furthermore tends to decrease chain mobility, increase strength and toughness, and increase the glass transition temperature ($T_g$).

## 4.7 Polymerization and Long-Chain Hydrocarbons

In polymer chemistry, polymerization is defined as a process of reacting with monomer molecules via a variety of reaction mechanisms that vary in complexity due to the functional groups present in the reacting compounds [62]. An example of alkene polymerization is shown in Fig. 4.30. Here, each styrene monomer's double bond reforms as a single bond plus a bond to another styrene monomer.

The chain growth polymerization (which is also called addition polymerization) involves the linking together of molecules incorporating double or triple carbon–carbon bonds. The unsaturated monomers have extra internal bonds that can break or link up with other monomers to form a repeating chain containing only carbon (C) atoms.

The chain growth polymerization can be cationic or anionic. In the former case, a cationic initiator transfers charge to a monomer which then becomes reactive. The reactive monomer then can react with other monomers to form a polymer [63]. This type of polymerization is being used in the production of polyisobutylene and poly

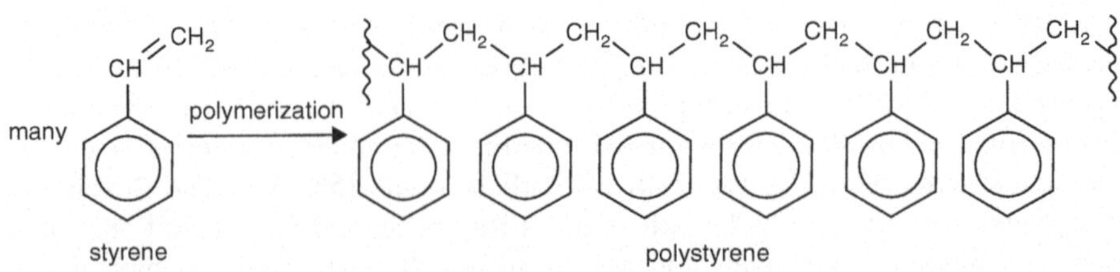


**Fig. 4.30** Reaction mechanisms of polymerization of a typical alkene group (Courtesy: H. Padleckas, WIKI)

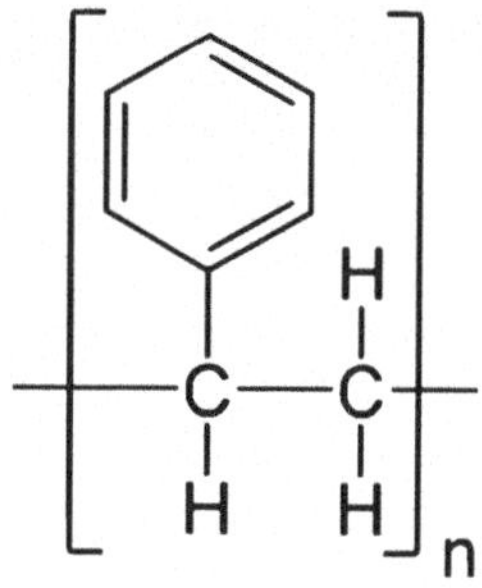

**Fig. 4.31** Structure of polystyrene

(N-vinylcarbazole) (PVK) [64]. Anionic polymerization, on the other hand, involves the polymerization of vinyl monomers with strong electronegative groups [65, 66]. It is used in the production of polydiene synthetic rubbers, solution styrene/butadiene, and styrene thermoplastic elastomers.

Polystyrene (Fig. 4.31) is a synthetic aromatic long-chain hydrocarbon made from the monomer styrene, a liquid petrochemical [67]. It is one of the most widely used plastics (for making compact disc (CD) and DVD cases). It is solid at room temperature but flows if heated to 100 °C, which is its glass transition temperature ($Tg$). Its molecular formula is $(C_8H_8)_n$, and Fig. 4.31 represents its molecular structure. The melting point of polystyrene is approximately 240 °C, density is between 0.96 and 1.04 g/cm$^3$, thermal conductivity is 0.033 W/(m.K), dielectric constant is 2.6 (1 kHz to 1GHz), and refractive index is 1.6. During polymerization, the carbon–carbon pi bond (in the vinyl group) is broken, and a new carbon–carbon single (sigma) bond is formed, attaching another styrene monomer to the chain. Styrofoam is a trademarked brand of a closed-cell extruded polystyrene foam, manufactured by Dow's Chemical Physics Lab in 1941.

Polymers that contain only a single type of repeat units are known as homopolymers, e.g., polystyrene, whereas polymers containing a mixture of repeat units are known as copolymers, e.g., ethylene-vinyl acetate. Some biological polymers are composed of a variety of different but structurally related monomer residues, for example, polynucleotides such as DNA (*deoxyribonucleic acid*). It can be well said that DNA is a long-chain polymer made from repeating units called nucleotides [68].

Figure 4.32 shows the double helix structure of DNA. Most DNA molecules are double-stranded helices, consisting of two long biopolymers made of simpler units called nucleotides. Two polynucleotide (polymer) chains are held together by weak thermodynamic forces to form a DNA molecule [69]. DNA molecules are circular in nature, and the geometry of the deoxyribose sugar (has five carbons and three oxygens) – backbone of DNA – forces the chains into a right-handed double helix [70].

There are several techniques that are used to study the properties of a polymer. Techniques such as *wide-angle X-ray scattering*, *small-angle X-ray scattering*, and *small-angle neutron scattering* are used to determine the crystalline structure of

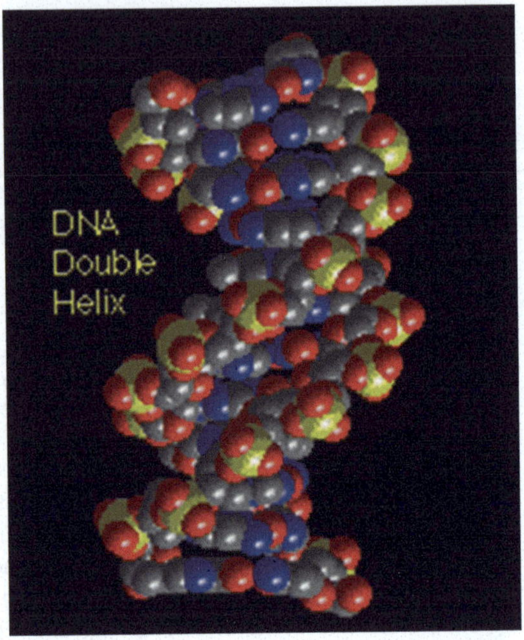


**Fig. 4.32** Double helix structure of DNA (Courtesy: Brian0918 & #153 WIKI)

polymers. However, the term crystalline in protein or polynucleotide is defined in terms of a conventional unit cell composed of one or more polymer molecules with cell dimensions of hundreds of angstrom or more.

The degree of crystallinity of a polymer can vary from zero (amorphous) to highly crystalline structure [71–73]. The amorphous one is transparent in appearance, while more crystalline structured polymer will be opaque in appearance due to light scattering (Fig. 4.33).

The average molecular weight of a polymer is determined from gel formation chromatography. Differential scanning colorimetry is used to determine the glass transition temperature ($Tg$) of the polymer, and thermal stability of a particular polymer can be determined by thermogravimetry method. Another polymer characterization technique is automatic continuous online monitoring of polymerization reactions (ACOMP) which provides real-time characterization of polymer reactions.

The susceptibility of a polymer depends on its structure. Epoxies and chains containing aromatic functionalities are especially vulnerable to ultraviolet (UV) light, and the polyesters are susceptible to degradation by hydrolysis [73–75]. Polymers containing unsaturated backbone are susceptible to ozone cracking, while carbon-based polymers are susceptible to thermal degradation. The degradation of polymers to form smaller molecules may proceed by random scission, as for example in polyethylene, or by specific scission as in the case of alpha methylstyrene.

**Fig. 4.33** Spherulite structure of a special type of polymer when viewed between crossed polarizers in an optical microscope (Courtesy: WIKI)

It has been observed that the polymeric mixtures are less miscible than mixtures of small-molecule materials. This is because of the fact that the driving force for mixing is the entropy, not the interaction energy. Furthermore, the phase behavior of polymer solutions and mixtures is more complex than that of small-molecule mixtures [76–79]. Small-molecule solutions exhibit only an upper critical solution temperature phase transition, at which phase separation occurs with cooling. On the other hand, polymer mixtures commonly exhibit a lower critical solution temperature phase transition, at which phase separation occurs with heating.

Heating, exposed to ultraviolet light, may degrade or change the properties of a polymer. These changes may be in molecular weight, tensile strength, color, or shape. The degradation of polymers to form smaller molecules may proceed by random scission or specific scission. Polyethylene when heated to 450 °C degrades to form a mixture of hydrocarbons. Although these changes are undesirable, in some cases, such as biodegradation and recycling, they may be intended to prevent environmental pollution. Degradation can also be useful in biomedical settings. For example, a copolymer of polylactic acid and polyglycolic acid is employed in hydrolyzable stitches that slowly degrade after they are applied to a wound [80–82].

There is another class of polymer that is designed to change properties upon exposure to light (UV, deep UV, or extreme UV). These polymers can be identified as photon-sensitive (or radiation-sensitive) polymers or *photoresists*. The story of *photoresists* is largely one of polymer modification chemistry. To process a thorough understanding of the chemistry of radiation-sensitive polymers, we need a development engineer and a scientist. The engineering tools and the scientific background required are the essential parts to develop the micro (now nano) lithography to produce statistically good controlled feature size of the integrated circuits [83–85].

## 4.8 Innovation of a Special Type of Polymer: The Photoresists

*Patterning the word with a special polymer was the theme at the beginning of the nineteenth century*. Indeed, pattern generation by photolithographic technique is based on the application of a polymer film in the desired configuration of much greater complexity. It has been possible only through the film made out of an organic compound (materials that contain carbon as a backbone) whose structure and solubility change when exposed to light, particularly in the ultraviolet (UV) region. The first materials found to have this property were natural products such as fish glue, asphaltum, sugars, or gelatin synthesized with dichromates [85, 86].

*Transistor was probably the most important invention of the twentieth century and the story behind the invention is one of clashing egos* [85]. Indeed, the first transistor invented at Bell Lab, Murray Hill, NJ was few inches long (Fig. 4.34a), and in 1984 the size reduced to 1.0 μm and could accommodate almost 1,000000 transistors in an integrated circuit (IC) [83]. The continuing miniaturization of the device size following Moore's law remains unabated, and the recent device (IC) can accommodate 58 million transistors fabricated by 130 nm node technology. The successful new technology introduced by IBM Research, NY is called G-5 process (Fig. 4.34b) [83].

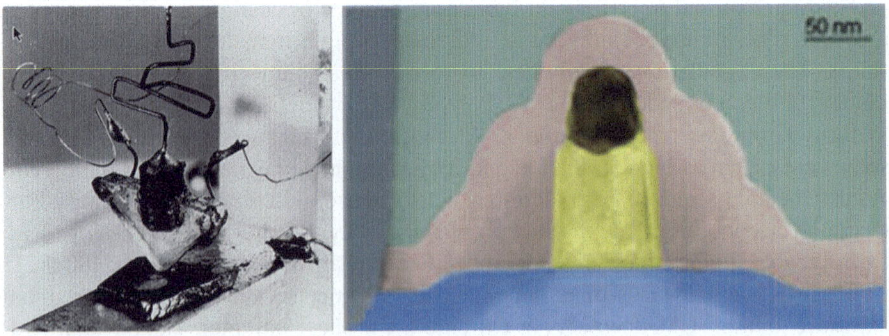


**Fig. 4.34** (**a**) The first transistor invented at Bell Lab in 1947 and (**b**) the modern twenty-first century transistor (130 nm node technology) (IBM)

On December 1, 2014, Transparency Market Research group indicates that photoresists (g-line and l-line, KrF, ArF Dry, ArF immersion) and photoresist ancillaries (antireflective coatings, photoresist developers, edge bead removers, and others) market for semiconductors and integrated circuits (ICs), printed circuit boards (PCBs) will grow at a rate of 6.2% between 2014 and 2020 [87].

The *carbon*-based organic chemical called the photoresist (a polymer with high molecular weight) is the most important member of the 32-nm device technology. The road map started with 365 nm lithography (from Greek *lithos* meaning *stone*, and *graphein* meaning *to write)*, or photolithography, a process used in modern microfabrication to pattern paths of a thin film or bulk of a substrate, with g-line ultraviolet light [84, 87]. The modern technology has developed alternatives to chemically amplified 193 nm photoresist materials that will be able to achieve the requirements associated with 32 nm device technology. In one of the polymers, the random copolymer Poly(styrene-co-4-hydroxystyrene) is prepared by radical polymerization of styrene, and 4-acetoxy styrene by hydrolysis of acetoxy group to OH [88, 89].

Figure 4.35 shows the molecular structure of SU-8, a special type of polymer developed by Microchem, Newton, MA. With the rapid growth of semiconductor industry, photolithography has advanced from UV technology to excimer laser technology to create structures from 1 μm to 45 nm and below in 2010 [90].

By far, the most important change in molecular structure and molecular weight due to exposure to UV light is observed in radiation-sensitive polymers [91–93].

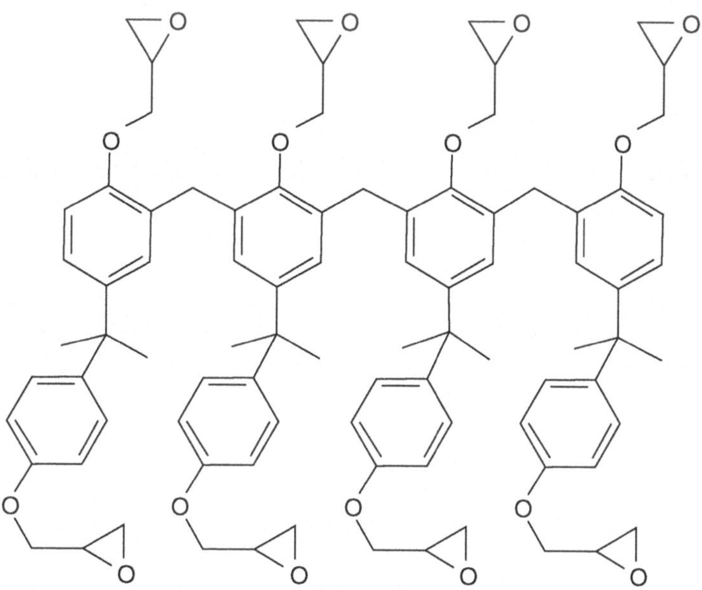


**Fig. 4.35** Molecular structure of SU-8 (Courtesy: Shaddack, WIKI)

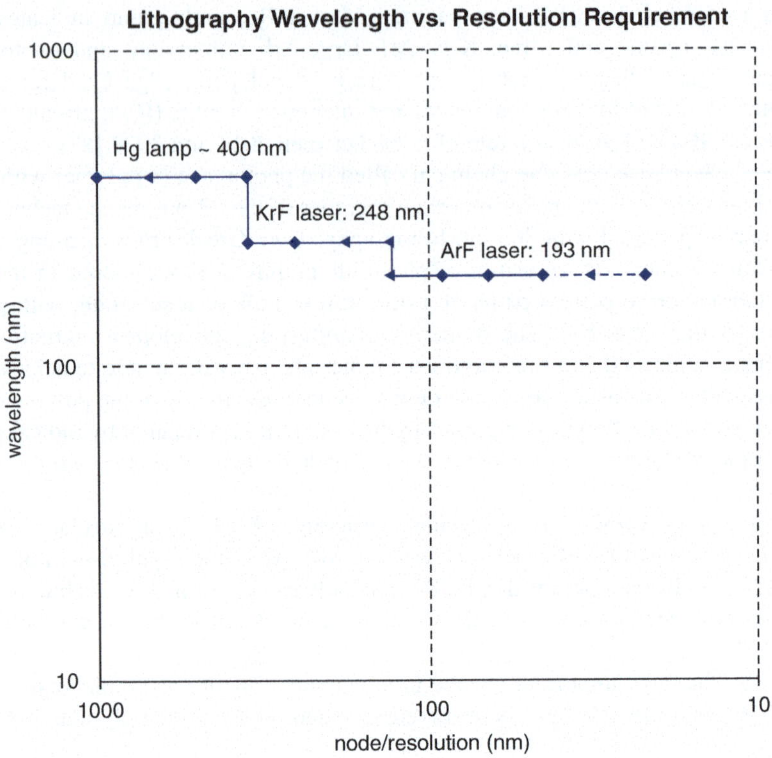


**Fig. 4.36** One of the evolutionary paths of photolithography (Courtesy: Guiding light, Creative Comm. WIKI)

These polymers are at the heart of the miniaturization of electronic devices that had brought revolution in integrated circuit (IC) technology in the year 1974. The continuing miniaturization shows little sign of abating from 1975 onward. Currently, devices containing over $5 \times 10^8$ transistors are manufactured with minimum circuit features of 22 nm. The circuit elements are fabricated by a series of processes known collectively as lithography. During the past several years, the term nanolithography has been used increasingly to denote the trend toward decreasing feature sizes. Figure 4.36 shows one of the evolutionary paths of photolithography that shows decreasing trends of nodes (wavelengths).

There are two types of photoresists (polymers), we call now as resists available in the market for semiconductor device processing. If the *resist* upon exposure to radiation increases resist solubility, and the ultimate illuminated resist forms a positive image of the mask applied to create fine structures, the resist will be called a *positive resist*. On the other hand, if the *resist* loses solubility when illuminated and forms a negative image of the photomask, it will be identified as *negative resist*. Both the resists belong to the polymer (a *carbon*-based organic material) family and contain a basic structural material called resin, a sensitizer, for complete

**Fig. 4.37** A 3D model of ethylene, the simplest alkene (Courtesy: WIKI)

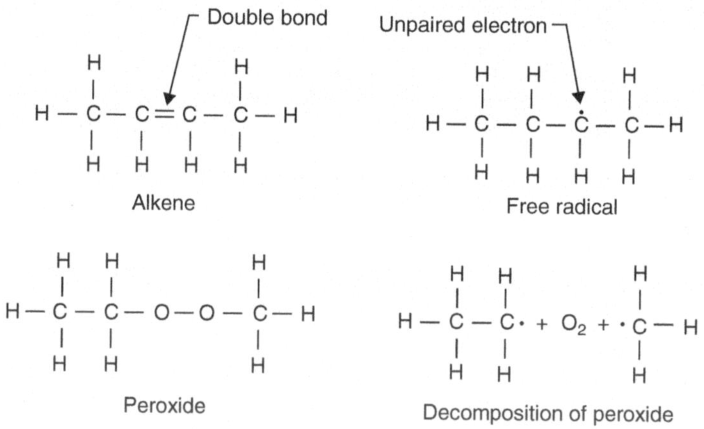


**Fig. 4.38** Chemical concepts pertinent to negative resist

polymerization upon exposure, and a solvent. However, the structures formed upon irradiation and developments are different with the two different photoresists [84].

Organic chemistry is the chemistry of carbon compounds, and an important class of carbon compounds is the hydrocarbons. Alkene is an unsaturated hydrocarbon, which has one or more carbon–carbon double bonds. Polymers from alkene monomers are referred to in a general way as polyolefins or in rare instances as polyalkenes [94–96]. The simplest alkene is ethylene ($C_2H_4$) (Fig. 4.37). The free radical (a free radical is any chemical species containing a single electron available for bonding) polymerization of alkenes (dialkene isoprene) gives rise to polyisoprene – a form of rubber, which has a high industrial value at great economy.

The isoprene is cyclized by encouraging some of the double bonds to react to form a closed loop and form the resin of the negative resist (Fig. 4.38).

The sensitizer of a negative resist is a six-sided carbon ring called bis-aryl diazide (BAD) such as 4–4′-diazidostilbene. The BAD species R is sandwiched between two azide substituents, $N_3$. On exposure to UV light, the compound loses

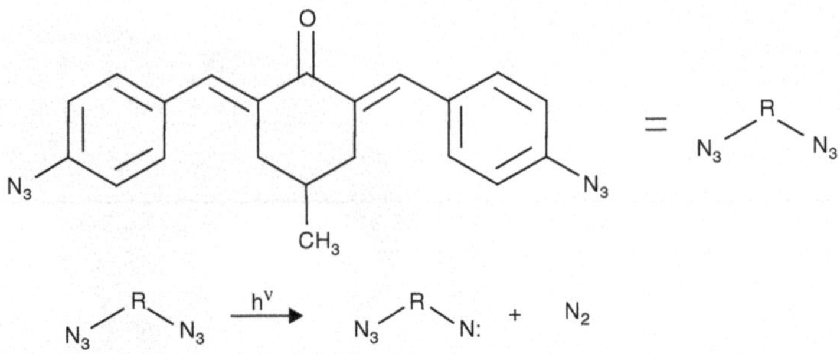


**Fig. 4.39** Bis-aryl diazide photosensitizer for negative resist

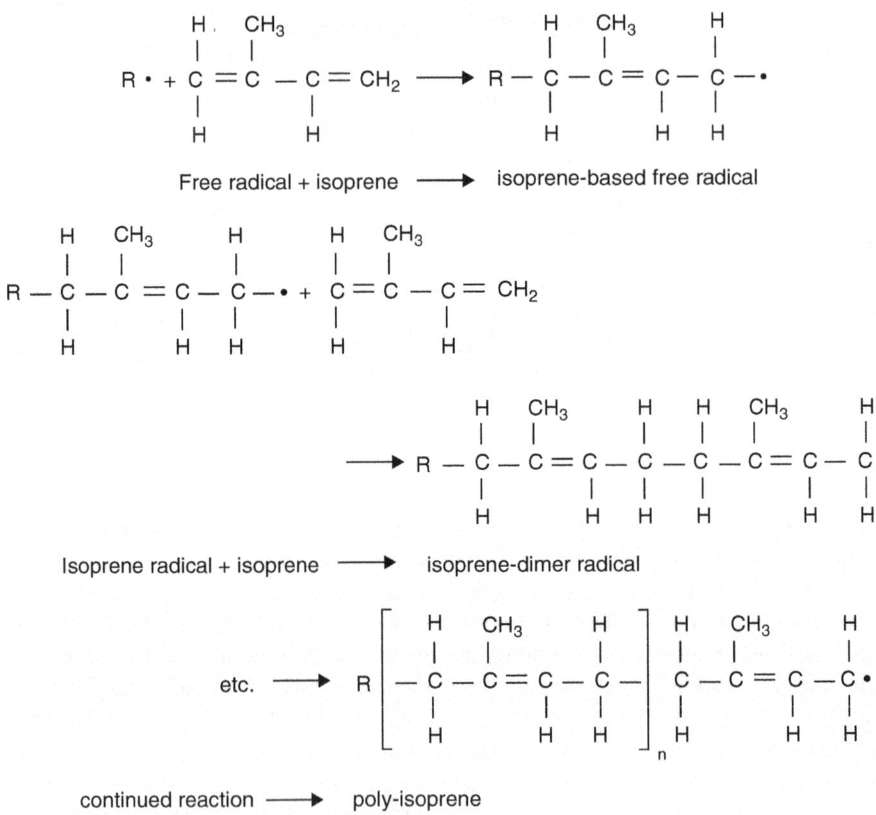


**Fig. 4.40** Free radical polymerization of isoprene

$N_2$ from each end and forms free radicals that can react with an alkene site in the resin (Fig. 4.39) [97–99].

The free radical polymerization of alkenes is a classic example of chain reaction. The chains made up of numerous units based on the original alkene are called polymers of the alkene (Fig. 4.40). Chain propagation continues indefinitely, until

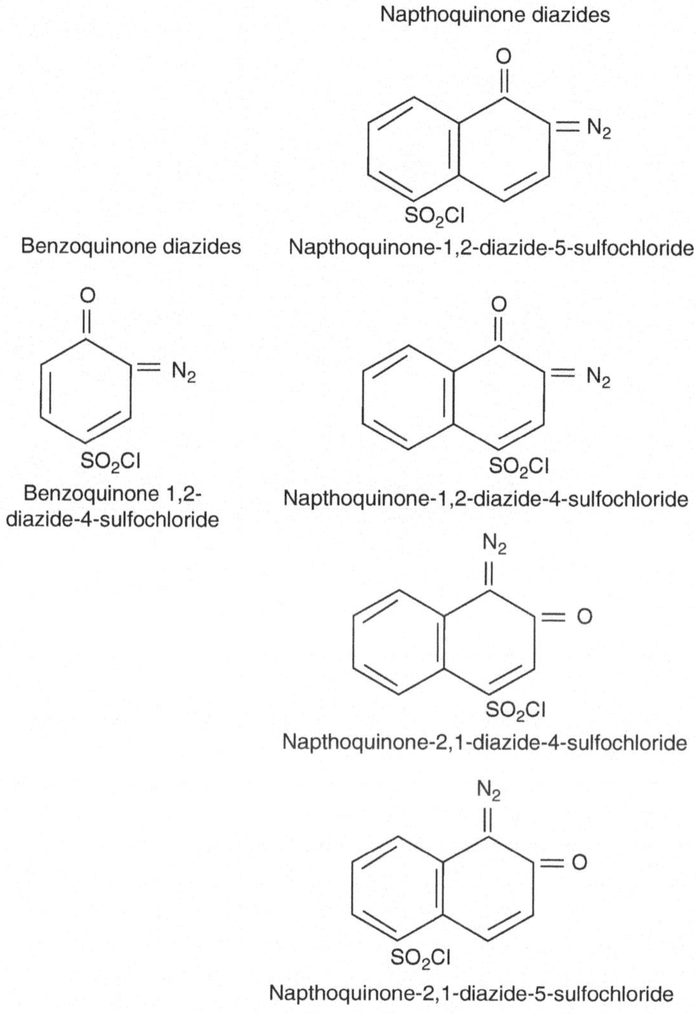


**Fig. 4.41** Some positive resist sensitizers

the odd electron is deactivated, either by reaction with the reactor wall, an impurity, or another free radical. Figure 4.40 illustrates these concepts using the dialkene isoprene, the building block of natural synthetic rubber.

On the other hand, some of the sensitizers used in positive resist resin are chemical compounds of sulfur chloride, together with either benzoquinone or napthaquinone (Fig. 4.41).

A positive resist is an $N_2$ group bonded hydrocarbon ring adjacent to a double bonded oxygen (ketone) group. Exposure to UV light leads to the dissociation of the $N_2$ group with the postulated formation of a five-carbon ring and a ketene: a=C=O

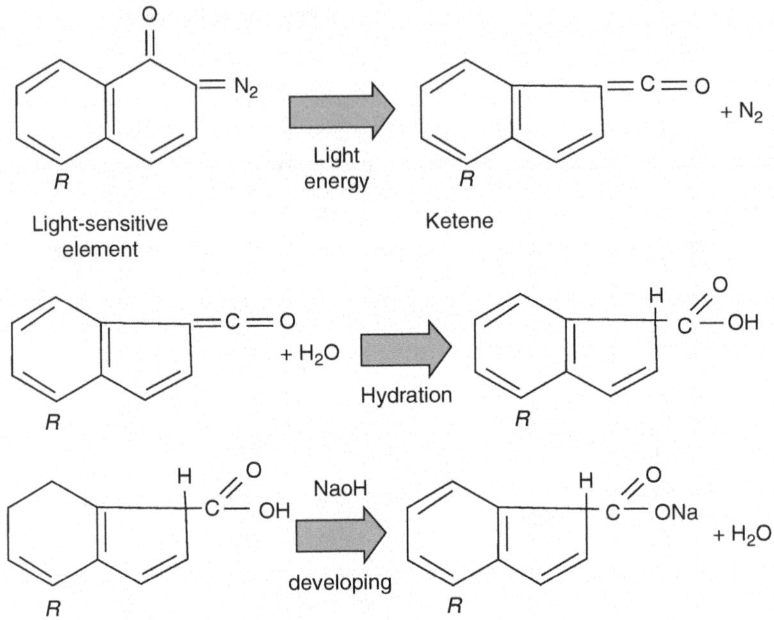


**Fig. 4.42** Irradiated positive resist after development

group. In the presence of moisture, ketene reacts to form carboxylic acid. Finally, in the presence of an alkaline material (NaOH), it forms a soluble ester (Fig. 4.42).

The semiconductor industry has continually pushed lithography technology to create smaller features and to achieve improved yields, and ultimately created an ongoing chemical imperative for the innovation of new photoresists. In the development of lithographic technology, a principal metric of advances has been the employment of shorter wavelengths of light in the process. Shorter wavelengths, paired with new photoresists capable of interacting with them, have allowed smaller features to be created. For the past three decades, lithographic processing has been carried out by using advanced, automated, and expensive manufacturing equipment known in the silicon community as process *tools* [100, 101].

In the beginning of the twenty-first century, the new process *tools* of making 45–22 nm circuit features has been implemented due to the development of extreme ultraviolet (EUV) nanolithography with the working emission of 13.5 nm [102, 103].

In organic chemistry, there are two effective ways of creating free radicals: exposure to ultraviolet (UV) light, or the addition of peroxides. Figure 4.43 shows the structure of a novolak resin and the additive in positive photoresist.

**Carbon Films from Pyrolyzed Photoresist** Carbon is a versatile material and has a wide range of applications in almost every sphere of our life. It shows reasonable electrical conductivity and good corrosion resistance that have made carbon as a better choice in the electrode industry. The thin carbon films produced from

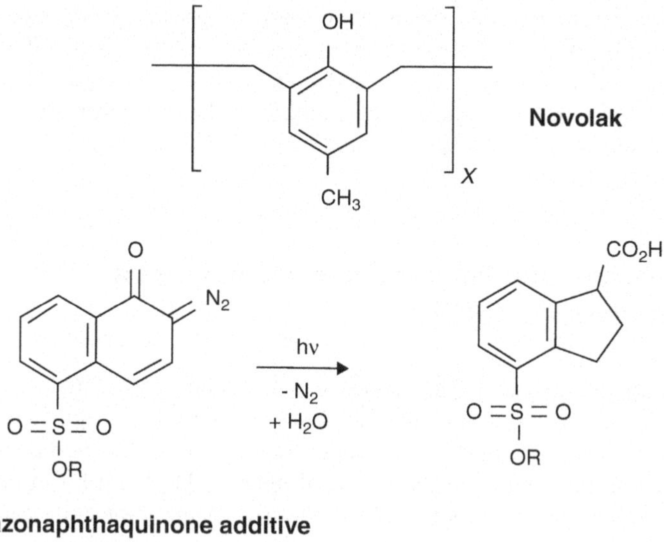


**Fig. 4.43** Structure of a novolak resin and the additive in positive photoresist

pyrolyzed photoresist are also used in microelectromechanical system (MEMS) devices [104, 105]. Carbon has been manufactured from coal tar pitches, phenol formaldehyde, polyfurfuryl alcohol, and polyvinyl alcohol. The yield of carbon is different from different materials, and the yield depends on the way it is manufactured [106]. Recently, thin carbon films have been used extensively as electrode materials [107, 108]., The photoresists can be carbonized by heating in an inert atmosphere for 60–90 min at different temperatures (starting from 700 to 1100°°C). The resist is spun on a silicon wafer and heated to 10° per minute. The chemical composition of the carbon films that are derived from pyrolyzed precursors is strongly dependent on the temperature at which carbonaceous films are formed. As the resist is heated in an ambient or inert atmosphere, it loses the solvent, and a -gradual weight loss is noticed. The weight loss is due to the loss of solvent, water, CO, and $CO_2$, followed by gradual thermal degradation of the aromatic carbon structure. The oxygen/carbon ratio determined from XPS is low ('1% for 1100 °C pretreatment), which increased more slowly upon exposure to air than that of the glassy carbon treated under identical conditions [109–111]. A novolac resin (Plenco, Plastic Engineering Company, Sheboygan, WI) shows hydrogen to carbon (H/C) ratio to decrease from 0.11 to 0.03 as the pyrolysis temperature increased from 800 to 1100 °C. Thus, the electrical resistivity of the carbonized photoresist changes due to the presence of hydrogen in the carbon structure. However, carbonaceous material produced from pyrolyzed photoresist at 1100 °C exposed to ambient air for 24 h does not show a measurable change in sheet resistance [107].

The silicon wafer with the carbonaceous polymer is cut into actual size and used for electrochemical measurements [107, 112]. Human neuroblastoma cells, or mouse teratocarcinoma cells, can be attached to the photoresist-derived carbon film [108]. The advantage of pyrolyzed photoresist carbon films is that they are highly conductive, with low surface oxygen and microstructure similar to glassy carbon. In addition to that, one can pattern the photoresist before pyrolysis and can form desired structured cell for a specific application [84, 113, 114].

## 4.9 Coordination Polymers and Metal Organic Frameworks (MOF)

Coordination polymers [115] and metal organic frameworks [116, 117] colloquially known as MOFs constitute an interdisciplinary field, with its origin in inorganic and coordination chemistry that has expanded rapidly in the last two decades, and is now attracting the interest of the chemical industry [118, 119]. Compounds comprising metal cluster vertices linked together by organic molecules can form one-, two-, or three-dimensional (3D) porous structures. Coordination polymers do not need to be crystalline. One can visualize the picture of metal clusters as joints where organic molecules are linkers. As a matter of fact, one can build a highly porous structure with lots of open space either by increasing the linker length or by increasing the size of the joint. Permutation and combination of the parameters such as the nature of metal ions and predesigned organic legends as interacting partners, their mole ratios, solvent medium, and experimental conditions (classical or hydrothermal synthesis) can lead to isolation of materials. With a wide range of architecture and with different levels of nano to meso porosities, it can form a potential framework of different materials, for delivery of medication into the human body or storage of voluminous quantities of gas molecules [120].

Briefly, when a single chain of polyethylene is cross-linked in 3D, it loses its identity as single chain of polyethylene or many individual chains of a polymer, the family it belongs to. The coordination polymers, on the other hand, either in solution or in polycrystalline state, show polynuclear entities, and when synthesized in a crystallized form become polymerized to form the coordination polymer [121].

Coordination polymers (CPs) and coordination networks (CNs), metal organic frameworks (MOFs), and organic–inorganic hybrid solids (OIHSs) are some of the terminologies that were used to identify the dimensionality, nature of interactions, and permanent porosity of the individuals belonging to a certain group. First generation of MOFs due to their permanent porosity and robustness were investigated mainly for fuel gas storage, separation, purification, and heterogeneous catalysis. However, hydrogen storage applications in energy conservative technologies and clean energy are considered to be the greatest achievement.

A hydrogen/air polymer electrolyte membrane fuel cell is fueled from an onboard reformer that can convert these fuels into hydrogen-rich gas mixtures. However, sulfur is found in all carbon-based fuels. The carbon monoxide (CO) released during the operation is removed immediately to avoid poisoning

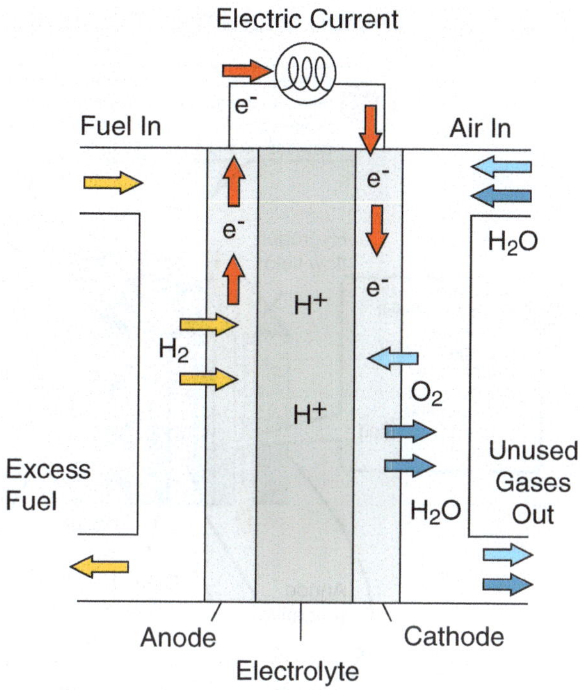


**Fig. 4.44** A proton-conducting fuel cell (Courtesy: R. Dervisoglu, WIKI)

the fuel cell catalyst. The catalyst material is platinum powder dispersed on carbon, and the ink is prepared by dissolving the powders in alcohol. The ink is then applied to the surface of the solid membrane made out of an organic polymer, usually poly (perfluorocarbon) acid [122]. Figure 4.44 shows a proton-conducting fuel cell.

## 4.10 Fuel Cells

Fuel cells are made up of three different segments, namely anode, the cathode, and an electrolyte in between. The electrolyte is a nonconductor of electricity and it allows ions to pass through it but not the electrons. The electrolyte substance usually defines the type of the fuel cell. The anode catalyst is made of very fine platinum powder, and is sufficiently reactive in bonding H and O intermediates, that facilitates the electrode process and effectively helps in releasing the intermediate to form the final product. The cathode catalyst turns the waste chemicals into water or carbon dioxide ($CO_2$). A fuel cell produces voltage from 0.6 to 0.7 V at full rated load. The voltage can be low due to some losses, e.g., ohmic loss, activation loss, or mass transport loss.

The cells that are to be operated at low temperatures are proton exchange membrane fuel cells (PEMFCs) and phosphoric acid fuel cells (PAFCs). In

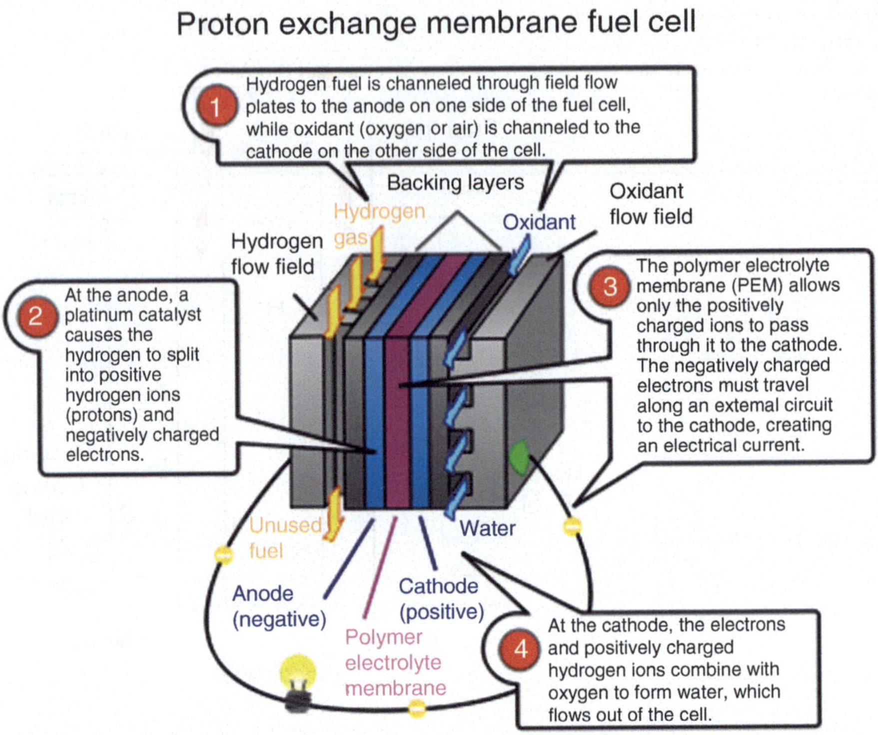


**Fig. 4.45** A high-temperature proton exchange membrane fuel cell (Courtesy: Jafet WIKI)

PEMFC cell, hydrogen is introduced on the anode side where it dissociates into proton and electrons. Protons are conducted through the membrane to the cathode, but electrons are forced to travel in an external circuit supplying power.

Figure 4.45 shows a high-temperature proton exchange membrane fuel cell (PEMFC). The cell is also called a solid polymer electrolyte fuel cell which works on the principle of proton exchange mechanism. The different components that are shown in Fig. 4.45 are bipolar plates, electrodes, catalyst, membrane, and the necessary hardware [123, 124]. There are two high-temperature fuel cells that work between 500 and 650 °C. They are: (a) solid-oxide fuel cells (SOFC), where a ceramic material called yttria-stabilized zirconia is used as the electrolyte [125] and (b) molten carbonate fuel cells (MCFCs), where lithium potassium carbonate salt is used as the electrolyte [126].

The *polymer-electrolyte* membrane cell is a low-temperature operated fuel cell, where the electrolyte is a solid polymer, and the trade name of the polymer is Nafion. It has a Teflon-like fluorocarbon backbone, and has hundreds of repeating –$CF_2$-CF-$CF_2$- units in length. A side chain –O-$CF_2$-CF-O-$CF_2$-$CF_2$- connects the molecular backbone to the third region and the ion clusters consisting of $SO_3^-$ negative ion. The $SO_3^-$ ion moves when hydrated, and the hydrogen ion ($H^+$) is available at the cathode when it moves through the ion-conducting membrane. The

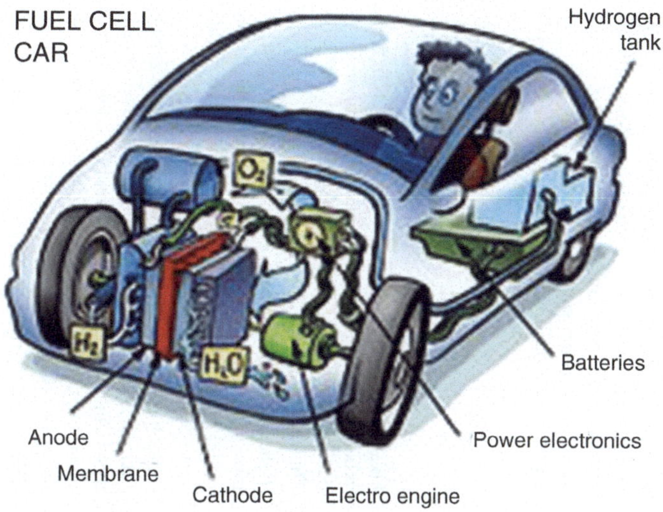


**Fig. 4.46** Configuration of components in a fuel cell car (Courtesy: WIKI)

reactions are favorable at a temperature around 80 °C. The final products of the overall cell reaction are electric power, water, and excess heat. The temperature of the cell stack is maintained at 80 °C [127, 128].

The other low-temperature fuel cell which uses phosphoric acid as a nonconductive electrolyte to transport hydrogen ions from the anode to the cathode is called phosphoric acid fuel cell (PAFC). These cells can work in temperatures between 150 and 200 °C. Since the invention, several companies have developed different types of cells: Honda (Honda FCX Clarity), Toyota (Toyota FCHV-adv), and Mecedes-Benz (F-Cell) [129]. Figure 4.46 shows the configuration of components in a fuel cell car. In 2012, fuel cell industry revenues exceeded $1 billion market value worldwide, with Asian Pacific countries shipping more than ¾ of the fuel cell system worldwide [130].

In most of the fuel cells, *carbonaceous material* is one of the ingredients of the electrolyte. In 2011, one of the published research papers documented the use of relatively inexpensive doped carbon nanotube instead of expensive metal like platinum (Pt) [131]. As for example, in molten carbonate fuel cells (MCFCs), lithium potassium carbonate is used as the electrolyte. Not only that; in hydrocarbon fuel cell, methanol and diesel are used. In addition to these fuel cells, polymer membranes are used in microbial fuel cells, regenerative fuel cells, and direct formic acid fuel cells [132–134]. The fuel cell is one of the fields where carbon derivative materials are used. A comparative analysis of different fuel cells is given in Table 4.1.

**Table 4.1** Comparison of different fuel cells

| Fuel cell Type | Common electrolyte | Operating temperature | Typical stack size | Efficiency | Applications | Advantages | Disadvantages |
|---|---|---|---|---|---|---|---|
| Polymer electrolyte membrane (PEM) | Perfluoro sulfonic acid | Typically ~80 °C | <1 kW–100 kW | 60% transportation 35% stationary | 1. Backup power<br>2. Portable power<br>3.Distributed generation<br>4.Transportation<br>5. Specifically vehicles | Solid electrolyte reduces corrosion and management problems, low temperature operation, and reactions start quickly | Expensive catalysis, sensitive to fuel impurities and waste heat |
| Alkaline fuel cell (AFC) | Aqueous solution of potassium hydroxide soaked in a matrix | 90–100 °C | 10–100 W | 60% | Military and space | Alkaline solution helps cathode reaction faster, leading to higher efficiency | Sensitive to $CO_2$ in fuel and air, electrolyte management |
| Phosphoric acid fuel cell (PAFC) | | 150–200 °C | 400 kW module | 40% | Distributed generation | Higher temperature operation, increase tolerance to fuel impurities | Expensive Pt-catalyst, start-up time is high, low current and power yield |
| Molten carbonate fuel cell (MCFC) | Solution of lithium, and sodium or potassium carbonates soaked in a matrix | 600–700 °C | 300 kW–3 MW module | 45–50% | Electric utility and distributed generation | High efficiency, fuel flexibility, variety of catalysts can be used, can manage combined heat and power (CHP) | High-temperature corrosion, breakdown of cell components, long start-up time, low power density |
| Solid oxide fuel cell (SOFC) | Yttria-stabilized zirconia | 700–1000 °C | 7 kW–2 MW | 60% | Auxiliary power, electric utility, distributed generation | High efficiency, fuel flexibility, can use a variety of catalysts, solid electrolyte, hybrid/GT cycle, suitable for CHP | High-temperature corrosion, breakdown of cell components, long start-up time |

Adapted from WIKI

## 4.11 Polymer Electrolytes

The classic *polymer electrolytes* are *ion polymer complexes* that contain a Lewis base (e.g., ethylene oxide unit $-OCH_2CH_2$-) to solvate the lithium (Li) salt. The distinguishing feature of polymer electrolytes is that ion movement occurs without long-range displacement of the solvent and the conductivity, which is coupled to the micro-Brownian motion of segments of the polymer chains above the glass transition temperature ($T_g$), rises significantly with increasing amorphicity of the polymer electrolyte. These ion–polymer complexes were discovered in 1973 by Fenton, Parker, and Wright. Since then, these complexes have been prepared with many different monatomic ions and with different polymeric legends. Polymer electrolytes offer pronounced advantages over conventional liquid electrolytes because of their versatile shape, mechanical strength, and stable contact between the electrode and electrolyte interface. As a result, these electrolytes are being used in *electrochromic* and *temperature-sensitive printed electronic devices, biomedical sensors, supercapacitors, solar cells,* and *batteries* [135].

One class of polymer electrolytes consists of polar pyridine units in the backbone and polyethylene oxide (PEO) in the side chains. Recently, solid polymer electrolytes have been prepared by blending the PEO-based polymers with lithium (Li) salt and plasticizer. The ion conducting pathway can be controlled in these complexes by blending PEO powder with PEO5 (cast PE with high content of Li salt (EO: Li = 5) at room temperature, followed by hot pressing at 60 °C. The enhancement in ionic conductivity is believed to be due to the formation of a controlled 3D network of the ion-conductive amorphous phase (Fig. 4.47) [136].

A comprehensive research of crystalline PEO6:LiXF6 ($X$ = P, phosphorus, As, Arsenic, or Sb, antimony) polymer electrolytes has been carried out by Bruce et al. [137, 138]. In 1:6 complexes, two PEO chains fold to form cylindrical tunnels inside which $Li^+$ ions are located and coordinated by the ether oxygen atoms. The channels provide pathways for the movements of cations (Fig. 4.48).

The cast films of these polymer electrolytes display entangled orientation of helices, which have a favorable disposition in the parallel-to-the-casting plane (denoted as XY). As a result, the longitudinal conductivity is much higher than that in the perpendicular direction. The most interesting phenomenon that has been observed with these casted polymer electrolytes is that mechanical stretching could increase the conductivity in the parallel direction. Figure 4.49 shows longitudinal DC conductivity (a) and orthogonal AC conductivity (b) at near ambient temperature of as-cast and stretched LiX:P $(EO)_{16–20}$ polymer electrolytes.

Upon stretching, most helices gain preferable orientation along the stretched direction. Scanning electron micrograph (SEM) shows clear grain structure for the pristine PE samples, whereas a fibrous structure is observed in lengthwise direction upon stretching (Fig. 4.50).

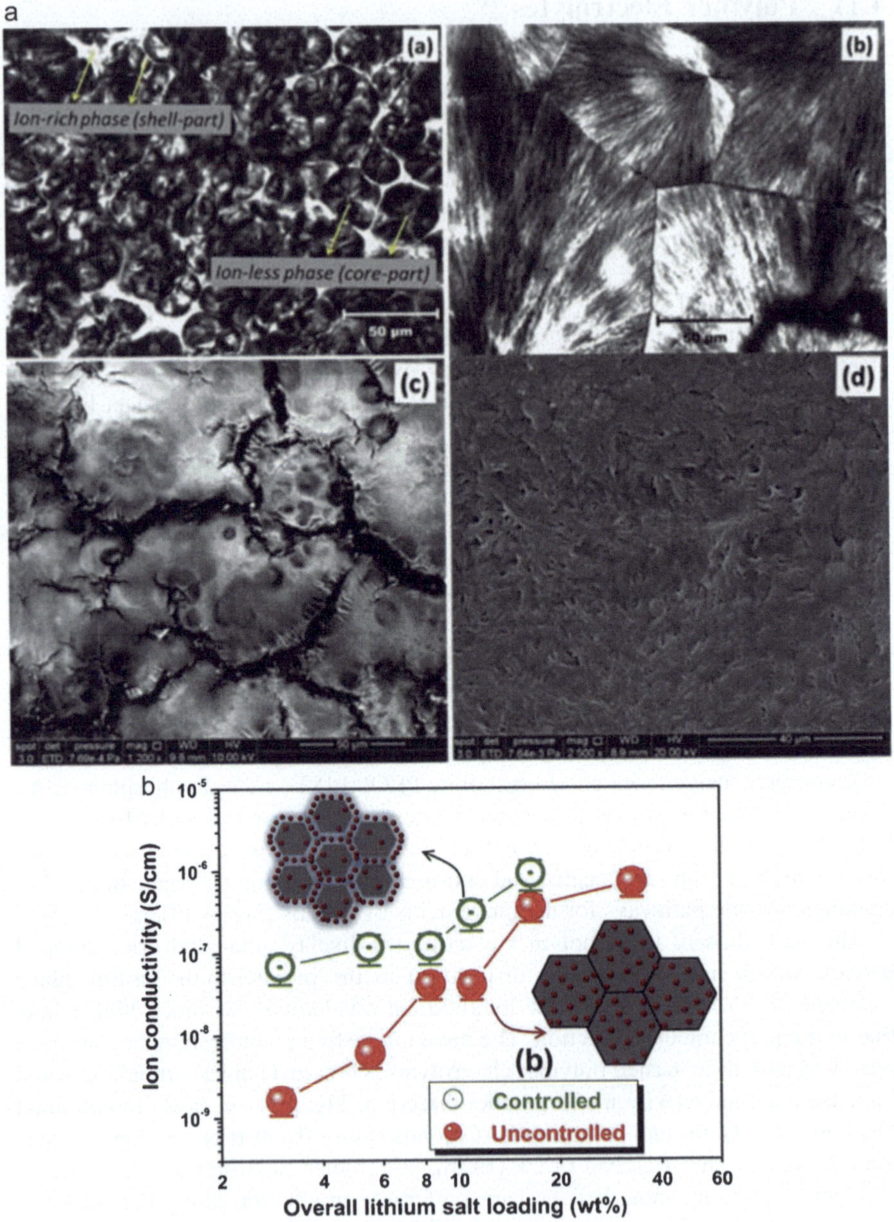


**Fig. 4.47** (**a**) and (**b**) Polar light microscopic images;(**c**) and (**d**) SEM images. The figures on the top and bottom of b show the controlled and typical uncontrolled electrolyte samples, respectively (Reproduced with permission of D. Goldnitsky et al. [135]).

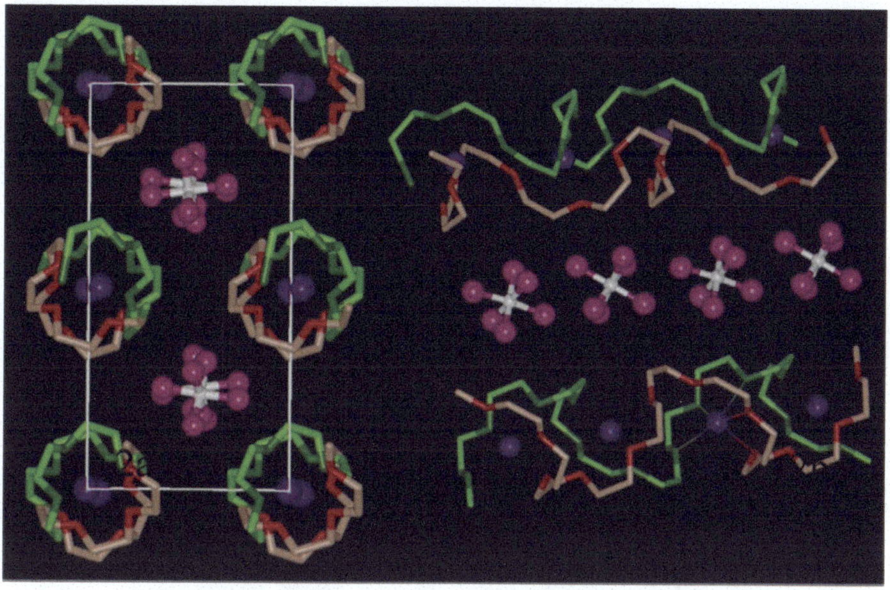

**Fig. 4.48** Structure of the polymer electrolyte PEO6:LiAsF6; *Left*: view of the structure along the chain axis showing rows of Li + ions perpendicular to the page; *Right*: view of the structure showing the relative position of the chains and their conformation (Reproduced with permission of D. Goldnitsky et al. [135])

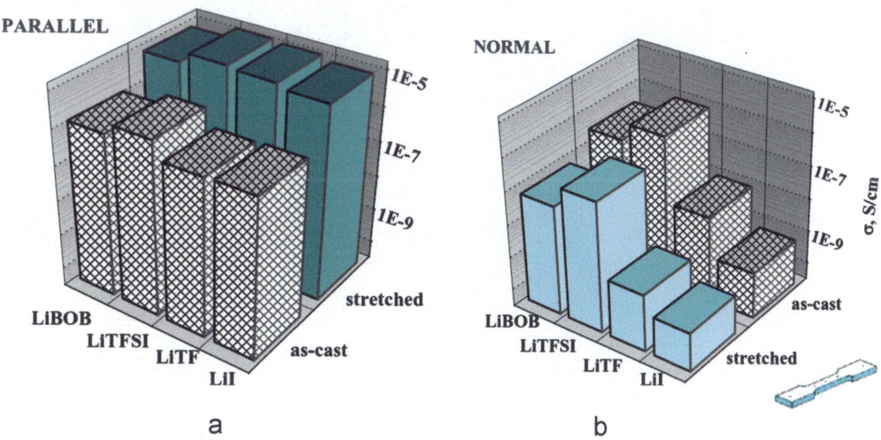


**Fig. 4.49** (**a**) Longitudinal DC conductivity and (**b**) AC conductivity at near ambient temperature of as-cast and stretched LiX:P(EO)$_{16–20}$ polymer electrolytes (Reproduced with permission of D. Goldnitsky et al. [135])

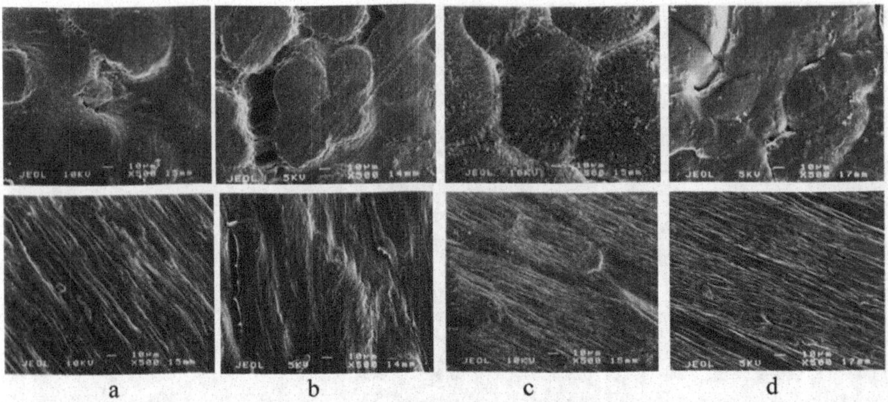

**Fig. 4.50** SEM micrographs of the as-cast (upper images) of the polymer electrolytes; lower images show when the films are stretched. (**a**) $LiCF_3SO_3$, (**b**) $LiAsF_6$, (**c**) LiI, (**d**) LiB $[(OCO)_2]_2$ (Reproduced with permission of D. Goldnitsky et al. [135])

**Fig. 4.51** Three-dimensional topography AFM images of (**a**) as-cast and (**b**) stretched LiI:P $(EO)_{40}$ polymer electrolyte (Reproduced with permission of D. Goldnitsky et al. [135])

Three-dimensional topography images of the as-cast and stretched films are taken by AFM. From Fig. 4.51a, it is clear that the films consist of grains of several micron size oriented randomly, and Fig. 4.51b shows the fibrous-type film upon stretching. The width of the fibers varied from 40 to 80 nm [139].

## 4.12 Summary

The term polymer is derived from the ancient Greek word and refers to a molecule whose structure is composed of multiple repeating units of *carbon*, and mostly hydrogen and/or nitrogen and oxygen. It is the major driving force behind most of the research that tries to design plastics fit for electronic applications. Each polymer type has already found widespread applications in many optical and electronic devices such as batteries, displays, plastic wires, optical signal processing, information storage, solar energy conversion, and in molecular biochemicals that shape the life of the living organisms (DNA, RNA, protein, peptides, glycogen, etc.), and in molecular electronics.

The modern concept of polymers is as covalently bonded macromolecular structures. Historically, products arising from the linkage of repeating units by covalent bonds have been the primary focus of polymer science. Polyisoprene of latex rubber and polystyrene of Styrofoam are examples of polymeric natural/biological and synthetic polymers, respectively. In biology, all biological macromolecules like proteins (polyamides), nucleic acids (polynucleotides), and polysaccharides are purely polymeric, or composed of mainly polymer components, e.g., isoprenylated/lipid-modified glycoproteins, where small lipidic molecule and oligosaccharide modifications occur on the polyamide backbone of protein.

Polymers are of two types: *natural* and *synthetic*. Natural occurring polymers are cotton, starch, wool, silk, and natural rubber. *Natural rubber* is a mixture of hydrocarbons with the empirical composition $C_5H_8$. *Natural rubber*, or the *latex rubber*, that comes from the latex of *Hevea brasiliensis* is mainly poly-cis-isoprene.

Heating, exposed to ultraviolet light, may degrade or change the properties of a polymer. These changes may be in molecular weight, tensile strength, color, or shape. The degradation of polymers to form smaller molecules may proceed by random scission or specific scission. Polyethylene when heated to 450 °C degrades to form a mixture of hydrocarbons. Although these changes are undesirable, in some cases, such as biodegradation and recycling, they may be intended to prevent environmental pollution. Degradation can also be useful in biomedical settings. For example, a copolymer of polylactic acid and polyglycolic acid is employed in hydrolysable stitches that slowly degrade after they are applied to a wound.

There is another class of polymer which changes its properties upon exposure to light (UV, deep UV, or extreme UV). The polymer can be identified as photon-sensitive (or radiation-sensitive) polymers or *photoresists*. The story of *photoresists* is largely one of polymer modification chemistry. To process a photoresist and to understand the chemistry of radiation-sensitive polymers, we need a development engineer and a scientist. The engineering tools and the scientific background are required to develop the nanolithography and to produce statistically good controlled feature size of the integrated circuits.

Another important use of polymers is the polymer-electrolyte membrane cell, which is a low-temperature operated fuel cell, where the electrolyte is a solid polymer (the trade name of the polymer is Nafion). In most of the fuel cells, carbonaceous material is one of the ingredients of the electrolyte. In 2011, inexpensive doped carbon nanotube was used in a fuel cell instead of expensive metal like platinum (Pt) [131]. In molten carbonate fuel cells (MCFCs), lithium potassium carbonate is used as an electrolyte.

The last but not the least polymer that has been attracted the attention of the research scientists is polymer electrolyte. The classic polymer electrolyte comprises organic macromolecules (usually polyether polymer) that are complexed with inorganic salts. The high-conductive polymer electrolyte has been used in electrochromic and temperature-sensitive printed electronic devices, biomedical sensors, supercapacitors, solar cells, and batteries. Polymer electrolyte is generally a semicrystalline material most commonly derived from archetypal polyethylene oxide (PEO).

## References

1. R.J. Young, *Introduction to Polymers* (Chapman & Hall, London, 1987)
2. U.W. Gedde, *Polymer Physics* (Springer Sci, Heidelberg), pp. 1–3
3. A. Rudin, P. Choi, *The Elements of Polymer Science and Engineering*, 3rd edn. (Academic Press, San Diego, 2013)
4. W.B. Jensen, Ask the historian: the origin of the polymer concept. J. Chem. Educ. **88**, 624 (2008)
5. M. Rubinstein, R.H. Colby, *Polymer Physics* (Oxford University Press, New York, 2003), p. 6
6. W.S. Johnson, W.H. Stockmayer, H. Taube, *Biographical Memoirs, Paul Flory*, vol 82 (The National Academic Press, Washinton DC, 2002)
7. R. Haag, Supermolecular drug delivery system based on polymeric core-shell architecture. Angew. Chem. Int. Ed. **43**(3), 278 (2004)
8. H. Cho, T.C. Lai, K. Tomoda, G.S. Kwon, Polymeric micelles for multidrug delivery in cancer. AAPS Pharm. Sci. Tech **16**(1), 10 (2015)
9. H.-J. Schnieder, *Applications of Supramolecular Chemistry* (CRC Press, Boca Raton, 2012)
10. Y. Zhang, Polymers in therapeutics and nanomedicine. Mater. Matt. **9**(3), (2014). Sigma-Aldrich, Pub. Millwaukee, Wisconsin
11. D. Sutton, N. Nasongkla, E. Blanco, J. Gao, Functionalized micellar system for cancer Targetted drug delivery. Pharm. Res. **24**(6), 1029 (2007)
12. Y. Matsumura et al., Phase I clinical trial and pharmacokinetic evaluation of NK911, a micelle-encapsulated doxorubicin. Br. J. Cancer **91**, 1775 (2004)
13. J.M. Shaw, Overview of Polymers for Electronic and Photonic Applications, in *Polymers for Electronic and Photonic Applications*, ed. by C.P. Wong (Academic, Boston, 1993), pp. 1–65
14. M. Tyagi, D. Tyagi, Polymer Nano-composite and their applications in electronic industry. Int. J. Electron. And Electric. Engineer **7**(6), 603 (2014)
15. C.A. Schalley, K. Beizai, F. Vogtle, On the way to Rotaxane-based molecular motors: studies in molecular mobilities and topological chirality. Am. Chem. Soc **34**(60), 465 (2001)
16. P.N. Taylor et al., Insulated molecular wires: synthesis of conjugated polyrotaxanes by Suzuki coupling in water. Agnew. Chem. Int. Ed **39**(19), 3456–3460 (2000)
17. C. Romuald, A. Arda, C. Clavel, J. Jimenez-Barbero, F. Coutrot, Tightening and loosening a pH –sensitive double Lasso molecular machine readily synthesized from an end-activated ($C_2$) daisy chain. Chem. Sci. (RSC Pub.) **3**, 1851 (2012)
18. H. Lodish et al., *Molecular Cell Biology*, 5th edn. (WH Freeman and Co., New York, 2004)

19. A. Gutteridge, J.M. Thornton, Understanding nature's catalytic toolkit. Trends Biochem. Sci. **30**(11), 622 (2005)
20. C. Scholz, Poly (amino acid) block copolymers, for drug delivery and other biomedical applications. Mater. Matt **9**(3), 73 (2014)
21. R. Riva, C. Jerome, Chitosan, a versatile platform. Mater. Matt **9**(3), 95 (2014.) Sigma-Aldrich Pub
22. I.M. Verma, N. Somia, Gene therapy—promises, problems, and propects. Nature **389**(6648), 239 (1997)
23. R. Riva et al., Chitosan derivatives in drug delivery and tissue engineering, in *Chitosan for Biomaterials*, ed. by R. Jayakumar et al. (Springer, Heidelberg, 2010), pp. 19–44
24. F.P. Guengerich, Cytome P450s and other enzymes in drug metabolism and toxicity. AAPS J. **8**(1), E101 (2006)
25. C.M. GGrisham, H.G. Reginald, *Biochemistry* (Saunders College Pub, Philadelphia, 1999)
26. H. Buc, T. Strick, S. Neidle, et al., *RNA Polymerase as Molecular Motors* (Royal Society Of Chemistry, Cambridge, 2009)
27. C.C. Richardson , R. D. Kornberg et al, Annual Rev. Biochem., Annual Reviews, Palo Alto, CA and C.C. Richardson , I.R. Lehman, and R. D. Kornberg Deoxyribonucleic acid Phosphate Exonuclease fron E-Coli II, J. Biol. Chem. 239, 251 (1964)
28. Genes and disease, Nat. Center for Biotechnology Information (US), Bethesda (MD), (1998)
29. S.K. Nair, T.L.Claderone, D.W. Christianson, C.A. Fierke, Alternating the mouth of a hydrophobic pocket. Structure and Kinetics of Humancarbonic anhydrase II mutants at residue Val-121. J. Bilogical Chem, **266**(26), 17320. Nat. Center for Biotechnology Information (US), Bethesda (MD), (1991)
30. Entrez Gene: CA2 carbonic anhydrase II, Nat. Center for Biotechnology Information (US), Bethesda (MD)
31. S.J. Weinstein et al., Null association between prostate cancer and Serum Folate, Vitamin B (6), Vitamin B (12), and homocysteine. Cancer Epidemiol Biomak. Prev **12**(11Pt 1), 1271. PMID 1452294 (2003)
32. V. Renugopalkrishnan et al., Rational design of thermally stable proteins: relevance to bionanotechnology. J. Nanosci. Nanotechnol. **5**(11), 1759 (2005)
33. K. Hult, P. Berguland, Engineered enzymes for improved organic synthesis. Curr. Opin. Biotechnol. **14**(4), 395 (2003)
34. R. Li, *Forensic Biology*, 2nd edn. (CRC Press, Boca Raton, 2015), p. 291
35. W.K. Purves, D.E. Sadova, G.H. Orians, H.C. Hailer, *Life: The Science of Biology*, 7th edn. (Sinauer Assoc. Inc. Pub, Sunderland, 2004), p. 48
36. T. Feizi, W. Chai, Oligosaccharide microarrays to decipher the glycol code. Nat. Rev. Mol. Cell Biol. **5**(7), 582 (2004)
37. L.H. Sperling, *Introduction to Physical Polymer Science* (Wiley, Hoboken, 2006)
38. J.C. Paterson Jones, M.G. Gillard, J. van Staden, The biosynthesis of natural rubber. J. Plant Physiol. **136**(3), 257 (1990)
39. M.J. McCoy, M.S. Matthew Wolter, K.E. Anderson, A safety Professional's review of natural rubber latex testing in work place. Am. J. SH & E Res **5**(3), 1 (2008)
40. M.D. Joesten, J.L. Hogg, M.E. Castellion, *The World of Chemistry Essentials* (Thomas Brooks Cole Pub, Bemont, 2007), p. 337
41. H.P. Affek, D. Yakir, Natural abundance carbon isotope comparison of isoprene reflects incomplete coupling between isoprene synthesis and photosynthetic carbon flow. Plant Physiol. **131**, 1727 (2003)
42. T.D. Sharkey, A.E. Wiberby, A.R. Donhue, Isoprene emission from plants: why and how. Ann. Bio. **10**(1), 5 (2008)
43. W.H. Carothers, Studies on polymerization and ring formation. I. An introduction to general theory of condensation polymers. J. Am. Chem. Soc. **51**(8), 2548 (1929)

44. T. Desmond, W. Obrecht, W. Wolfgang, G. Wachholz, R. Engehausen, Robber, 3 synthetic rubbers: introduction and overview, in *Ulmann's Encyclopedia of Industrial Chemistry*, (Wiley-VCH, Weinham, 2011)
45. A.K. Naskar et al., Tailored recovery of carbons from waste tires for enhanced performance as anodes in Lithium ion batteries. RSC Adv. **4**(72), 38213 (2014)
46. B. Kuhlman et al., Design of a novel globular protein fold with atomic level accuracy. Science **302**(5649), 1364 (2003)
47. H.M. Berman, The protein data bank: a historical perspective. Acta Crystallogr. **A-64**, 88 (2008)
48. J.W. Watson, *DNA, The Secret of Life* (Alfred A. Knopf pub, New York, 2004)
49. B. Alberts, A. Johnson, J. Lewis, M. Raff, K. Roberts, P. Walter, *Molecular Biology of the Cell* (Garland Science Pub, New York, 2002)
50. A. Smith, K. Shaw, Discovering the relationship between DNA and protein production. Nature Education **1**(10), 112 (2008)
51. S. Brenner et al., An unstable intermediate carrying information from genes to ribosomes for protein synthesis. Nature **190**, 576 (1961)
52. E. Fahey et al., Update of the LIPIDS MAPs comprehensive classification system of lipids. J. Lipid Res. **50**, S9 (2009)
53. Y. Chen, E.E. Kely, R.P. Masluk, C.L. Nelson, D.C. Cantu, P.J. Reilly, Structural classification and properties of Ketoacyl synthesis. Protein Sci. **20**(10), 1659 (2011)
54. R. Johnson, Proteomics: profiling lipited proteins. Nat. Chem. **7**, 456 (2015)
55. A.J. Lusis, P. Pajukanta, A treasure trove for lipoprotein biology. Nat. Genet. **40**(2), 129 (2008)
56. V. Kumar et al., Three dimensional CryoEM reconstruction of native LDL particles to 16ÅResolution at physiological body temperature. PLoS One **6**(5), e18841 (2011)
57. H. Garrett, M. Grishman, *Biochemistry*, 4th edn. (Brooks Cole, MA, 2008)
58. K. Bloch, D. Rittenberg, On the utilization of acetic acid for cholesterol formation. J. Biol. Chem. **145**, 625 (1942)
59. N. Kresge, R.D. Simon, R.L. Hill, The biosythetic pathway for cholesterol. J. Biol. Chem. **280** (2005)
60. J. T. Smith, Squalene potential Chemopreventive agent. Expert Opinion on Investigational Drugs. **9**(8), 1841 (2000) and also Gulf war Syndrome, U. Virginia archive, July 14, (2004)
61. M. Rubinstein, R.H. Colby, *Polymer Physics* (Oxford University Press, Oxford, GB, 2003)
62. J. Clayden, N. Greeves, S. Warren, *Organic Chemistry* (Oxford University Press, Oxford, London, 2000), p. 1450
63. G. Odian, *Principles of Polymerization* (Wiley, Hoboken, 2004)
64. K. Matyjaszewski, *Cationic Polymeriztions: Mechanisms, Synthesis, and Applications* (Marcel and Dekker, Inc., New York, 1996)
65. H. Hsieh, R.P. Quirk, *Anionic Polymerization, in Encyclopedia of Polymer Science and Technology* (Wiley, New York, 2003)
66. H. Hsieh, R. Quirk, *Anionic Polymerization: Principles and Practical Applications* (Marcel Dekker, New York, 1996)
67. J. Maul, B.G. Frushour, J.R. Kontoff, H. Eichenauer, K.-H. Ott, C. Shade, *Polystyrene and Styrene Copolymers, in Ullmann's Encyclopedia of Industrial Chemistry* (Wiley-VCH, Weinheim, 2007)
68. W. Saenger, *Principles of Nucleic Acid Structures* (Springer, New York, 1984)
69. A.G. Leslie, S. Amott, R. Chandrasekaran, R.L. Ratliff, Polymorphism of DNA double helix. J. Mol. Biol. **143**(1), 49 (1980)
70. J.Darnell, H. Lodish, D. Baltimore, *Molecular Cell Biology* (Scientific American, Washington, DC, 1986), p. 83
71. S. Kavesh, J.M. Schultz, Meaning and measurement of crystallinity in polymer. Poly. Eng. Sci **9**(5), 331 (1969)
72. G. Odian, *Principles of Polymerization*, 3rd edn. (Wiley, New York, 1991), p. 27

73. G.W. Ehrenstein, R.P. Theriault, *Polymeric Materials: Structure, Properties, Applications* (Hanser Verlag Pub., Munich), p. 67
74. B. Tipton, Prevention of Environmentally Induced Degradation of Carbon/Epoxy Composite Materials, M.S. Thesis, Mech. Mater. And Aerospace Eng. Univ. of Central Florida, FL., (2000)
75. K.H. Wu, K. Fa Cheng, C.C. Yang, C.P. Wang, C.I. Liu, Thermal and optical properties of epoxy/Siloxane Hybrimer based on sol-gel derived phenyl Siloxane. Open J. of Comp. Mater **5**, 49 (2015)
76. R.E. Young, in *Unsaturated Polyester Technology*, ed. by P.E. Bruins (Gordon and Beach, New York, 1976), p. 315
77. N. Eric, The Complex Phase Behavior of Aqueous Solutions of Water-Soluble Polymers, Ph. D. Thesis, Department of Chemistry, Katholieke Universiteit Leuven, Belgium, Leuven, (April 2014)
78. F. Castro-Marcano, A.M. Catano-Burrera, C.M. Collina, Phase behavior of polymer solutions from macroscopic properties: application to the perturbed chain statistical associating fluid theory equation of state. Ind. Eng. Chem. Res. **50**(2), 1046 (2011)
79. C. Wohlfarth, *Hand Book of Liquid-Liquid Equilibrium Data of Power Solutions* (CRC Press Pub, Boca Raton, 2008)
80. C.E. Astete, C.M. Sabliov, Synthesis and characterization of PLGA nanoparticles. J. Bio. Mater. Sci. Poly.ed **17**(3), 247 (2006)
81. E.Vey, A.F. Miller, M. Claybourn, and A. Saiani, In Vivbro Degradation of Poly (lactic-$CO_2$-Glycolic) Acid Random Co-polymers, The 6th. Int. Symp. On Polymer-Solvent Complexes, Aug. 29-1st Sept. (2006), Manchester, UK. Macromolcular Symp. 251 (1), 81, (2007)
82. T. Lin, C.K. Poh, W. Wang, Poly (lactic-co-glycolic acid) as a controlled release delivery device. J. Mater. Sci. Mater. Med. **20**(8), 1669 (2009)
83. T.K. Gupta, *Copper Interconnect Technology* (Springer, New York, 2009) and also T.K. Gupta, *Hand Book of Thick and Thin Film Hybrid Microelectronics* (Wiley, Hoboken, 2003)
84. T.K. Gupta, *Hand Book of Thick and Thin Film Hybrid Microelectronics* (Wiley, Hoboken, 2003)
85. J. Kosar, *Light Sensitive Systems* (Wiley, NY, 1965) and also Ira Flatow, Transistorized!, broad cast on PBS, and TV movie made by Ira Flatow, Executive Producer and Director, ScienCetral, in 1999, credit goes to Am. Physical Soc
86. R. Glang, L.V. Gregor, Generation of patterns in thin films, in *Handbook of Thin Film Technology*, ed. by L.I. Maissel, R. Glang (McGraw Hill, New York, 1970)
87. L.F. Thompson, C.G. Willson, J.M.J. Frechet (eds.), Am. Chem. Soc., (1984), Washington DC and Ms. A. K. Sheela, Assistant Manager, Transparency Market Research, State Tower, 90 State Street, Suite 700, Albany NY., 12207
88. J. Liu et al., Process research of high aspect ratio microstructure using SU-8 resist. Microsyst. Technol. **10**(4), 265 (2004)
89. S. Deokar, R.S. Ghadage, C.R. Rajan, S. Ponrothnam, Facile synthesis of poly(4-Hydroxy styrene) from polystyrene. J. Appl. Poly. Si **91**(5), 3192 (2004)
90. D. Basting et al., Historical review of excimer lase development, in *Excimer Laser Technology*, ed. by D. Basting, G. Marowsky (Springer, Heidelberg, 2005)
91. P.J. Wibawa, A. Agam, H. Nur, H. Sain, Changes in physical properties and molecular structures of polystyrene Nano sphere exposed with solar flux. AIP Conf. Proc **1341**, 54 (2011)
92. W. Zhou, *Nanoprint Lithography Resist, Chapt-3* (Springer, Hiedelberg, 2013), p. 99
93. S. Tgawa et al., Radiation photochemistry of onium salt acid generators I chemically amplified resist. Proc. SPIE **3999**, 204 (2000)
94. L.G. Wade, *Organic Chemistry*, 6th edn. (Pearson Prentice Hall Pub, Upper Saddle River, 2006), p. 279
95. G.P. Moss, P.A.S. Smith, D. Tavernier, Pure Appl. Chem. **67**(8–9), 1307 (1995)

96. Production and Growth is the Norm, Chem. Eng. News, ACS Nat. 232nd Meeting, San Francisco, Sept. 10–14 (2006), 84 (28), 59 (2006)
97. K. Sakurai, H. Nemoto, and A. Kumano, Radiation Sensitive Composition, U.S. Patent # 6348298, Feb. 19, (2002)
98. J. Saw, J. Gelorme, N. LaBianca, W. Conley, S. Holmes, Negative photoresists for optical lithography. IBM J. Res. & Devp **41**(1–2) (1997)
99. N. Zelentsova, S. Zelentsova, M. Abadie, E. Makareeva, *Photochemical Crosslinking of Low Molecular Weight Vinyl Containing Polysiloxane with Organic Azides* (Russian Univ. Scitific Res. 2000)
100. David C. Brock, Reflections on Moore's Law, Chapt-8, Chem. Heritage Foundation, Diane Pub. Co., Darby, PA, (2006)
101. B. Hoeneisen, C.E. Mead, Fundamental Limitations in Microelectronics. Solid State Electron. **15**(7), 819 (1972)
102. T. Kozava, H. Oizumi, T. Itani, S. Tagava, Latent image created by using small field exposure tool for extreme ultraviolet lithography. Jpn. Appl. Phys. Express **2**(7), 075006-1–075006-3 (2009)
103. T.K. Gupta, *Copper Interconnect Technology* (Springer, New York, 2009)
104. G.T. Teixidor et al., Carbon microelectromechanical systems as a substratum for cell growth. Biomed. Mater. **3**, 1 (2008)
105. S. Ranganathan et al., Photoresist derived carbon for microelectrochemical systems and electrochemical applications. J. Electrochem. Soc. **147**, 447 (2000)
106. G. Jenkins, K. Kawamura, *Polymeric Carbons-Carbon Fiber, Glass and Char* (Cambridge University Press, Cambridge, 1976)
107. J. Kim, X. Song, K. Kinoshita, M. Madou, R. White, Electrochemical studies of carbon films from Pyrolyzed photoresist. J. Electrochem. Soc. **145**(7), 2314 (1998)
108. H. Zhou, A. Gupta, J. Zou, J. Zhou, Photoresist derived carbon for growth and differentiation of neuron cells. Int. J. Mol. Sci. **8**(8), 884 (2007)
109. S. Ranganathan, R. McCreery, S.M. Majji, M. Madou, Photoresist derived carbon for microelectromechanical systems and electrochemical applications. J. Electrochem. Soc. **147**(1), 277 (2000)
110. H.Q. Ziang, S.B. Fang, Y.Y. Jiang, Carboneceous anodes for lithium ion batteries prepared from phenolic Resis with different cross-linking densities. J. Electrochem. Soc. **144**, L187 (1997)
111. V. Rehacek et al., Pyrolyzed photoresist film electrodes for application in electro-analysis. J. Electr. Eng. **62**(1), 49 (2011)
112. A. Sing, J. Jayaram, M. Madou, S. Akbar, Pyrolysis of negative photoresists to fabricate carbon structures for microelectrochemical systems and electrochemical applications. J. Electrochem. Soc. **149**(3), E-78 (2002)
113. A.J. Griffiths et al., *Introduction to Genetic Analysis*, 9th edn. (W.H. Freeman and Co., New York, 2008)
114. J.C. Liao et al., Use of electrochemical DNA biosensor for rapid molecular identification of Uropathogenes in clinical urine specimens. J. Clin. Microbiol. **44**, 561 (2006)
115. S.R. Batten, S.M. Neville, D.R. Turner, *Coordination Polymers: Design, Analysis and Applications* (RSC, Cambridge, 2009)
116. J.L.C. Rowsell, O.M. Yaghi, Metal organic frameworks: a new class of porous materials. Microporous Mesoporous Mater. **73**, 3 (2004)
117. J.R. Long, O.M. Yaghi, The pervasive chemistry of metal organic frame works. Chem. Soc. Rev. **38**, 1213 (2009)
118. M. Jacoby, Heading the market with MOFS. In Chem. Eng. News **86**, 13 (2008)
119. A.U. Czaja, N. Trukhan, U. Muller, Industrial application of metal organic frameworks. Chem. Soc. Rev. **38**, 1284 (2009)
120. A.D. Naik et al., Coordination polymers and metal organic frameworks derived from 1,2,4-Triazole amino acid linkers. Polymer **3**, 1750 (2011)

121. S.R. Batten et al., Terminology of metal-organic frameworks and coordination polymers. Pure Appl. Chem. **85**(8), 1715 (2013)
122. S. Thomas M. Zalbowitz, *Fuel Cells, Green Power* (Los Alamos Nat. Lab. Los Alamos, New Mexico)
123. B.K. Kakati, D. Deka, Development of low-cost advanced bipolar plate for P.E.M.Fuel cell. Fuel Cells **8**(1), 45 (2008)
124. S. Gottesfeld, Polymer Electrolyte Fuel Cells, Adv. in Electrochem. Sci., Eng. 5, (1997), Wiley-VCH, Germany
125. A.B. Stambouli, Solid oxide fuel cells (SOFCs), a review of an environmentally clean and efficient source of energy. Renew. Sustain. Energy Rev. **6**(5), 433 (2002)
126. M. Spinelli et al., Application of Molten Carbonate Fuel Cells in Cement Plants for $CO_2$ Capture and Clean Power Generation, Energy Procedia, 63, 6517 (2014) and also Molten Carbonate Fuel Cells, U.S. dept. Energy, 9th August, 2011
127. C.E (Sandy).Thomas et al., Integrated Analysis of Hydrogen Passenger Vehicle Transporation Pathways, National Renewable Energy Lab. March (1998)
128. Fuel Cell Handbook (4th ed.) U.S. Dept. of Energy Office of Fossil Energy, Fed. Energy Tech. Center, Nov. 1998
129. Hydrogen and Fuel Cell Vehicles Worldwide, Industrie Service-GmbH, Accessed on 2nd August, 2011
130. M. Millikin (ed.), Green Car Congress, Navigant: fuel cell industry passed $1 billion revenue mark in 2012, Green Car Congress, August 12th, 2013
131. S. Wang, D. Yu, L. Dai, Polyelectrolyte fuctionalized carbon Nano-tubes as efficient metal free electrocatalysts for oxygen reduction. J. Am. Chem. Soc. **133**(14), 5182 (2011)
132. M.H. Robson, K. Artyushkova, W. Patterson, P. Atanassov, M. Hibbs, Non-platinum carbon supported oxygen reduction catalyst ink evaluation basen on poly (Sulfone) and poly (Phenylene)-derived Ionomers in alkaline media. Electrocatalysis **5**(2), 148 (2014)
133. K. Kinoshita, *Carbon, Electrochemical and Physicochemical Properties* (Wiley, New York, 1998)
134. G. Wang, Y. Weng, C. Deryn, D. Xie, R. Chen, Preparation of alkaline anion exchange membranes based on fuctional poly (ether-amide) polymers for potential fuel cell applications. J. Membr. Sci. **326**(1), 4 (2008)
135. D. Golodnitsky, E. Strauss, E. Peled, G. Greenbaum, Review – on order and disorder in polymer electrolytes. J. Electrochem. Soc. **162**(14), A2551–A2556 (2015)
136. Y. Wang, B. Li, J. Liu, Q. Li, et al., Lithium and Lithium ion batteries for applications in microelectronic devices—A review. J. Power Sources **247**, 452 (2014)
137. Z. Gadjourova, Y.G. Andreev, D.P. Tunstall, P.G. Bruce, Ionic conductivity in crystalline polymer electrolytes. Nature **412**, 520 (2001)
138. Z. Stoeva, I. Martin-Litas, E. Staunton, Y.G. Andreev, P.G. Bruce, Ionic conductivity in crystalline polymer electrolytes. J. Am. Chem. Soc. **125**, 4619 (2003)
139. L. Gitelman, M. Israeli, A. Averbauch, M. Nathan, Z. Schuss, D. Golodnitsky, Modeling and simulation of Li-ion conduction in poly (ethylene Oxide). J. Computational Phys. **227**, 1162 (2007) and also L. Long, S. Wang, M. Xiao, Y. Meng, Polymer electrolytes for lithium polymer batteries. J. Mater. Chem. **A4**, 10038 (2016)

# Chapter 5
# Coal, the *Black Carbon*

## 5.1 Introduction

*Formation of Coal* The first thing that we could think of about *black carbon* is *coal*. Coals are complex heterogeneous solids that consist of a large polymeric matrix of aromatic structures commonly called the coal macromolecules. These macromolecules may vary widely in their chemical and physical properties [1]. Coal is composed primarily of carbon along with variable quantities of other elements, chiefly hydrogen, sulfur, oxygen, and nitrogen [2]. Indeed, coal (from old English term Col, meaning *mineral of fossilized carbon*) contains mainly carbon, the conversion of dead vegetation called *carbonization* [3]. Archeological evidence in China indicates surface mining of coal and household usage after approximately 3490 BC [4].
Figure 5.1 shows the formation of coal from peat. As geological processes apply pressure to peat over time, it is transformed successively into different types of coal. Elemental analysis of coal gives empirical formulas as follows [5]:

$$\text{Bituminus coal} : C_{137}H_{97}O_9NS \tag{5.1}$$

$$\text{Anthracite coal} : C_{240}H_{90}O_4NS \tag{5.2}$$

Carbon forms more than 50% by weight and more than 70% by volume of coal (this includes inherent moisture). Pure anthracite coal can even contain 95% of carbon [5]. According to a geologist, *coal* is defined as an organic *sedimentary rock* (usually occurring in *rock strat*a in layers, or veins called *coal seam*) that forms from the accumulation and preservation of plant materials, usually in a swamp environment (Fig. 5.1). To form the thick layer of plant debris, to produce a coal seam, the rate of the plant debris accumulation must be greater than the rate of decay. It has been estimated that 50 ft of plant debris is needed to make a five-feet thick coal seam, and it would require thousands of years to accumulate. Indeed, it is the most plentiful fuel in the fossil family and has the longest and, perhaps, the most



T. Gupta, *Carbon*, DOI 10.1007/978-3-319-66405-7_5

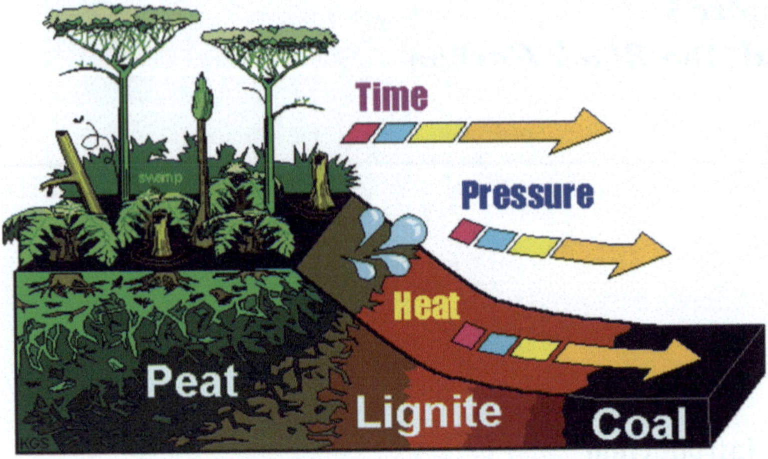


**Fig. 5.1** Formation of coal from peat (Photo courtesy: Geological Survey, Kentucky)

varied history. Coal has been used for heating since the cave man. Archeologists have also found evidence that the Romans in England used it in the second and third centuries (100–200 AD) [5].

## 5.2 Types of Coal and Its Structure

Coal is extracted from the ground by coal mining. Since 1983, the world's top producer of coal is China [6]. In 2011, China produced 3520 million tons of coal, which was almost 49.55% of the total coal mined throughout the world. Other large producers were the United States (993 million tons, mT), India (589 mT), European Union (576 mT), and Australia (416 mT). Based on the composition and properties, coals are assigned to a rank progression that corresponds to their level of organic metamorphism. Figure 5.2 shows the coal ranking system in the United States.

The *anthracite* coal is the highest ranked coal followed by *bituminous*, and *lignite*. Figure 5.3 shows the chemical structure of coal [7, 8].

Under high pressure and high temperature, dead vegetation is converted into coal. Thus, coals have different chemical compositions, and therefore different structures. Even within certain ranks of coal, such as *lignite*, *bituminous*, or *anthracite*, the structure may vary depending on the environment in which that particular rank of coal was formed. A typical chemical structure is illustrated in Fig. 5.3, where the hexagons (and the occasional pentagons) represent six (five carbon) atoms that are arranged in the so-called *aromatic structure*.

The *bituminous* coals are composed of three major *maceral* groups, which, in the common European terminology, are known as *vitrinite*, *exinite*, and *inertinite*. The term *inertinite* refers to components *inert* in *carbonization*. However, the word *inert*

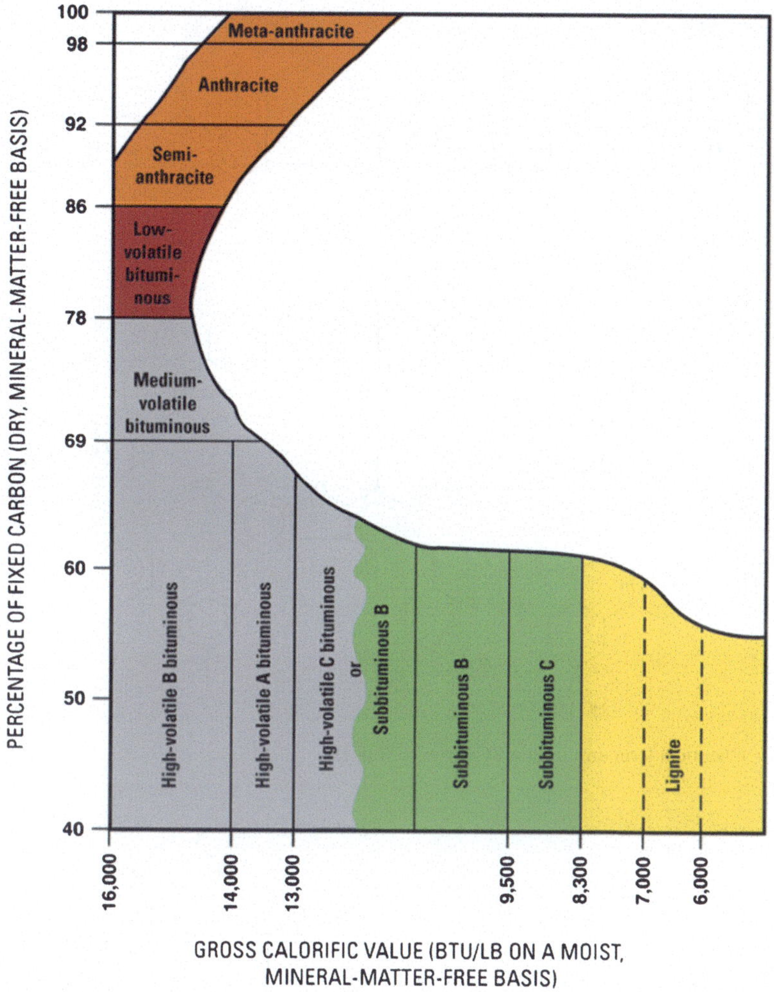


**Fig. 5.2** The coal ranking system in the United States (Photo courtesy: US Geological Survey, Adapted from WIKI)

should not be taken in the sense that the material does not undergo any chemical reactions on the changed environments. Rather, it means that in changed environments (especially at higher temperatures) it does not soften or swell and does contribute little to the volatile matter. However, vitrinite is quantitatively the most important maceral, accounting for probably 60–80% of most worked coals, and it is the material primarily responsible for the characteristic coking behavior of higher rank *bituminous coals* [9, 10]. Figure 5.4 shows the hypothetical segment of a typical molecule in bituminous coal.

*Carbonization* of coal is carried out in the United States almost entirely for the purpose of making coke. It is carried out entirely by heating mostly bituminous coal

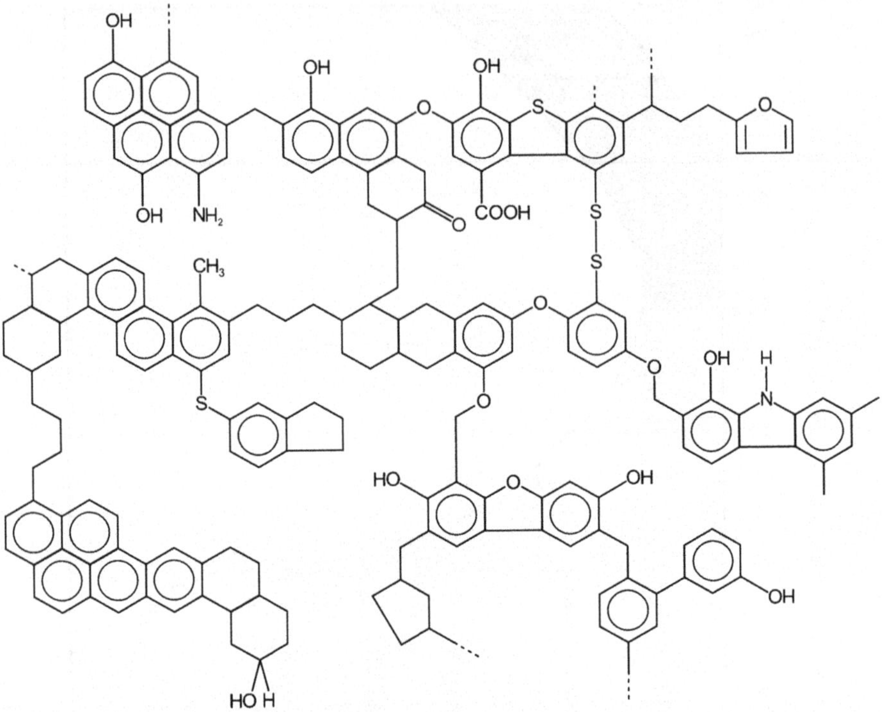


**Fig. 5.3** Chemical structure of coal (Courtesy: WIKI)

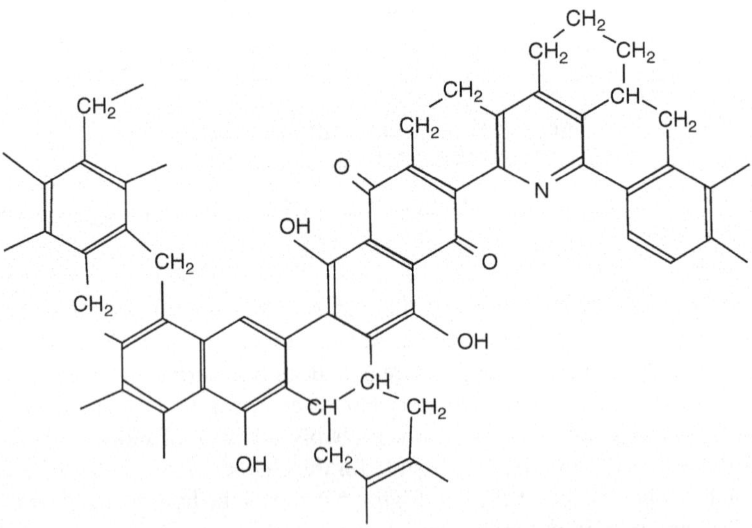


**Fig. 5.4** Hypothetical segment of a typical molecule in a bituminous coal (Courtesy: P.H. Given)

in the absence of oxygen. The first major change that has been observed during heating is softening of the volatile solids, which represents the lower molecular weight hydrogen-rich fraction of the coal. This softening is accompanied by the initiation of the decomposition of this fraction of hydrogen from hydro-aromatics, shortly followed by the dissociation of OH. These various changes, initiated by dissociation of H and OH, are more or less complete by about 500–550 °C. When the sample is further heated and the temperature reaches between 600 and 650 °C, enough energy is available to break a few aromatic carbon–carbon bonds, and the carbon skeleton can arrange and condense to larger polycyclic nuclei of lower H/C ratio. As a result, hydrogen gas is released. These changes also permit the pairing of the odd electrons in the free radical centers. The dramatic change in electrical conductivity results from the growth of the aromatic lamellae [11, 12].

Indeed, coals have many structural features such as aromaticity, various functional groups of oxygen, nitrogen, and sulfur, covalent and noncovalent cross-links, physical associations, and several surface structural features that determine the physical properties and reactivities [13, 14]. In fact, advanced solid-state $^{13}C$ NMR techniques (CP/MAS and dipolar dephasing), combined with carbon counting, have been used to determine the chemical structure of coal [15, 16].

As coal contains mainly carbon, the conversion of dead vegetation into coal is called *carbonization* [17]. For a smaller area, Hilt's law (which is a geological term) states that the more deeper the coal, higher is the rank (grade). The classification of coal is generally based on the percentage of volatile contents. However, the exact classification varies between countries. According to German classification, coal is classified as in Table 5.1 [18].

Based on composition and properties, coals are assigned to a rank progression that corresponds to their level of organic metamorphism. The *anthracite* coal is the highest ranked coal followed by *bituminous* and *lignite*. The peat variety, which is a mass of partially *carbonized* plant debris

is the lowest variety of *fossilized carbon*. Geologically, the largest most concentrated *anthracite* deposit in the world is found in northeastern Pennsylvania, United States. The highest grade anthracite coal can contain 95–98% of carbon, and the heat content can range from 22–28 million Btu (British Thermal Units) per short ton (between 26 and 33 MJ/kg) on a moist, mineral matter-free basis [19]. Figure 5.5 shows two varieties of anthracite coal (a) from near Bay City, Michigan and (b) from Ibbenburen, Germany.

In the year 2013, a group of scientists from Rice University, Houston, Texas, has been able to successfully produce graphene quantum dots (GQD) from anthracite and bituminous coals. Coke produced GQDs between 4 and 8 nm, and anthracite made stacked structures from 18 to 40 nm [20]. Unique coal structure has an advantage over pure $sp^2$-carbon allotropes for producing quantum dots [21]. These dots are water-soluble, easy to handle, nontoxic, and offer much promise in medical imaging. Figure 5.6 shows the nanostructure made from bituminous coal.

**Table 5.1** German classification of coal

| German classification | English designation | Volatiles % | C Carbon % | H Hydrogen % | O Oxygen % | S Sulfur % | Heat content kJ/kg |
|---|---|---|---|---|---|---|---|
| *Braunkohle* | Lignite (brown coal) | 45–65 | 60–75 | 6.0–5.8 | 34–17 | 0.5–3 | <28,470 |
| *Flammkohle* | Flame coal | 40–45 | 75–82 | 6.0–5.8 | >9.8 | ~1 | <32,870 |
| *Gasflammkohle* | Gas flame coal | 35–40 | 82–85 | 5.8–5.6 | 9.8–7.3 | ~1 | <33,910 |
| *Gaskohle* | Gas coal | 28–35 | 85–87.5 | 5.6–5.0 | 7.3–4.5 | ~1 | <34,960 |
| *Fettkohle* | Fat coal | 19–28 | 87.5–89.5 | 5.0–4.5 | 4.5–3.2 | ~1 | <35,380 |
| *Esskohle* | Forge coal | 14–19 | 89.5–90.5 | 4.5–4.0 | 3.2–2.8 | ~1 | <35,380 |
| *Magerkohle* | Nonbaking coal | 10–14 | 90.5–91.5 | 4.0–3.75 | 2.8–3.5 | ~1 | 35,380 |
| *Anthrazit* | Anthracite | 7–12 | >91.5 | <3.75 | <2.5 | ~1 | <35,300 |

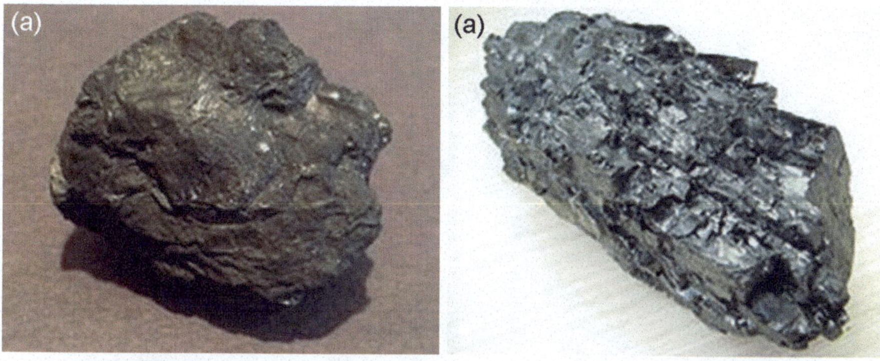


**Fig. 5.5** Two varieties of anthracite coal (**a**) from near Bay City, Michigan and (**b**) from Ibbenburen, Germany (Courtesy: WIKI)

**Fig. 5.6** Nanostructure made from bituminous coal (Courtesy, Rice Univ., Dr. Tour and his group)

## 5.3 Characteristics of Coal

Figure 5.7 shows the stacking layer structure of graphene quantum dots extracted from anthracite coal. *Density values* of coals as have been reported by different research scientists differ considerably from one scholar to another (even after correcting for the mineral matter) because of the experimental procedure each individual adopted. However, the measurements related to the displacement of helium (He), after the absorbed gas has been released from the coal sample, have been found to be the best method for determining the density of coal. Measurements show that the *density* can vary from 1.4 to 1.6 g/cm$^3$ depending on the percentage of carbon content in the coal sample [22].

The *specific electrical conductivity ($\sigma$)* of dry coals is very low (specific resistance can vary between $10^{10}$ to $10^{14}$ ohm-cm). However, the *thermal conductivity* (K) and *thermal diffusivity ($\alpha$)* (defined as thermal conductivity divided by density

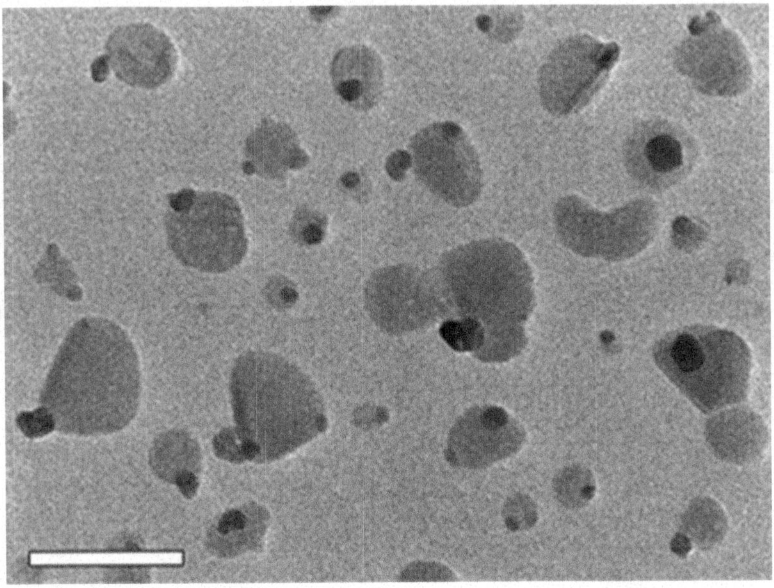

**Fig. 5.7** Stacking layer structure of graphene quantum dots extracted from anthracite coal (Coutesy, Rice Univ. Tour and his group)

and specific heat capacity) of coals depend on the *porosity* and *crack* structure. It has been found that the *thermal conductivity* (K) of coal varies from 0.23 to 0.35 W/(m.k). The *specific heat* of coal has been measured from the ratio of thermal conductivity and thermal diffusivity and was found to be around 1.25 J/ (g.k) at 20 °C [23]. Further studies show that it decreases as the temperature rises. On the other hand, the dielectric constant (€) value of coal is seen to be very much dependent on the structural change of the material during strong heating.

## 5.4 Uses of Coal

It is said that *coal keeps the light on*. Coal has always been *a boom and bust business*, but the recovery may never come, particularly in the Appalachian region where the high production cost operation has made its coal increasingly uncompetitive. More than 70 % of affected power plants have already installed scrubbers and other controls to meet EPA requirements, so that they have lower cost coal options than the higher cost Appalachian thermal coal.

Coal-fired electric power generation emits around 2000 pounds of carbon dioxide gas for every megawatt-hour power generated, which is almost double the amount of carbon dioxide gas released by natural gas fired electric plant. In the United States, higher carbon efficiency of natural gas generations is introduced, and the use of coal has been reduced. As a result, carbon dioxide emission has fallen rapidly, and the measured emission of carbon dioxide data as recorded in the first quarter of 2012 were the lowest of any recorded for the first quarter of any year since 1992 [24].

However, in the winter of 2013, the cold blast of arctic air that had settled over much of the New England cities of the United States, due to *polar vortex effect*, pushed the energy demand more than 24% compared to the usual energy consumption during winter [25, 26]. As a result, the electric companies burnt less expensive coal to meet the consumer demand. Again, during spring 2013, the air-conditioning season of North America, the peak power demand increased considerably and the power companies did not have any alternative other than the use of coal – the least expensive fuel to meet the energy crisis.

Stratospheric sudden warmings (SSWs) are a ubiquitous feature of wintertime flow in the northern hemisphere. Radiosonde method was discovered more than 60 years ago [27], and it had been used to provide routine observations at altitudes higher than 20 km above the surface of the earth. The events take their name from a rapid increase of temperature several tens of Kelvin in over a few days in the high northern latitudes. The event in early January 2013 is thus not typical, but, like all of these events, has unique dynamic characteristics in terms of development and interactions with the tropospheric flow. Figure 5.8 shows the development of a broad cyclonic flow, likely signifying the formation of a new polar vortex and the end of the SSW on January 22, 2013.

According to the US Energy Information Administration (EIA), 40% of America's electricity in the future is going to come from coal. According to the most recent data available for the first 11 months of 2013, coal accounted for about 39% of the 3.7 trillion kilowatt-hours of the electricity generated in the United States. Therefore, it will not be inappropriate to say that no matter how we dislike coal and its burning products, in case of emergency, it is the most abundant and cheapest fuel till date to meet our energy demands.

Another important use of coal is in the production of coke. Coke is produced by heating (~ 1000 °C) coal under controlled conditions in the absence of air. This drives off some of the volatile materials and concentrates the carbon content. It is generally used as a high carbon content fuel for metal processing. Coke made from coal is gray, hard, and porous (Fig. 5.10). The manufacturing process of coke is either by *hearth process* or it can be processed in a *beehive coke oven*, which is a fire brick chamber shaped like a dome, typically 4 meters wide and 2.5 meter high [28, 29]. Due to its high porosity, the specific gravity never reaches 0.77. The market value of coke depends on the percentage of ash and sulfur contents. Petroleum coke is a solid residue obtained in oil refining, which resembles coke but contains many impurities to be useful in metallurgical applications. Figure 5.9 shows a smokeless Coke-Oven Plant in South Wales, Australia.

*Coal petrology* is used to identify the macerals and the microlitho types present in coals [30]. The difference in resistance of macerals to breakage together with other properties is used in the Longwy-Burstlein or Sovaco process in the selection of coals for blending in optimum proportions for producing high-quality coke [31, 32]. Figure 5.10 shows the picture of gray, hard, and porous raw coke. The

**Fig. 5.8** Development of a broad cyclonic flow, likely signifying the formation of a new polar vortex and the end of the SSW on January 22, 2013) (Photo courtesy, NASA, United States)

coke must be very strong to resist the overburden in the blast furnace, which is why coking coal is so important in making steel using the conventional route.

Figure 5.11 shows how the stacked coal from the mine is being transported to the market. According to the Association of American Rail Roads (ARRs), coal made up about 39% of all freight tonnage. Coal also accounted for 19% of rail revenue in July 2014. For years, coal has been a cost-effective fuel of choice for electrical power generation because of the efficiency of the rail road industry to move it from mine to power plant at competitive rates. In the year 2015, about 70% of all coal delivered to these plants were moved by rail.

The choice of mining method is largely determined by the geology of the coal deposits. In general, coal is mined by two methods, either by *surface or open cast mining* or by *underground or deep mining*. The first method is economical when the coal seam is near the surface, and the method recovers a higher proportion of the coal deposit than *underground mining*.

The *underground mining* can be subdivided into two groups (a) room and pillar mining and (b) long-wall mining. Figure 5.12 shows the procedure of the two mining technologies. Unfortunately, deep mining of coal causes environmental

**Fig. 5.9** Coke-Oven Plant at South Wales, Australia (Photo courtesy: WIKI)

**Fig. 5.10** Gray, hard, and porous raw coke (Courtesy: WIKI)

disruption due to strip mining. As a result, environmental disruption caused by strip mining contributed to the decline in the use of coal.

In the United States, the first record of coal mining was in the year 1673–1674. As a matter of fact, by the late 1700s, coal was being mined in Coal Hill, now Mount Washington in Pittsburgh, Pennsylvania. In 1816, the city of Baltimore,

**Fig. 5.11** Stacked coal from coal mines being removed by tractors

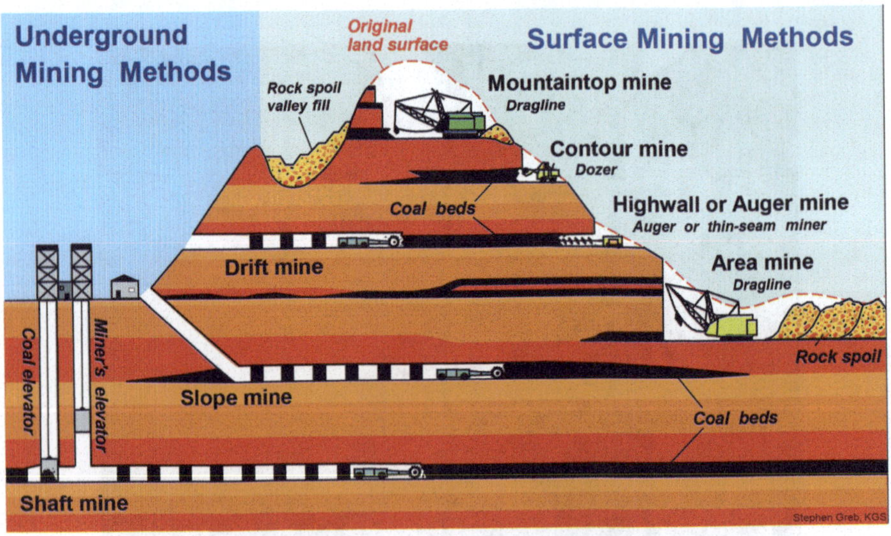


**Fig. 5.12** Coal-mining methods (Courtesy: University of Kentucky)

Maryland, began to light its streets with combustible gas made from coal. By the 1830s, coal was being used to make glass in Fayette County, Pennsylvania.

The Industrial Revolution played a major role in expanding the use of coal. During the first half of the 1800s, James Watt's invention of steam engine brought the Industrial Revolution that spread in the United States, and steamships and

**Fig. 5.13** Coal-fired steam engine, a revolutionary discovery of the power of coal

steam-powered railroads became the chief forms of transportation, where coal was used as the main fuel for their boilers (Fig. 5.13).

The roots of today's primary use of coal for electric power generation can be traced back to the time of Thomas Edison. In 1882, Edison built the first practical coal-fired electric generation station, supplying electricity to some residents of New York City. However, the firing of coal for the generation of electricity produced huge amounts of black soot and carbon dioxide gas in the environment. Figure 2.6 shows how burning of black coal forms black soot which is responsible for more than 30% warming in the arctic regions [33, 34]. It is a powder-like *amorphous carbon*.

Soot as an airborne contaminant in the environment has many different sources, all of which are results of some form of pyrolysis. They include soot from coal burning, internal combustion engines, power plant boilers, hog fuel boilers, ship boilers, central steam heat boilers, waste incineration, local field burning, house fires, forest fires, fireplaces, furnaces, etc. It is a powder-like form of amorphous carbon. The gas phase of soots contains polycyclic aromatic hydrocarbons (PAHs) [35].

*Amorphous carbon* is formed when a material containing carbon is burned without enough oxygen. Table 5.2 shows the physical properties of amorphous carbon.

**Table 5.2** Physical properties of amorphous carbon

| | |
|---|---|
| Atomic number | 6 |
| Density (g/cm$^3$) | 2.26 |
| Molecular weight (g/mol.) | 12.011 |
| Melting point (°C) | 3727 °C |
| Surface area (m$^2$/g) | 90–460 |
| Thermal conductivity (cal/s-cm- °C) | 0.056 |
| Hardness (at room temperature) | 2–2.9 mhos |
| pH | 3–10 |
| Specific heat at room temperature | 0.165 cal/g |
| Crystallography | Hexagonal |
| Particle shape | Irregular, spherical, needle, crystal |
| Particle size | 12–94 nm |

## 5.5 Carbon Black

The black soot coming out of the steam engine, shown in Fig. 5.13, is *gas black* or *channel black*, or *carbon black* (black carbon). Two carbon black manufacturing processes (furnace black and thermal black) produce nearly all of the world's carbon blacks, and the furnace black process is the most common process among the two. The furnace black process uses heavy aromatic oils as feedback. The carbon black produced in this process is conveyed through the reactor, cooled, and collected in bag filters in a continuous process. Thermally reduced graphene exhibits a close relationship with amorphous carbon [36].

Carbon black has been used in various applications for electronics. As a good conductor of electricity, carbon black is used as a filler mixed in plastics, elastomer, films, [37] adhesives, paints, and rubber products [38–40]. However, morbidity studies of US carbon black workers for over 20 years show that exposure to carbon black at 1.0 mg/m$^3$, over a 40-year career, could result in an additional 27 ml decrement in FEV1 (forced expiratory volume in 1 s) [41, 42].

Now in the twenty-first century, most of the coal-fired engines are gone. Though electric engines have replaced most of them, still some diesel locomotives exist. Soot, especially diesel exhaust pollution, accounts for over one quarter of the hazardous pollution in the air. Figure 5.14 shows emission of soot from a large idling diesel truck. Long-term exposure to urban air pollution containing soot increases the risk of coronary heart disease, according to a major study published in *New England Journal of Medicine* [43].

Different types of soot contain different amounts of black carbon (BC). The definition and measurement methods of atmospheric black carbon (BC) have long been subjects of scientific controversy. The recent discovery of light-absorbing carbon that is not black makes it imperative to reassess and redefine the components that make up light-absorbing carbonaceous matter (LAC) [44].

Generally, the blacker the soot, more is the global warming. BC warms climate in two ways. When suspended in air, BC absorbs sunlight and generates heat in the

**Fig. 5.14** Emission of soot from a large idling diesel truck (Courtesy:WIKI)

atmosphere, which warms the air and can affect regional cloud formation and precipitation patterns. The light-absorbing characteristics of BC have been successfully measured and compared with other light-absorbing materials. The measurements are based on the refractory nature of the material (inertness at high temperatures). Thermal–optical process is known to be by far the most commonly used method to measure the ECa (apparent elemental carbon, refractory) and OCa (apparent organic carbon, non-refractory) components of ambient and source aerosols.

Light-absorbing technique, a more reliable method, is adopted here to measure the light-absorbing capacity of the suspended particles. A laser beam at a specific wavelength is passed through a particle in an air volume. The light after it is refracted through the particle is deposited onto a filter and is measured. The intensities of the incident light and the light after it is passed through the particles are measured, and the amount of light absorbed by the particle is calculated. The measurements are repeated with different suspended particles, and the average of all the measurements is calculated. In isolated areas with high soot production, as much as 50% of surface-warming greenhouse gases may be masked by atmospheric brown clouds [45]. Beside particulate soot, greenhouse gases and solar activity can change the global temperatures [46, 47]. Ninety percent of the global warming is occurred in ocean, and 10% is caused by the air [48]. Figure 5.15 shows the observed (1900–2009) and projected (2010–2070) rise in global temperature.

A model of soot formation from coal tar which included oxidation and agglomeration is expected to predict soot behavior in both oxidizing and nonoxidizing environments based on predicted tar yields from the CPD (chemical percolation devolatilization) coal devolatilization model [49]. When soot is deposited on ice, it

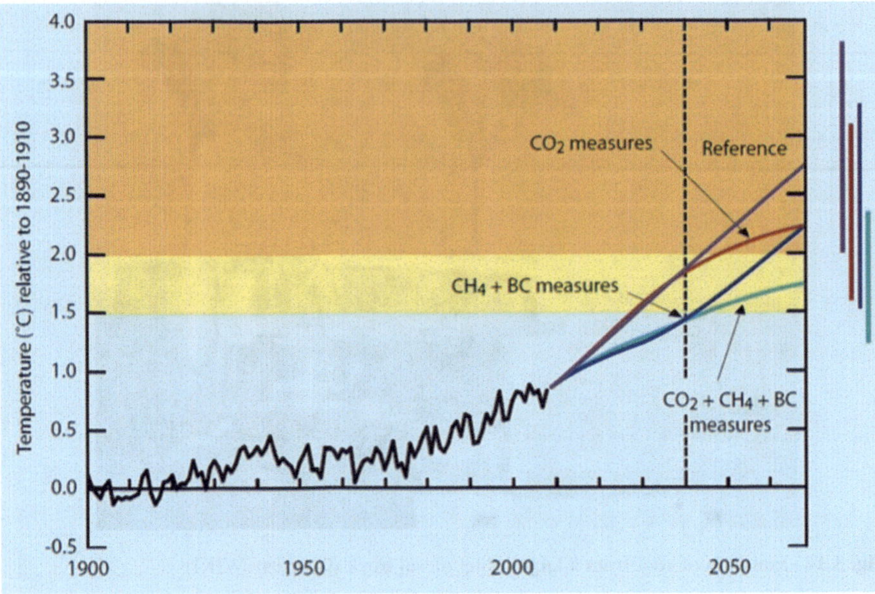


**Fig. 5.15** Observed (1900–2009) and projected (2010–2070) rise in global temperature (Adapted from the published research of R. Santoro, Cornell Univ, NY)

absorbs sunlight and generates heat. As a result, it warms both the air above and the ice below, thus accelerating melting (Fig. 5.16). Recent studies suggest that BC is responsible for more than 30% of recent warming [33].

Different types of soot contain different amounts of BC. The more the percentage of carbon in BC, more is the absorption of sunlight and more is the warming. Fossil fuel and the biofuel soot are blacker than biomass burning. However, according to some scientists, reduced BC emission represents a win–in scenario, because it should have an immediate cooling effect on earth's climate.

Figure 5.16 shows the retreat of glaciers in the regions of Himalayas. Studies show that plateau-type glaciers retreat faster than valley-type glaciers. According to a study by the Kathmandu-based International Center for Integrated Mountain Development in 2011, there are 54, 000 individual glaciers in the Hindu Kush Himalayan regions stretching from Afghanistan in the West to Yunnan in the Southwest China in the east. The glaciers in these regions show signs of shrinking, thinning, and retreating.

Future climate change and associated impacts will vary from region to region around the globe. The effects of an increase in global temperature include rise in sea levels and a change in the amount and pattern of precipitation, as well as a probable expansion of subtropical deserts. To prevent the unwanted global warming, United Nations Framework Convention on Climate Change (UNFCCC) has adopted some policies that include mitigation by emissions reduction, adoption to its effects, building systems resilient to its effects, and possible future climate engineering [50, 51].

**Fig. 5.16** Glacier retreat in the regions of Himalayas (Photo courtesy: Ed King)

Figure 5.17 shows the processing of coal to generate electricity. Fine powdered coal is burnt in a pulverized coal combustion system at a high temperature. The hot gases and heat energy produced during the firing convert water into steam. The high-pressure steam is passed into a turbine. The condensed water then returns to the boiler to be heated again. The electricity thus generated is transformed into higher voltages and transmitted through power line grids. Modern pulverized coal combustion (PCC) system with new designed combustion technologies having higher efficiency allows more production of electricity by burning less coal, which ultimately reduces less carbon dioxide ($CO_2$) emission per megawatt (MW). In 1999, world gross carbon dioxide emissions from coal usage were 8,666 million tons of carbon dioxide ($CO_2$) [52]. Coal plays an important role in electricity generation in spite of the facts that coal burning produces carbon emission, depletion of the ozone layer, etc. Coal-fired power plants currently fuel 41% of global electricity (Source IEA, 2011). Table 5.3 shows the percentage of coal used in different countries to produce electricity.

Another useful utilization of coal is the production of coal gas by gasification (UCG). It is an industrial process which converts coal into product gas (Fig. 5.18).

UCG is an in situ gasification of coal which is carried out in nonmined coal seams using injection of oxidants and bringing the product gas to the surface. It is

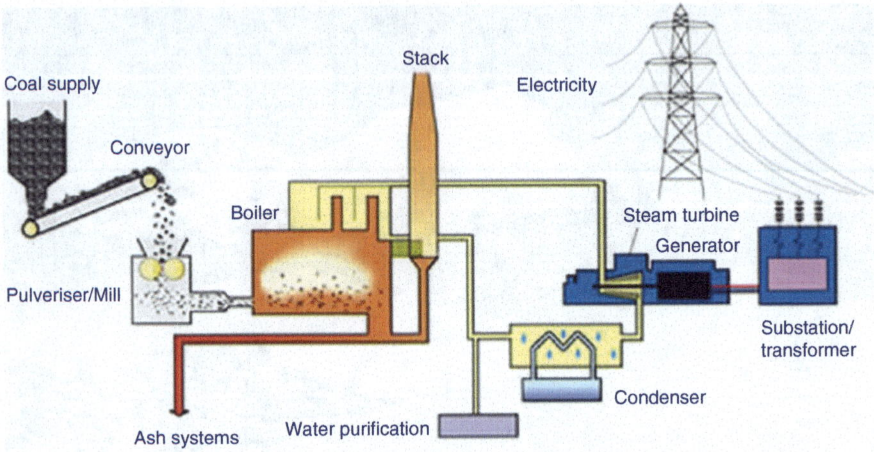


**Fig. 5.17** Processing of coal to produce electricity (Courtesy: Earthlyissues, United States)

mostly a mixture of methane, carbon monoxide, hydrogen, and water vapor. Historically, early technology was used to produce coal gas or town gas for municipal lighting and heating (Fig. 5.19).

The coal industry has many forces working against it as the world moves toward cleaner energy. Figure 5.20 shows coal's long goodbye because it faces three hurdles such as market forces, government regulations, and moral arguments. It is nicely said that *if coal were a medical patient, its prognosis would be creeping toward the critical list* [52].

The energy department's arm, the Energy Information Administration, forecasts in its latest annual energy outlook that *US coal production remains below its 2008 level through 2015*. And that is without weaving in the impact of Clean Power Plan, because it has not yet taken effect. However, clean coal technology – a collection of technologies – is being developed to mitigate the environmental impact of coal energy generation. Much of the challenge is in commercializing the technology so that coal use remains economically competitive despite the cost of achieving low, eventually near-zero, emissions. Alongside nuclear power and harnessing renewable energy sources, one hope for this is via *clean coal* technologies, such as carbon capture and sequestration (CCS). Few major developments were seen in 2013, and policies necessary to facilitate the transition from demonstration to development are still largely missing. For its low-carbon 2DS (2 degree scenario), *the rate of capture and storage must increase by two orders of magnitude* by 2025.

An early example of a coal-based plant using (oxy-fuel) carbon capture technology is a Swedish company Vattenfall Schwarze Pumpe power station located in Spremberg, Germany, built by German firm Siemens, which went online in September 2008. Figure 5.21 shows the operation process of an oxy-fuel CCS fossil fuel power plant.

**Table 5.3** Coal in electricity generation

| Countries | United States | South Africa | Australia | Israel | Greece | Poland | Kazakhstan | Czech Republic | China | India | Morocco | Germany |
|---|---|---|---|---|---|---|---|---|---|---|---|---|
| % of coal used | 45 | 93 | 78 | 58 | 54 | 87 | 75 | 51 | 79 | 85 | 51 | 41 |

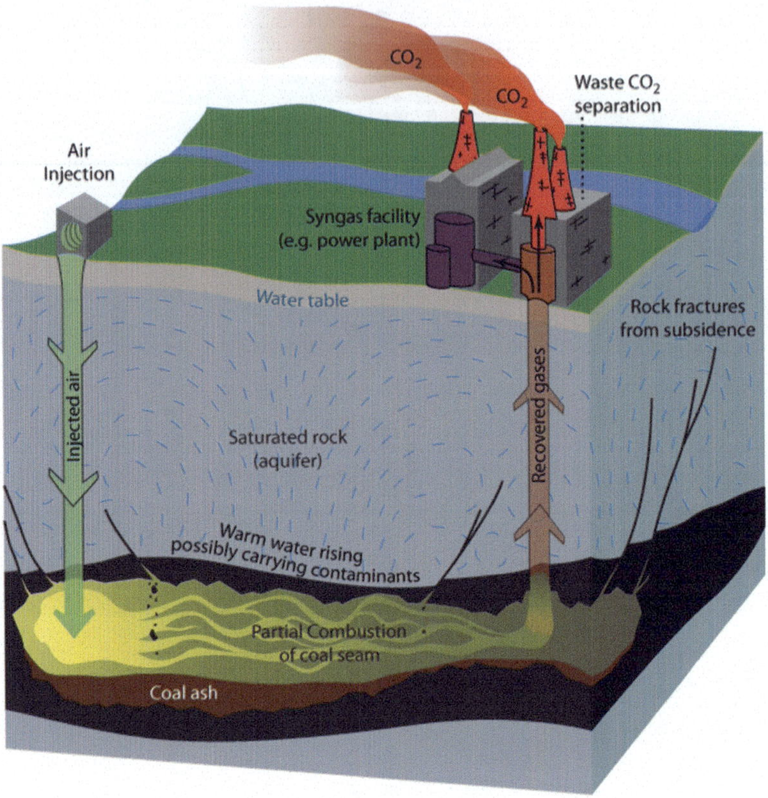


**Fig. 5.18** The process technology involved in gasification of nonmined coal (Photo courtesy: Bretwood, Creative Comm. Wikipedia)

## 5.6 Coal Products

Figure 5.22 shows different products that are obtained when coal is subjected to destructive distillation at high temperatures between 1273 and 1473 K. In high-temperature pyrolysis process, coal is heated in the absence of air, and the major products are coke, coal gas, coal tar, ammonium sulfate, and oil. The light oil by-products contain benzene, toluene, and xylene, and the middle-grade oil generally contains phenol, cresol, naphthalene, and anthracene oil. Figure 5.23 shows the different by-products produced from coal.

Figure 5.24 shows a typical equipment to produce coal gas, in Seattle's (United States) Gas Works Park. It is the only surviving plant in the United States. Coal gas is a flammable gaseous fuel made from coal and supplied to the user via piped distribution system. Originally, coal gas is created as a by-product in the coking process. It had been developed during the nineteenth and early twentieth centuries. Manufactured gas can be made either by carbonization or gasification.

**Fig. 5.19** Municipal lighting using coal gas or town gas (Courtesy: Julo, WIKI)

**Fig. 5.20** Coal's long goodbye (Photo courtesy: Wikipedia)

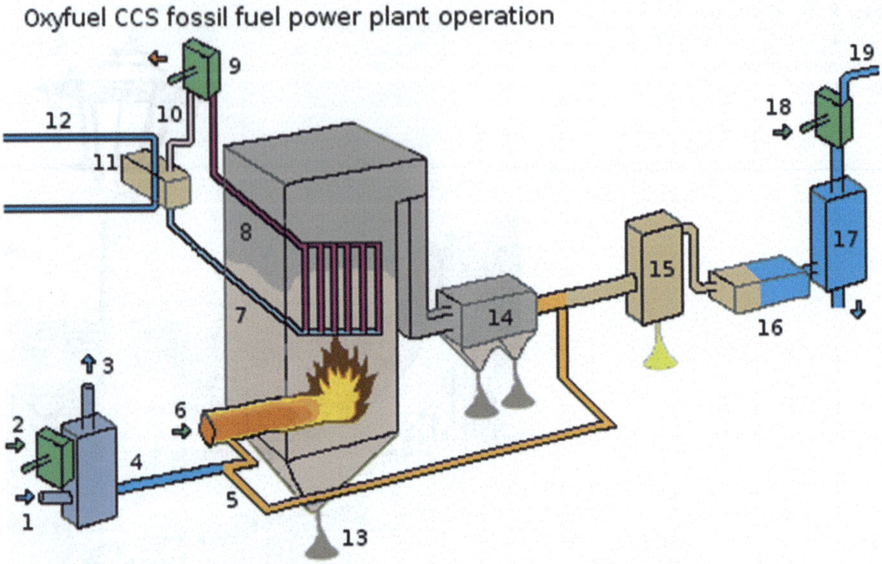


**Fig. 5.21** Operation processes of an oxy-fuel power plant (Credit: KVDP, Wikipedia) *1* = air inlet, *2* = mechanical energy is supplied, *3* = nitrogen outlet, *4* = oxygen outlet, *5* = recycled exhaust gas inlet, *6* = fuel inlet (coal), *7* = cold water pipe, *8* = steam inlet pipe, *9* = steam turbine, *10* = steam outlet pipe, *11* = steam condenser, *12* = cooling pipe of steam condenser, *13* = bottom ash, *14* = fly-ash removal, *15* = Sulfur + gypsum removal, *16* = cooler, *17* = water condenser (water removal), *18* = mechanical energy is supplied ($CO_2$ compressor), *19* = $CO_2$ outlet

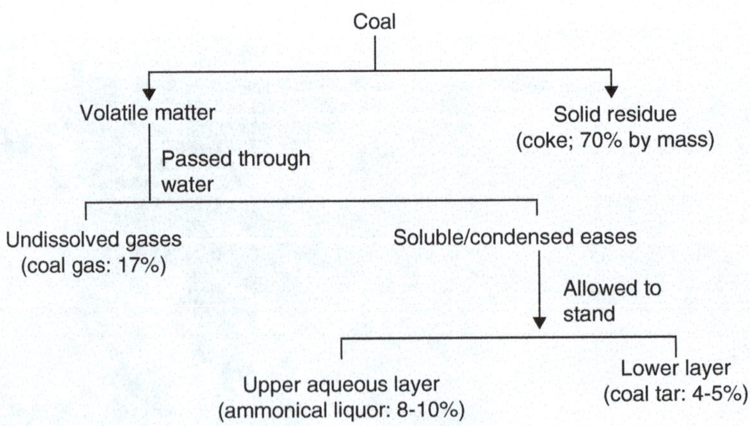


**Fig. 5.22** Different products that are obtained when coal is subjected to destructive distillation at high temperatures

Carbonization refers to the devolatilization of an organic feedstock to yield gas and char. Gasification, on the other hand, is the process of subjecting feedstock to chemical reactions that produce gas [53].

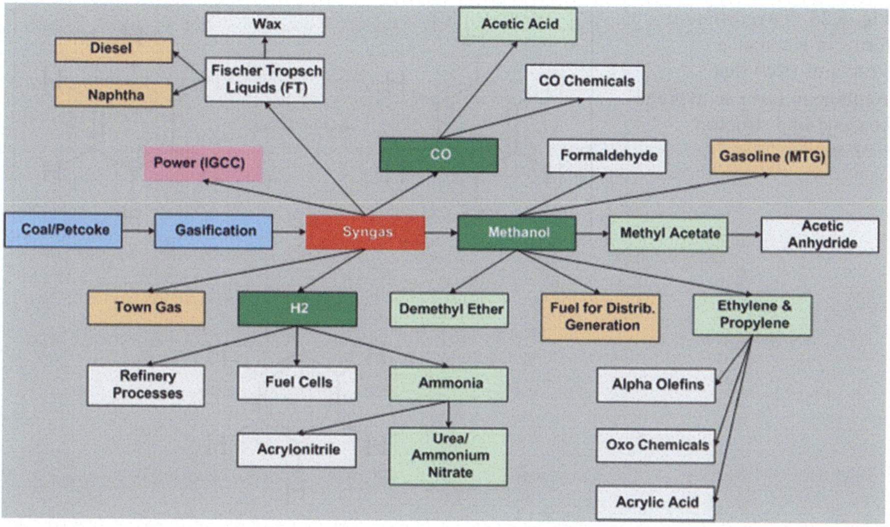


**Fig. 5.23** Different by-products produced from coal (Courtesy: Nattional Energy Technology Laboratory, WIKI)

**Fig. 5.24** Equipment for making coal gas (Courtesy: WIKI)

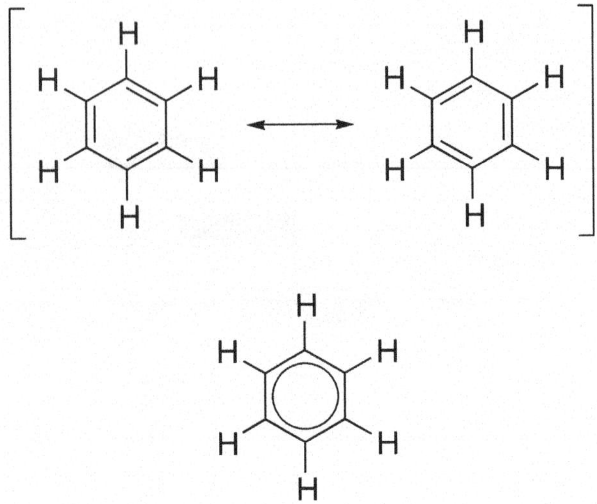

**Fig. 5.25** Two different forms of resonance structures (*top*) that combine to form an average benzene-like structure (*bottom*)

The most useful carbon-containing material is the black coal. About 79% of our annual coal production is burned to produce electricity. The remaining 21% are mostly aromatic class of carbons and hydrocarbons. Coal reserves are vast compared to the other fossil fuels. One ton of bituminous coal (soft coal) yields about 1500 pounds of coke, 8 gallons of coal tar, which can be distilled to yield aromatic compounds (discussed below), and 10,000 cubic feet of coal gas, used as fuel.

Aromatic compounds are originally named because of their fragrant properties [54]. They are unsaturated hydrocarbons with ring structures that exhibit special properties due to their aromatic nature, and they possess an unusual stability. They are often represented by resonance structures containing single and double bonds. However, the bonding is stronger than expected for a conjugated structure and is more accurately depicted as delocalized electron density shared between all the atoms in the ring, as it is seen in benzene structure. Indeed, aromatic compounds have cyclic structures in which each ring atom is a participant in a $\pi$ bond, resulting in delocalized $\pi$ electron density on both sides of the ring, and are planar unlike cycloalkanes.

Figure 5.25 shows two different resonance forms of benzene ($C_6H_6$) (top) that can combine and produce an average structure [55, 56]. It is a common practice to show a standard resonance diagram with a double-headed arrow, which indicates that the two structures are not distinct entities, but merely hypothetical possibilities.

Aromatic compounds contain phenyl group, and they are aromatic in nature. It is a special chemical property that is possessed by some organic compounds, and it describes a way in which a conjugated ring of unsaturated bond or empty orbital exhibits strong stabilization.

Aromatic compounds are generally nonpolar and immiscible with water. They are usually used as solvents for other nonpolar compounds. Because of the $\pi$ electrons, aromatic compounds can act as Lewis bases and can participate in

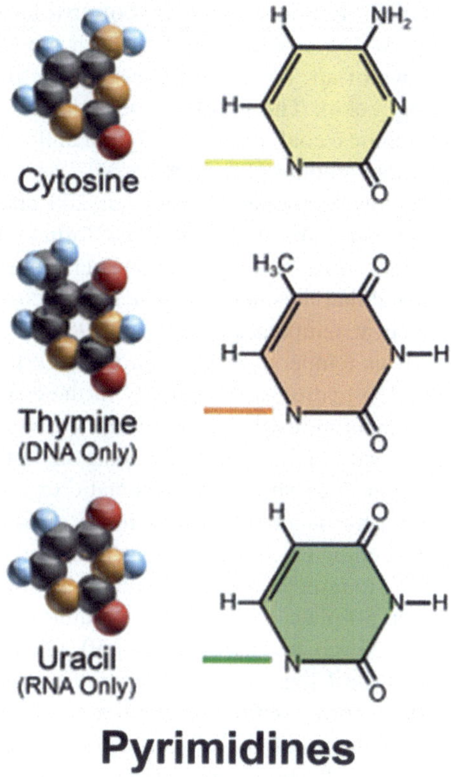


**Fig. 5.26** Three of the five nucleotides (adenine, thymine, cytosine, guanine, and uracil) that make up the sequence of the genetic code in DNA and RNA (Courtesy: BruceBlaus, WIKI)

electrophilic substitution reactions. In biochemistry, aromatic compounds play important roles. The 20 basic building blocks of proteins can contain one of the four aromatic amino acids, namely, *histidine*, *tryptophan*, *phenylalanine*, and *tyrosine*. Moreover, the five nucleotides (*adenine*, *thymine*, *cytosine*, *guanine*, and *uracil*) that make up the sequence of the genetic code in *DNA* and *RNA* are *aromatic purines* or *pyrimidines* [57, 58]. The molecule *heme* contains an aromatic system with 22 $\pi$ electrons. *Chlorophyll* also has a similar aromatic system. The overwhelming majority of aromatic compounds of carbon need not be hydrocarbons.

Figure 5.26 shows three of the five nucleotides (*adenine, thymine, cytosine, guanine, and uracil*) that make up the sequence of the genetic code in *DNA* and *RNA*.

## 5.7 Pyrolysis of Coal

The macromolecules of coal consist of clusters of aromatic carbon linked to other aromatic structures by bridges. There are other attachments to the aromatic structures that do not form bridges. Due to the large variety of functional groups that make up the bridge, structures of coal have a large distribution of bond strengths [59]. As the coal is heated at high temperatures (from 200 to 1000 °C) in the absence of air, tar and light gases (volatiles) are driven out of the coal (which is mostly coke). The pyrolysis behavior of coal depends on the type of coal used. Bituminous coal produces significantly more tar than the low-rank coals and moderate amount of light gases.

At low temperatures between 200 and 250 °C, gases like CO and $CO_2$ are formed, and at ~300–500 °C tar formation occurs. During heating (from 300 to 450 °C), cross-linking reactions start and continue as the temperature rises. It is believed that the chemical structure of coal affects the pyrolysis process. Pyrolysis of coal at temperatures 750 and 950 °C with 100% nitrogen gas indicates that pyrolysis temperature has a profound effect on the structures of the evolved tars [60]. The hydrogen aromaticity of the coal tars is seen to increase dramatically as the temperature of the pyrolysis increased. The pyrolysis of coal can also be performed in a high-pressure controlled profile (HPCP) drop tube reactor [61].

Figure 5.27 shows the schematic of a HPCP drop tube reactor [62]. The drop tube reactor is a laminar flow furnace with a controlled wall temperature.

The crude tar thus formed is processed in a cast iron vessel and heated gently to 200°°C to remove water and light oils. The light oils are then fractionated on either 30- or 50-plate low hold-up column to give benzole (up to 100 °C), toluole (100–125 °C), and xylole (125–150 °C). The residue from this distillation is further treated until the vapor temperature reaches ~360 °C. Table 5.4 shows different products from continuous distillation of crude coal tar [63–66].

Naphtha normally refers to the number of flammable liquid mixtures of hydrocarbons, and is used primarily as feedstock for producing high-octane gasoline. The second material recovered from coal tar distillation is phenanthrene, a polycyclic aromatic hydrocarbon (PAH), composed of fused benzene rings. It is more stable than its linear isomer anthracene. It is found as a natural compound in a mineral called creosote. The next PAH is fluoranthene, where naphthalene and a benzene unit are connected by a five-membered ring. It is the most abundant coal tar constituent [67–69]. Thermodynamically, fluoranthene is not as stable as pyrene, because its electrons cannot resonate throughout the complete structure like the corresponding ones in pyrene [70, 71]. Pyrene, the next PAH extracted from coal tar distillation, consists of four fused benzene rings resulting in a flat aromatic system. It is used in making dyes and dye precursors. It is also produced in a wide range of combustion conditions. The fluoranthene, the next candidate, also belongs to PAH family. It forms white crystals that exhibit characteristic aromatic odor similar to that of naphthalene ($C_{10}H_8$). It is a combustible chemical and has a violet fluorescence. Polyfluoranthene is electrically conductive and has been investigated for use as a luminophore in organic light-emitting diodes (OLED). The next family member of PAH that is extracted from coal tar during distillation is anthracene. It is used to make red dye [72, 73]. It is a colorless solid and exhibits blue fluorescence under UV light. It is used as a scintillator for detectors of high-energy photons,

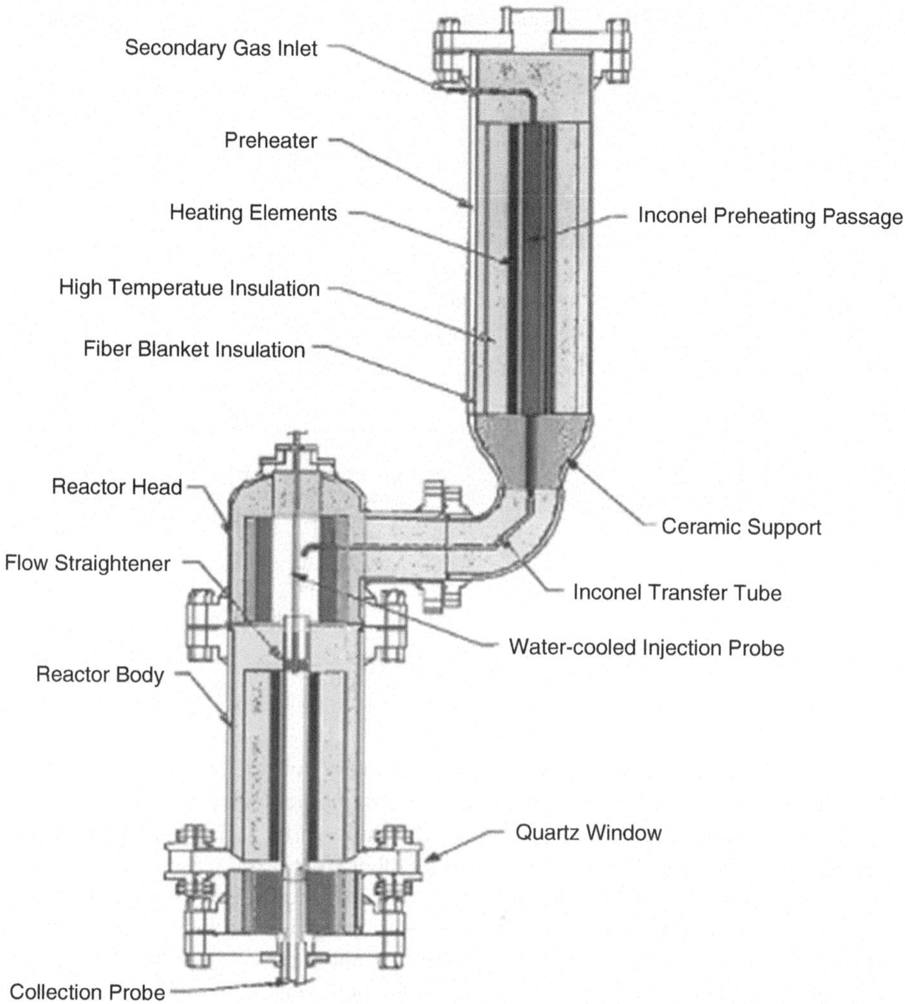


**Fig. 5.27** Schematic of a high-pressure controlled profile (HPCP) drop tube reactor (adapted from E.M. Hambly, Brigham Young University, Utah, U.S.A)

α-particles, and electrons. Polyvinyl toluene is mixed with anthracene to produce plastic scintillators [74–77]. Chrysene, one of the most abundant constituents of coal tar pitch, consists of fused benzene rings. It is also found in creosote, a chemical used to preserve wood. Carbazole ($C_{12}H_9N$) (Fig. 5.28), the next constituent of the coal tar is an aromatic heterocyclic organic compound. It has a tricyclic structure, consisting of six-membered benzene rings fused on either sides of a five-membered nitrogen-containing ring.

**Table 5.4** Different products from continuous distillation of crude coal tar

| | |
|---|---|
| (a) Naphthalene | 10% |
| (b) Phenanthrene | 5% |
| (c) Fluoranthene | 3.3% |
| (d) Pyrene | 2.1% |
| (e) Fluorene | 2.0% |
| (f) Chrysene | 1.8% |
| (g) Anthracene | 1.5% |
| (h) Carbazole | 1.4% |
| (i) 2-Methylnaphthalene | 1.0% |
| (j) Dibenzofuran | 1.0% |
| (k) 1-Methylnaphthalene | 1.0% |

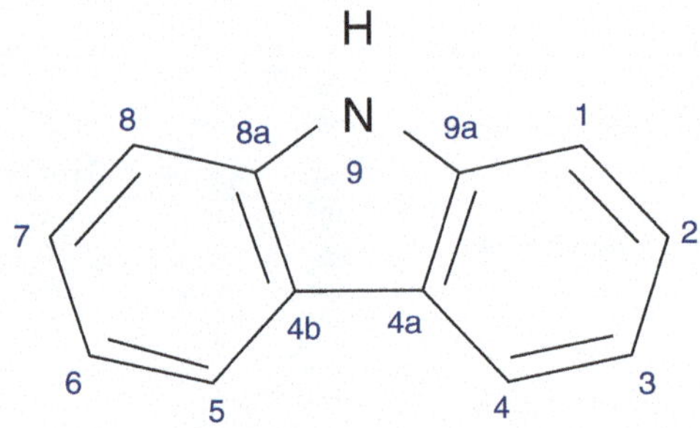


**Fig. 5.28** Molecular structure of carbazole (Courtesy: Edgart, WIKI)

2-methylnaphthalene is present in coal tar in very considerable amounts. In the high-temperature tar, the ratio of naphthalene to monomethylnaphthalene is 4:1; the lower the cooking temperature, the more this ratio shifts in favor of methyl and dimethylnaphthalenes. The next extracted member from coal tar is dibenzofuran. It is a heterocyclic organic compound. It has two benzene rings fused to a central furan ring. It is a very stable compound and has been tried to use as a heat transfer agent. It is the skeletal substance of morphine. It undergoes electrophilic reactions. The last PAH of the coal tar family is 1-methylnaphthalene. It has no cetane number and was previously used as the lower reference for cetane number (CN). However, due to the expense and handling difficulty it was replaced in its capacity by isocetane with a CN of 15. Ignition delay of diesel fuels is characterized by the cetane number (CN). Cetane or hexadecane is assigned CN = 100, and isocetane is arbitrarily assigned CN = 15. The cetane number is calculated as CN = {(% cetane) + (0.15% isocetane)} [78].

The coal tar, during continuous distillation, in the temperature range between 200 and 250 °C, produces naphthalene which is almost 65% of the tar-acid free oil; the remaining 35% is made up of thionaphthalene and ß-methyl naphthalene. A naphthalene molecule can be viewed as the fusion of a pair of benzene rings. Two

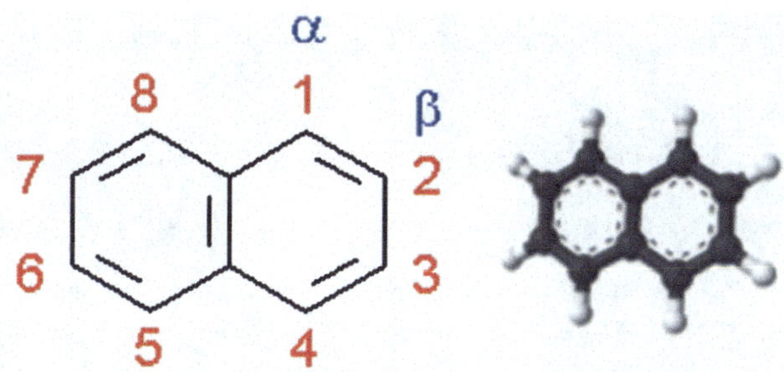


**Fig. 5.29** Ball and stick model of two fused benzene rings representing naphthalene

**Fig. 5.30** Typical polycyclic aromatic hydrocarbons (Courtesy: WIKI)

isomers are possible for mono-substituted naphthalene corresponding to substitution at α or ß position. Naphthalene is used mainly as a precursor to other chemicals. It also provides an excellent stabilizing medium for poorly soluble aromatic compounds [79].

Figure 5.29 shows the ball and stick model of two fused benzene rings representing naphthalene. The fused ring structure of carbon (C) and hydrogen (H) is a common bonding structure of polyaromatic hydrocarbons (PAHs) or polynuclear aromatics. PAHs are lipophilic, meaning they mix more easily with oil than water. Natural crude oil and coal deposits contain significant amounts of PAHs, arising from the chemical conversion of natural product molecules, such as steroids, and aromatic hydrocarbons [80].

Figure 5.30 illustrates the typical polycyclic aromatic hydrocarbons. Clockwise from left: benz[e]acephenanthrylene, pyrene, and dibenz[a,h]anthracene. PAHs are

**Fig. 5.31** Extremely bright stars illuminate a mist of PAHs in this Spitzer image (Spitzer, WIKI)

abundant in the universe, seem to have been formed as early as a couple of billion years after the Big Bang, and are associated with new stars and exoplanets [81]. A controversial hypothesis suggests that the convection currents cause carbon and hydrogen in the nebulae's core to get caught in stellar winds, and radiate outward. Witt and his team inferred that since they discovered PAHs, which may have been vital in the formation of early life on earth – in a nebula, by necessity they must originate in nebulae [82, 83]. Figure 5.31 shows two extremely bright stars that illuminate a mist of PAHs in the Spitzer image.

The presence of fullerenes – the buckyballs – has been reported by Garcia-Hernandez and his coworkers in other nebulae [84]. Some astronomers thought that the seeds for life on earth had been possible because of the presence of fullerene in the outer space [85]. The studies of the NASA scientists, in September 2012, revealed that PAHs subjected to interstellar medium (ISM) (the matter that exists in the space between the star systems in a galaxy) conditions are transformed through hydrogenation, oxygenation, and hydroxylation to more complex organics – a step along the path toward amino acids and nucleotides, the raw materials of proteins and DNA, respectively. As a result of these transformations, the PAHs lose their spectroscopic signature which could be one of the reasons for the lack of PAH detection in interstellar ice grains, dense clouds, and the upper molecular layers of protoplanetary disks [86]. In the year 2013, Gudipati and coworkers detected complex organic materials at the lower atmosphere of Titan, the largest moon of Saturn. Their further studies revealed that these complex organics can develop amino acids and nucleotides, the basic ingredients of RNA [87].

## 5.8 Summary

The first thing that we could think of about *black carbon* is *coal*, meaning *mineral of fossilized carbon*. Coals are complex heterogeneous solids that consist of a large polymeric matrix of aromatic structures commonly called the coal macromolecules. Archeological evidence in China indicates surface mining of coal and household usage after approximately 3490 BC.

Based on the composition and properties, coals are assigned to a rank progression that corresponds to their level of organic metamorphism. The *anthracite* coal is the highest ranked coal followed by *bituminous*, and *lignite*. Measurements show that the *density* of coal can vary from 1.4 to 1.6 g/cm$^3$ depending on the percentage of carbon content in the sample. The *specific electrical conductivity ($\sigma$)* of dry coals is very low (specific resistance can vary between $10^{10}$ and $10^{14}$ ohm-cm). However, the *thermal conductivity* (K) and *thermal diffusivity ($\alpha$)* (defined as thermal conductivity divided by density and specific heat capacity) of coals depend on the *porosity* and *crack* structure. It has been found that the *thermal conductivity* (K) of coal varies from 0.23 to 0.35 $Wm^{-1}\ k^{-1}$. The *specific heat* of coal has been measured from the ratio of thermal conductivity and thermal diffusivity and was found to be around 1.25 J $m^{-1}.k^{-1}$ at 20 °C.

Coal plays an important role in electricity generation in spite of the facts that coal burning produces carbon emission, depletion of the ozone layer, etc. Coal-fired power plants currently fuel 41% of global electricity. As the coal is heated at high temperatures (from 200 to 1000 °C) in the absence of air, tar and light gases (volatiles) are driven out of coal (which is mostly coke). The pyrolysis behavior of the coal depends on the type of coal used. At low temperatures, between 200 and 250 °C, gases like CO and $CO_2$ form, and at ~300–500 °C tar formation occurs. Bituminous coal produces significantly more tar than the low rank coals and moderate amount of light gases. The coal tar, during continuous distillation, in the temperature range from 200 to 250 °C, produces naphthalene which is almost 65% of the tar-acid free oil; the remaining 35% is made up of thionaphthalene and ß-methyl naphthalene.

In the first quarter of 2016, US coal exports (14.2 million short tons) decreased 7.5% from the fourth quarter of 2015. The import of coal on the other hand continued in the United States from Colombia (81.2%), Canada (8.9%), and Indonesia (7.4%) (Report of US Energy Information Administration, August 16, 2016) (Fig. 5.32).

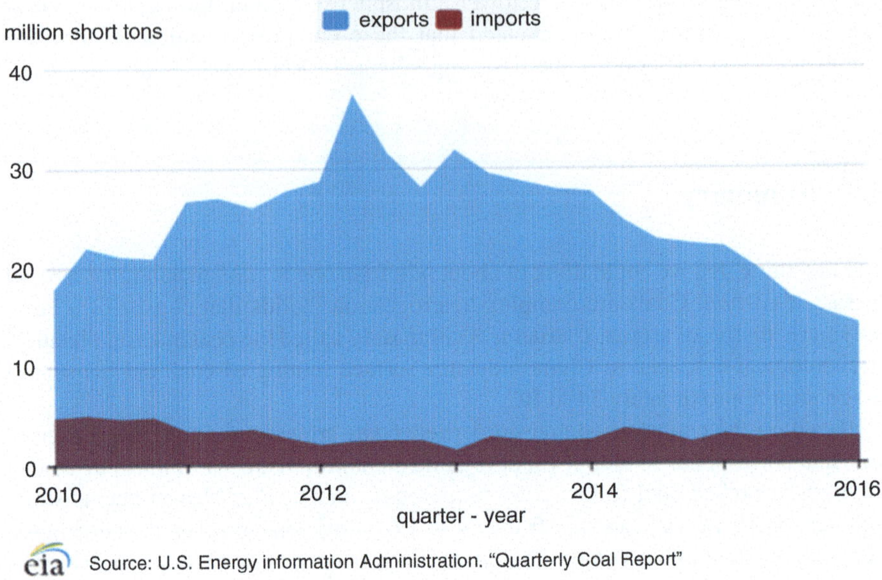


**Fig. 5.32** US coal exports and imports (Courtesy: U.S. Energy Information Administration)

## References


1. D.W.V. Krevelen, *Coal: Topology, Chemistry, Physics, and Constitution* (Elsevier, New York, 1981)
2. O. Charon, S.G. Kang, K. Graham, A.F. Sarofim, J.M. Boer, Variation in coal composition, DOE Grant, RA22-86PC90751, (1986), Chem. Eng. Dept., MIT, Cambridge, MA 02139, USA
3. T.N. Taylor, L.E. Taylor, M. Krings, *The Biology and Evolution of Fossil Plants* (Academic, Cambridge, 2009)
4. J. Dodson et al., *Use of Coal in Bronze Age in China* (Sage, Thousand Oaks, 2014)
5. B.H. Bowen, M.W. Irwin, Coal Characteristics CCTR Basic Facts File # 8, Indiana Center for Coal Technology Research, Perdue Univ., W. Lafayette, IN. and also US Department of Energy, A brief History of Coal, 1000 Independence Avenue, SW/ Wasington, DC., Feb. 12, (2013)
6. BP statistical review of world energy, 2012, British Petroleum. Retrieved 18th August, 2011
7. J.W. de Leeuw, C. Largeau, A review of macromolecular organic compounds that comprise living organisms and their role in kerogen, coal, and petroleum formation, in *Organic Geochemistry: Principles and Applications*, ed. by M. Engel, S. A. Macko (Springer, New York, 1993)
8. J.C. Crelling, Chemical structure of *bituminous coal* and its constituting maceral fractions as revealed by flash pyrolysis. Energy Fuel **6**(2), 125 (1992)
9. P.H. Given, Chemical Structure of Bituyminous coal, in *Coal*, ed. by W. francis (Arnold, London, 1961)
10. P.H. Given, An essay on the organic geochemistry of coal, in *Coal Science*, vol. 3, ed. by M. L. Gorbaty, J. W. Larsen, I. Wender (Academic, Orlando, 1984), p. 65
11. M.L. Gorbaty, J.W. Larsen, I. Wender, *Coal Science* (Academic, New York, 1982), p. 52
12. S. Norman, K.M. Thomas, Fundamental aspects of coal structural changes in the thermoplastic phase. Fuel. **77**(8), 829 (1998) and P.H. Given, M.E. Peover, W.F. Wyss, Fuel. **39**, 323 (1960)

13. F. Derbyshire et al., Molecular structure of coals: a debate. Fuel **68**(9), 1091 (1989)
14. J.H. Shinn, From coal to single stage and two stage products: a reactive model of coal structure. Fuel **63**(9), 1187 (1984)
15. T.H. Fletcher, R.J. Pugmire, *Chemical Structure Changes of Coal, Char, and Tar During Devolatilization, DOE and NSF Joint Funded Program* (University of Utah and Brigham Young University, Utah, 1994)
16. A.M. Orendt et al., in *Advances in Coal Spectroscopy*, ed. by H. L. C. Meuzelaar (Plenum Press, New York, 1992), pp. 215–254
17. T.N. Taylor, E.L. Taylor, M. Krings, *Paleobotany, The Biology and Evolution of Fossil Plants* (Academic Press, Amsterdam/Boston, 2009)
18. Eberhard Lindner: Chemie fur ingenieure: Linder Verlag Karlsruhe, S.258
19. 1879. American Cyclopedia article Anthracite, source:WIKI
20. R. Ye et al., Coal as an abundant source of grapheme quantum dots. Nat. Commun. **4**, 2943 (2013)
21. M. Hook, W. Zittel, J. Schindler, K. Aleklett, Global coal production outlooks based on a logistic model. Fuel **89**, 3546 (2010)
22. V. Bondarenko, I. Kovalevs'ka, K. Gonushevych, *Progressive Technologies of Coal* (CRC Press, E.H. Leiden, 2014), p. 225
23. J.M. Herrin, Thermal conductivity of US coal. J. Geogr Res **101**(B-11), 25381 (1996)
24. R. Nuwer, A Twenty Year Low in U.S. Carbon Emission, *Green*, The New York Times, August 17, 2012 and also, Ron Flanary, Coal's life continues to deam, Trains. **75**(11), 8 (2015)
25. L. Coy, S. Pawson, GEOS-5 Analyses and Forecasts of the Major Stratosheric Sudden Warning of January 2013, GMAO-Res. Nat. Aeronaut. Space Admin. (NASA, US) (2013)
26. A. Simmons et al., ECMWF analyses forecasts of stratospheric winter polar Votex break up in September (2002) in southern hemisphere related events. J. Atmos. Sci. **62**, 668 (2005)
27. R. Scherhag, Die Explosionsartgen Stratospharenerwamungen des Spatwinters 1951.52. Ber. Deutsch. Wetterdienst (US Zone) **6**, 51 (1952)
28. S.H. Beaver, Coke manufacture in Great Britain: a study in industrial geography. Trans. & Papers, Inst. Brit. Geogr. **17**, 133 (1951), Royal Geographical Soc. Pub
29. T.A. Wertime, *Coming of the Ages of Steel* (University of Chicago Pub, Chicago, 1962)
30. C.R. Ward, I. Suarez-Ruiz, Introduction to applied coal petrology, in *Applied Coal Petrology*, ed. by I. Suarez-Ruiz, J. C. Crelling (Elsevier, Amsterdam, 2008)
31. E. Burstlein, La Preparation Selective er Petrographique des Carbons en vue de Leur Cokegaction. Chal. Ind. **30**, 351 (1954)
32. N. Berkowitz, *An Introduction to Coal Technology* (Academic, New York, 1979), p. 28
33. D. Shindell et al., Climate response to regional radiative forcing during the twentieth century. Nat. Geosci. **2**, 294–300 (2009)
34. S.C. Graham, J.B. Homer, J.J. Rosenfeld, The formation and coagulation of soot aerosols generated by the pyrolysis of aromatic hydrocarbons. Proc. R. Soc. Lond. A **344**, 259 (1975)
35. R. Ruthann, Polycyclic aromatic hydrocarbons, phthalates, and phenols, in *Indoor Air Quality Handbook*, ed. by J. Spengleer, J. M. Samet, J. F. McCarthy (McGraw Hill, New York, 2001), pp. 34.1–34.2
36. C.H.A. Wong, A. Ambrosi, M. Pumera, Thermally reduced Graphenes exhibiting a close relationship to amorphous carbon. Nanoscale **16**(4), 4972 (2012)
37. T.K. Gupta, Effect of temperature on electrical conduction of carbon black filled polyimide. IEEE Trans. Compon, Hybrids Manuf. Technol. **12**(4), 696 (1989)
38. Y. Zhang et al., Effect of carbon black and silica fillers in elastomer blends. Macromolecules **14**, 7056 (2001)
39. M. Kluppel, The role of disorder in filler reinforcement of elastomers on various length scales. Adv. Polym. Sci. **164**, 1 (2003)
40. J.E. Mark, B. Erman, F.R. Eirich, *Science and Technology of Rubber* (Academic, New York, 1994), p. 188, and also *Mass Production Technology for Plant Oil-based Carbon Black*, (Mitsubishi Chemical, Japan, 2013)

41. Carbon Black User's Guide, Safety, Health, and Environment, A Publication by an International Carbon Black Association, June 2004
42. K. Gardiner, Effects on respiratory morbidity of occupational exposure to carbon black: a review. Arch. Environ. Health **50**(10), 44 (1995)
43. K.A. Miller et al., Long-term exposure to air pollution and incidence of cardiovascular events in women. N. Engl. J. Med. **356**, 447 (2007)
44. M.O. Andreae, A. Gelencser, Black carbon or brown carbon? The nature of light absorbing carbonaceous aerosols. Atmos. Chem. Phys. Discuss **6**, 3419–3463 (2006)
45. V. Ramanathan et al., Atmospheric Brown Clouds : impact on South Asian climate and hydrological cycle. Proc. Natl. Acad. Sci. **102**(15), 5326 (2005)
46. Jessica Blue, What is Natural Green House Effect?, National Geographic, 27 May 2013
47. W.J. Randel et al., An update of observed stratospheric temperature trends. J. Geophys. Res. **114**(D2), D02107 (2009)
48. R. Bailey, *Global Warming and Other Eco Myths* (Ronald Bailey and Competitive Enterprise Institute, Roseville, 2009)
49. A.L. Brown, T.H. Fletcher, Modeling soot derived from pulverized coal. Energy Fuel **12**, 745 (1998)
50. S. Gupta et al., Climate Change and Other related Policies, IPCC AR4 WG3 (2007)
51. United Nations Framework Convention on Climate Change, *Compilation and Synthesis of Fifth National Communications. Executive Summary*, Geneva, Switzerland, (2011)
52. International Energy Annual, 2006, International Energy Agency (IEA) Report, 9 rue de la Federation, 75739 Paris Cadex 15, France, and J.H. Cushman Jr., Coal Faces Three Hurdles and Steady Decline, Projection Show, inside climate news, May 18, 2015
53. M.R. Beychok, Coal Gassification and the Phenolsovan Process, Am. Chem. Soc. 168th Nat. Meet. Atlantic City, Sept (1974)
54. K.-G. Fahlbusch et al., Flavors and Frgrance, in *Encyclopedia of Industrial Chemistry*, (Wiley, Hoboken, NJ, 2003)
55. A.W. Hofmann, On insolinic acid. Proc. R. Soc. Lond. **8**, 1 (1855)
56. P. von Rague Schleyer, Introduction: delocalized pi and sigma. Chem. Rev. **105**(10), 3433 (2005)
57. T.L. Gilchrist, *Hetro Cyclic Chemistry* (Longman, New York, 1997)
58. D.J. Brown, R.F. Evans, W.B. Cowden, M.D. Fenn, *The Pyrimidines* (Wiley, New York, 1994)
59. P.R. Solomon, M.A. Serio, Evaluation of coal pyrolysis kinetics, in *Fundamentals of the Physical Chemistry of Pulverized Coal Combustion*, ed. by J. Lahaye, G. Prado (Martinus Nijhoff, Dordrecht, 1987), pp. 126–151
60. T.H. Fletcher, M.S. Solum, D.M. Grant, S. Critchfield, R.J. Pugmire, 26th International Symposium On Combustion, The Combustion Inst, Pittsburgh, PA, p. 1231, (1990)
61. C.R. Monson, G.K. Germane, A high pressure drop-tube facility for coal combustion studies. Energy Fuels **7**, 928 (1993)
62. C.R. Monson, Ph.D. Dissertation, Mechanical Engineering Department, BRIGHAM University, Utah, USA
63. T.-C.L. Chang, C. Karr Jr., Gas-liquid chromatographic analysis of aromatic hydrocarbons boiling between 202 C and 280 C in a low temperature coal tar. Anal. Chem. Acta **24**, 343 (1961)
64. J.G. Speight, *The Chemistry and Technology of Coal*, 3rd edn. (CRC Press, Boca Raton, 2013), p. 719
65. H. Ratte, J. Geller, Conitinuous Distillation of Coal Tar, US Patent # us 2868713A, Jan 1959
66. K.D. Bartle, D.W. Jones, in *Analytical Methods for Coal and Coal Products*, vol. 11, ed. by C. Karr Jr. (Academic, New York, 1978), pp. 103–160
67. S.A. Wise et al., Determination of polycyclic aromatic hydrocarbons in a coal tar standard reference material. Anal. Chem. **60**(9), 887 (1988)
68. L.S. Lee, P.S.C. Rao, I. Okuda, Equillibrium Partioning of polycyclic aromatic hydrocarbons from coal tar into water. Environ. Sci. Technol. **26**(11), 2110 (1992)

69. I.M. Fanasov, A.V. Kepman, A.N. Seleznev, V.A. Avdeev, Determination of polycyclic aromatic hydrocarbons in coal tar pitch. J. Analyt. Chem. **64**(4), 361 (2009)
70. J.W. Rice, J. Fu, E.M. Suuberg, Thermodynamics multicomponent PAH mixtures and development of tar like behavior. Ind. Eng. Chem. Res. **50**(6), 3613 (2011)
71. S.S. Deshpande, *Food Technology* (Marcel Dekker, New York, 2002), p. 286
72. I.B. Beriman, *Handbook of Fluorescence Spectra of Aromatic Molecules* (Academic, New York, 1971)
73. F. Rpdriguez, J.C. Burillo, L.F. Adrados, J.F. Tijero, Recovery of anthracene from coal tar by solvent extraction. Sep. Sci. Technol. **24**(3–4), 275 (1989)
74. D. Patranabi, *Sensors and Transducers*, 2nd edn. (PHI Pub, Delhi, 2002), p. 224
75. R.C. Powell, L.A. Harrah, Non-linear responses of poly (venyl toluene) plastic scintillators at high excitation dose. J. Chem. Phys. **55**, 1878 (1971)
76. K.D. Bakes, Evaluating the Response of Polyvenyl Toluene Scintillators Used in Portal Detectors, Ph.D. Thesis, Airforce Institute of Technology, Dept. of Eng. Phys., Wright-Patterson Airforce Base, OH. Mar 2008
77. D.J. Mitchell, C. Brusseau, Neutron Counting and Gamma Ray Spectroscopy with PVT Detectors, Sandia Rept. SAND 2011–4361, June (2011), Sandia Nat. Lab. Albuquerque, New Mexico
78. A.W. Date, *Analytical Combustion with Thermodynamics: Chemical Kinetics and Mass Transfer* (Cambridge University Press, New York, 2011)
79. K. Komatsua, Y. Murataa, N. Sugitaa, K. Takeuchib, T.S.M. Wan, Use of naphthalene as a solvent for selective formation of the 1:1 diels-alder adduct of $C_{60}$ with anthracene. Tetrahedron Lett. **34**(52), 8473 (1993)
80. G.M. Roy, *Activated Carbon Applications in Food and Pharmaceutical Industries* (CRC Press, Roca Baton, 1995), p. 125
81. D.M. Hudgins, C.W. Bauschlicher Jr., L.J. Allamandola, Variations in the peak position of 6.2μm interstellar emission feature:a tracer of N in the interstellar polycyclic aromatic hydrocarbon population. Astrophys. J. **632**, 316 (2014)
82. S. Battersby, Space Molecules Point to Organic Origins, New Scientists, 11 Dec 2009
83. G. Mullas, G. Malloci, C. Joblin, D. Toublanc, Estimated IR and phosphorescence Emision fluxes for specific polycyclic aromatic hydrocarbons in the red rectangle. Astron. Astrophys. **446**(2), 537 (2006)
84. D.A. Garcia-Hernandez et al., Formation of fullerenes in H-containing platary nebulae. Astron. J. Lett. **724**(1), 139 (2009)
85. N. Atkinson, Buckyballs Could be Plentiful in the Universe, Universe Today, 10 Oct 2010
86. S.M. Gudipati, R. Yang, In situ probing of radiation induced processing of organics in astrophysical ice analogs---novel Lser desorption laser ionization time of flight mass spectroscopic studies. Astron. J. Lett. **756**(1), L24 (2012)
87. S.M. Gudipati, R. Jacovi, I.C. Tamburelli, A. Lignell, M. Allen, Photochemical activity of Titan's long altitude condensed haze. Nat. Commun. **4**, 1648 (2013)

# Chapter 6
# Graphite: Carbon the Gray

## 6.1 Introduction

*Graphite* The gray color graphite is an allotrope of carbon, known since antiquity and has been named from the Greek verb graphein. The name was given by Abraham Gottlob, in the year 1789. The old European name is plumbago, meaning *black lead*. The name lead (*black lead*) used to mean that graphite is being used in *lead pencils* and should not be mixed up with the *metal lead* (*Pb*, as we can see in the periodic table of the elements). These two are completely different materials, because *lead* (Pb) is a *metal* whereas *graphite* is a *semimetal* [1–3]. Indeed, the term *graphite* designates mineral planar sheets of carbon atoms, with each atom bound to three neighbors in a noncompact, honeycomb structure, stacked regularly, with three-dimensional order [4].

Figure 6.1 shows the filling of the electronic density of states in various types of materials when the density of states is plotted against the energy of the electrons.

A semimetal is a material with a very small overlap between the bottom of the conduction band and top of the valence band. The semimetallic state is very similar to metallic state, but in semimetals both electrons and holes contribute to electrical conduction. Because of its semimetallic nature, graphite has been used since 1900, in batteries, in steel making, and for making crucible.

Graphite is considered to be the highest grade of black coal, above anthracite, and alternatively called meta-anthracite, although it cannot be used as a fuel like coal. It is the most stable state carbon under standard conditions. It has a hexagonal (dihexagonal) crystal structure, with dipyramidal symmetry [5–7]. It appears iron black to steel gray in daylight and deep blue in transmitted light. The Mohs scale hardness is between 1 and 2 depending on the porosity. Density is in between 2.09 and 2.23 g/cm$^3$, and it is soluble in molten nickel (Ni).

It is estimated that the world reserves of graphite exceed 800 million tons. The United States imports more than one-half of graphite from China, Brazil, Canada, and Sri Lanka. Austria possesses large reserves of amorphous and flake graphite.



T. Gupta, *Carbon*, DOI 10.1007/978-3-319-66405-7_6

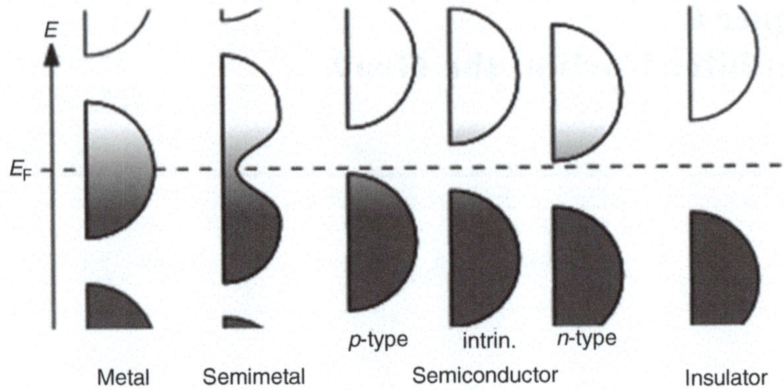


**Fig. 6.1** Filling of the electronic density of states in various types of materials (Courtesy: WIKI)

**Fig. 6.2** (**a**) A lump of graphite mined from the ore (Photo credit: W. Gray Theodore); (**b**) Partial picture of lead pencils – one of the earlier uses of graphite (Courtesy: Dmgerman, WIKI)

Figure 6.2a shows a picture of a lump of graphite mined from the ore, and b represents a partial picture of a lead pencil, which is made from graphite.

Figure 6.3 shows a graphite plant in Passmore, British Columbia, Canada. The plant produces high-grade carbon flake (0.94–0.98% graphite carbon) graphite for the worldwide market.

## 6.2 Graphite Mining, Beneficiation, and Milling

Graphite is mined as ore either by pit method or by underground method. The type of mining depends on how deep the graphite is in earth. For deep deposits, underground mines are dug and graphite is extracted. The mining process is similar

**Fig. 6.3** A graphite plant (Courtesy: Eagle Graphite, Passmore, BC, Canada)

to coal mining, extracting from open pit and underground mines. Graphite usually needs beneficiation, meaning varieties of processes to extract the highest purity of the mineral [8]. In graphite, beneficiation is carried out by handpicking the pieces of gangue (rock) and hand screening the product or by crushing the rock and floating out of the graphite.

Repeated grinding and floating is practiced to obtain a pure highest carbon containing graphite. Acid leaching is also performed to extract pure graphite from the floating pieces of graphite. In the United States, American Graphite Technologies (AGT) is a mineral exploration and technology development company that has 100% ownership in the United States and Canada [9–12].

Depending on the origin of ore, three principal varieties of natural graphite are found. These are crystalline, amorphous, and lump graphite. Crystalline flake graphite occurs as isolated flat plates with hexagonal edges, while the amorphous variety occurs as a result of thermal metamorphism of coal having finer particles [13]. The last variety of the graphite which is known as lump graphite appears to be grown due to massive platy intergrowth and is hydrothermal in origin.

Graphite occurs naturally in metamorphic rocks, such as gneiss, marble, and schist, as a result of the reduction of the sedimentary carbon compounds during metamorphism [9]. The highly oriented ordered pyrolytic graphite (HOPG) has an angular spread of $1^{\circ}$. Some of the physical properties of graphite are included in Table 6.1 for better understanding of graphite – the gray carbon.

**Table 6.1** Physical characteristics of graphite

| | |
|---|---|
| Apparent density | 1.829 g/cm$^3$ |
| Atomic volume | 5.315 cm$^3$/mol |
| Sublimation point at 1 atm. | 400k |
| Triple point | 4200k |
| Boiling point | 450k |
| Heat fusion | 46.84 kJ/mol |
| Sp. Heat at 25 °C | 0.690–0.719 kJ/kg.k |
| Thermal conductivity at 25 °C: W/mk | ab direction: 398<br>c direction: 2.2 |
| Electrical resistivity: ohm.m | ab direction: 3000 $\times$ 10$^{-6}$<br>c direction: 2.5–5.0 $\times$ 10$^{-6}$ |
| Modulus of elasticity | 1.1 $\times$ 10$^4$ N/mm$^2$ |
| Comprehensive strength | 0.0069 MPa |

*Flake graphite* is the most geologically common variety of natural graphite that has been mined in the United States, Central America, South America, Canada, Africa, Germany, Ukraine, Russia, and China. Flake graphite is formed in either metamorphic or igneous geological environments. However, as igneous rocks are rare, most of the flake graphite is deposited in metamorphic rocks formed as *granulite facies* as classified by the metamorphic petrologists. Flake graphite can begin its geological life as either an organic or inorganic carbon. The flake graphite is released by crushing the enclosing country rock. Reduced ore is then fed to a mill where the ore is crushed further. Milling may be done in water slurry. The graphite/ore–rock slurry is then fed to froth floatation cells. Mixing is followed by air injection to form bubbles. The buoyancy induced by the air bubbles causes the individual flakes to float on the top of the bath where they are skimmed off. Subsequent to floatation, the flake is washed, dried, and finally packaged for shipment. Commercial flake graphite sizes are available in the ranges between 50 and 100 mesh, with a purity varying from 80 to 99+ percent carbon. Figure 6.4 shows a graphite crystal with a growth spiral-aligned with the prism edges (sample collected from Crestmore, CA, USA).

Graphite has a layered structure almost similar to lead iodide ($PbI_2$). Carbon atoms are arranged in the honeycomb lattice of each layer with a separation of 0.142 nm. The porosity of the best variety of graphite has been found to be about 16% of the volume of a bulk piece of graphite with open or closed pores. Graphite has two varieties, e.g., alpha (hexagonal) and beta (rhombohedral). Both the varieties are very similar in their physical properties except for the way the graphene layers are stacked.

Animated view of the unit cell in the three layers of graphene is slightly different than the unit cell of graphite. Figure 6.5a shows the arrangements of atoms in a unit cell, and b shows the ball and stick model of graphite (two graphene layers). In nature, graphite is found usually in association with feldspar, mica, quartz, pyroxene, rutile, pyrites, and apatite.

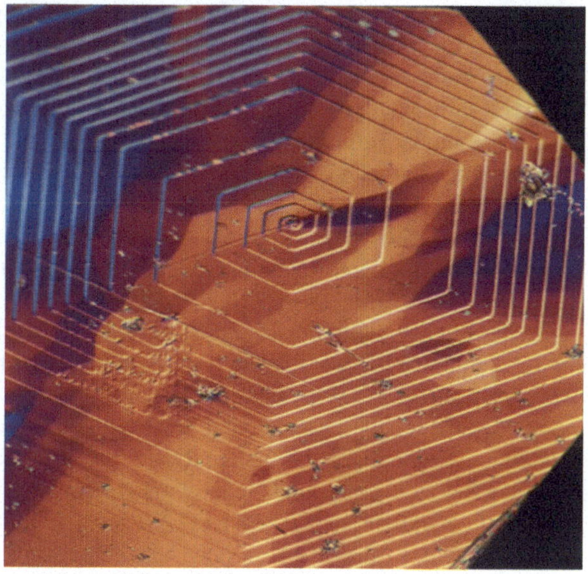

Fig. 6.4 A graphite crystal with a growth spiral-aligned with the prism edges (Photo courtesy J.A. Jaszczak, MTU)

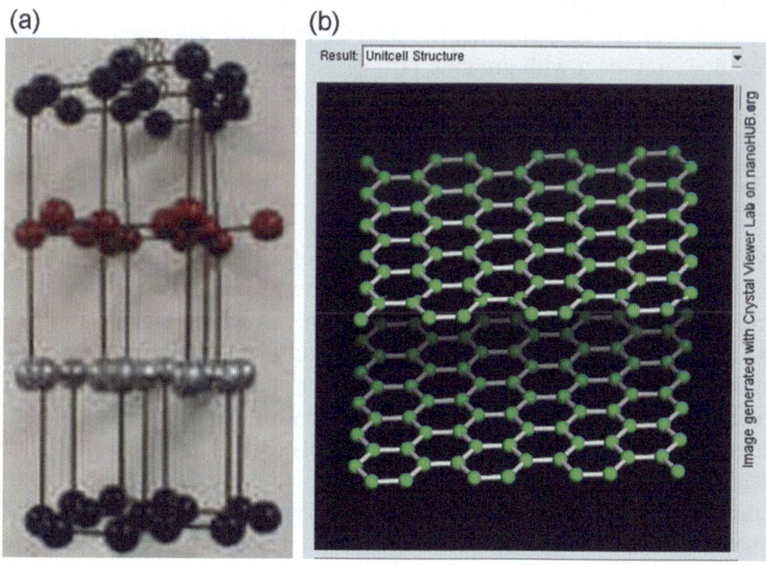


Fig. 6.5 (a) Graphite's unit cell; (b) Ball and stick model of graphite (Courtesy: S.R. Mehrotra and G. Klemeck, WIKI)

As the temperature of a piece of graphite is increased (up to about 2500 °C), its compressive strength increases. The temperature also has a pronounced effect on the electrical resistivity of graphite. As the temperature increases from room temperature to about 700 °C, the electrical resistivity decreases, but above

700 °C, the resistivity increases with temperature. Graphitization temperature also has an effect on electrical resistivity. The higher the graphitization temperature, the lower the electrical resistivity becomes, as measured at room temperature. Table 6.1 shows the physical characteristics of graphite.

The hexagonal (alpha) variety of graphite may be either flat or buckled. The alpha variety can be transformed to beta variety by mechanical treatment, while the beta form can be reverted to alpha by the application of heat (~1300 °C) [11, 14]. The acoustic and thermal properties of graphite are highly anisotropic. The phonons propagate quickly along the tight-bound planes, but travel slowly from one plane to another.

Graphite reacts with metals and forms carbides of the metal. As for example, it can react with the metals of group IV, V, and VI, like molybdenum (Mo), titanium (Ti), vanadium (V), and tantalum (Ta), and forms hard carbides of the respective metals. It also reacts with iron and forms iron carbide ($Fe_3C$). Alkali metals like potassium (K), calcium (Ca), strontium (Sr), and barium (Ba) also react with graphite. Halogen compounds react with graphite at different temperatures. Nitric acid ($HNO_3$) and sulfuric acid ($H_2SO_4$) react with graphite and produce oxides of nitrogen ($NO_2$), sulfur ($SO_2$), carbon ($CO_2$), and water ($H_2O$). Besides these, nitric acid also produces traces of graphite oxide ($C_7H_2O_4$), mellitic acid ($C_6$ $(CO_2H)_6$), and hydrocyanic acid (HCN). Similarly, with sulfuric acid, it produces traces of benzoic acid ($C_6H_5CO_2H$) and mellitic acid $C_6$ $(CO_2H)_6$. Hydrofluoric acid (HF), however, does not react with graphite [15].

## 6.3 Graphitization

*Graphitization* is defined as the transformation of a turbostratic graphite material (i.e., *carbon*) into a well-ordered graphite structure, when carbon is heated at high temperatures in excess of 2500 °C [16, 17]. The structural change from coal to graphite is accompanied by a large increase in electrical and thermal conductivities. The density also increases and approaches to that of a single crystal.

Figure 6.6 shows graphitization of coal at a very high temperature. Coal tar pitch graphitizes readily by the process called pyrolysis in the absence of air. The pyrolysis behavior of coal tar pitch is a strong function of coal rank, such as lignite, bituminous, and anthracite. As the temperature rises during pyrolysis process, the bonds between the aromatic clusters in the coal tar macromolecule break creating fragments that are detached from the macromolecule [18, 19].

Figure 6.7 shows the atomistic model of a diamond–graphite interface formed during graphitization of a diamond surface. The degree of graphitization is a function of the ratio between the aromatic and aliphatic hydrogen and is known as hydrogen aromaticity (the ratio of aromatic hydrogen atoms to the total number of hydrogen atoms). Graphitization is accompanied by the weight loss of the starting material which is attributed to the removal of interlayer chemical species,

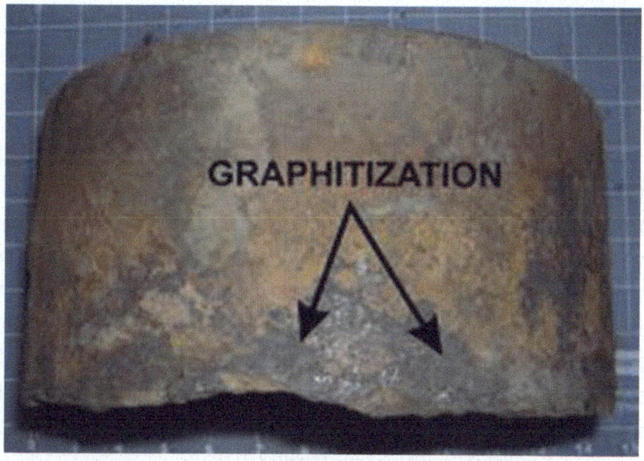


**Fig. 6.6** Graphitization of coal at a very high temperature (Courtesy: TE, CA)

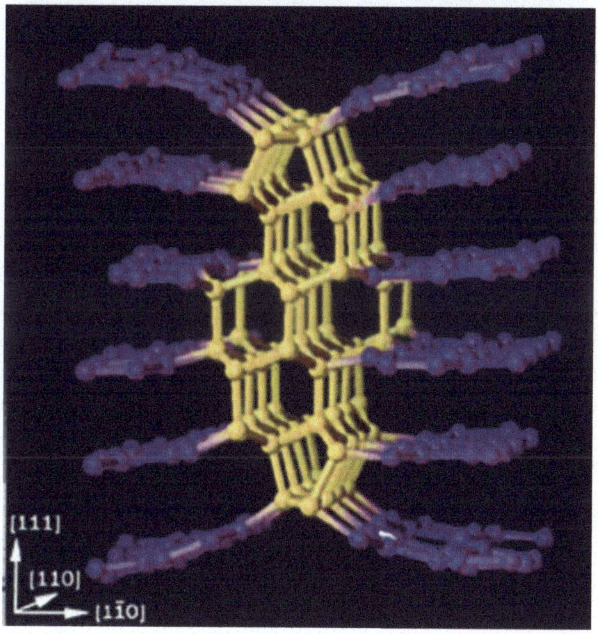


**Fig. 6.7** Atomistic model of a diamond–graphite interface formed during the graphitization of a diamond surface (Reproduced with permission, from Springer, Galli [60])

mostly interstitial carbon [20]. In 1951, R.E. Franklin first recognized the existence of graphitizable, nongraphitizable carbons, as well as intermediate graphitic carbon [21, 22].

It is now recognized that three-dimensional ordering of carbon occurs only after the graphene layers have reached a perfect two-dimensional structure. Defects are progressively removed as the heat treatment temperature increases. Thus, graphitization process depends on the transformation occurring within the layer itself. Hoska and Warren, and then Maire and Mering, postulate that pairs of nearest neighboring atomic layers order independently, rather than whole regions taking on the graphite structure [23, 24]. This ordering produces modulation in the two-dimensional hk-bands.

The word *aromatic* means a chemical compound having *aroma* or *odor*, and the word *aromaticity* in organic chemistry is defined as the way a conjugated ring of unsaturated bonds, lone pairs, or empty orbitals exhibits *stabilization* stronger than that would be expected by the *stabilization* of conjugation alone. It can be described also as the manifestation of cyclic delocalization and of resonance [25, 26]. As a matter of fact, the *quantum mechanical* origins of this *stability* or *aromaticity* was first modeled by Huckel, in 1931, and he was the first chemist to separate the bonding electrons into Sigma and Pi electrons [27, 28]. Polycyclic aromatic hydrocarbons (PAH) are the aromatic compounds contained in significant amounts in natural crude oil, coal, and coal tar. However, the aromaticity of PAHs depends on each individual's ring structures [29].

When coal is burnt, nitrogen is released in two stages. In the first stage, nitrogen is released as the volatiles in the tar and light gases. The second stage of nitrogen release occurs as the char is combusted. Chars are produced by carbonization of aliphatic hydrocarbons (compounds with an open-ended chain) and of most polymers. They do not undergo graphitization easily. However, when vitreous carbon is heated to about 1800 °C, it produces a material with interlayer spacing (d) of 0.36 nm and a crystallite size ($L_c$) of 1.5 nm. These results have been obtained from the broadening of the line along <110> and <002> directions of the X-ray diffraction (XRD) patterns.

It has been detected that when vitreous coal is heated up to 2750 °C, it does not change XRD line patterns. However, when petroleum coke is heated to 1800 °C, well-graphitized material with crystallite size produces interlayer spacing of 0.35 nm and $L_c$ value almost equal to 3 nm. After carbonization of the vitreous coal, the residual material is mostly a vitreous carbon which is a ribbon-like strained tangled aromatic molecules cross-linked by carbon–carbon covalent bonds with varying bond energies [30, 31].

## 6.4 Molded Graphite

The molded graphite is defined as a synthetic product of graphite. The manufacturing processes are similar to particle-in-binder (PIB) process, where carbon particles are mixed with organic binder which is subsequently carbonized and graphitized. The basic product was invented by E.G. Acheson, in the year 1896, and by then it was found in almost every corner of the industrial world and formed the base of the traditional graphite industry [32].

It is often difficult to obtain details of the specific process, particularly when most of the graphite manufacturers kept their manufacturing process as proprietary [33–35]. However, a great deal of information on the basic materials and processes is disclosed in the open literature. The selection of the appropriate precursor is the

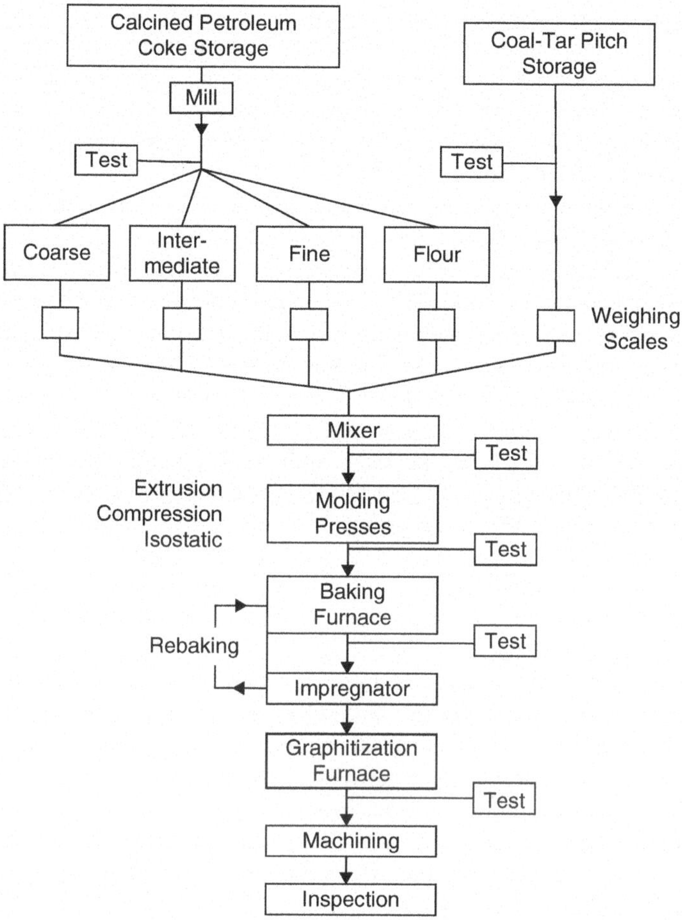


**Fig. 6.8** Flow diagram for the production of molded graphite

first and critical step in the manufacturing process. Figure 6.8 shows the flow diagram for the production of molded graphite.

Petroleum coke is a popular choice since it can rapidly be graphitized at ~2700 °C. The most common binder is coal tar pitch which is a hard, brittle, and glassy material, but the composition is also very complex. Other binders that are used for some specific applications are petroleum pitch and thermosetting. A typical production process flow diagram for molded graphite production is shown in Fig. 6.8 [36, 37]. Extrusion, compression, and isostatic pressing are used to mix the filler with the binder. Extrusion pressures are on the order of 7 MPa. The compression molding is performed in tungsten carbide dies, and the pressure varies from 28 to 280 MPa. Isostatic pressure on the mixture is applied from all directions through a rubber membrane in a liquid-filled chamber. The green-shaped material thus obtained is sent for heat treatment or carbonization.

The initial temperature of the furnace with an inert or reduced atmosphere is set to 550 °C. The temperature ramped to 1200 °C at an interval of every 15 min. Impregnation is followed by carbonization to minimize porosity and maximize density of the material. In special cases, to increase electrical conductivity, noncarbon impregnating materials such as silver (Ag) and lithium fluoride (LiF) are added during processing.

Next step is graphitization. Graphitization is a solid state transformation of thermodynamically (by heat treatment to 2800–3000 °C at atmospheric pressure) unstable carbon into graphite by thermal activation [38]. During this process, the material is heated up to 3000 °C. The temperature cycle is shorter than the carbonization cycle. Graphitization increases the resistance of the material to thermal shock and chemical attack. High-temperature (2800–3000 °C) processing during graphitization causes geometrical changes along with the release of the volatile gases very rapidly, which sometimes cause cracks and other structural defects. Irreversible expansion (puffing) isothermal heating with small amounts of metal/s can minimize these effects [39]. Graphitization of the anisotropic matrix carbon is characterized by the occurrence of fold-sharpening and polygonization [40]. The final product is heat-treated in a halogen atmosphere to remove further impurities from the product (like aluminum (Al), boron (B), calcium (Ca), iron (Fe), silicon (Si), titanium (Ti), and vanadium (V)). After purification, the product is machined to the final shape.

High-energy electrons or ions with large momenta can create interstitial vacancy pairs in graphite and may cause structural changes [41, 42]. As the temperature is raised too high (between 1200 and 1600 °C), vacancies aggregate and reconstruct, while the interstitials that aggregate between the basal planes form a new layer. The process is called plating out which leads to anisotropic changes in the dimensions of graphite grains and may cause the formation of carbon onions [43]. Density functional theory (DFT) calculations put the adatom on a basal plane above a host bond center, forming almost equivalent triangle of carbon atoms, perpendicular to, and slightly raised from, the basal plane.

Annealing at low temperature brings negligible change in the defect subsystem, and the defect centers can store substantial amounts of energy known as Wigner energy. Various experimental techniques have been used to study point defects concentration of quenched-in vacancy analysis.

The application of molded graphite is increasing day by day, and the technology for the production of molded graphite is continuously changing. The estimated world market for carbon and graphite products in 2009 was 1.82 billion US dollars, which is down from 2.79 billion US dollars [44], of which the molded carbon share was almost 40%. The major use of molded graphite in the metal-processing industry is in the production of electrodes. Molded graphite also has many applications in the processing of ferrous and nonferrous metals and alloys such as copper, copper–nickel, brass, bronze, and zinc.

In semiconductor industry, molded graphite is extensively used in crystal growing process. As a matter of fact, for the materials with high melting point, molded graphite crucibles are used as an alternative and cheap material to platinum (Pt),

**Fig. 6.9** (**a**) A molded graphite crucible; (**b**) Molded graphite electrodes

iridium (Ir), or alloy of Pt–Ir crucible. The cheap crucible-molded graphite not only can withstand very high temperature, but it also remains inactive chemically. In electrical applications, molded graphite is used for brushes in electric motors, current-collecting shoes for railroad conductor rails, heating elements for high-temperature furnaces, fuel cell components, and electrodes for lighting.

In mechanical engineering, high-density, low-porosity molded graphite is extensively used for bearings and seals, particularly for high-temperature applications where conventional liquid lubricants degrade rapidly. Starting from seal rings for gas turbine, to water pumps, automobiles, Freon air compressor for air-conditioning, washing machines, and dishwasher water pumps, all use molded graphite. Figure 6.9a shows a molded graphite crucible and Fig. 6.9b shows molded graphite electrodes.

## 6.5 Pyrolytic Graphite

It is also called the *pyrolytic carbon* or *pyrocarbon*. It differs from molded graphite in the sense that the precursor in this case is a gas instead of solid, like vitreous carbon, coal tar pitch, petroleum fractions, or polymers. Although it is produced in bulk form, its main use is in the form of coatings, deposited on molded graphite, carbon fibers, or porous carbon–carbon structures. The process involved in the production of pyrolytic graphite is known as chemical vapor deposition (CVD). It is a common method adopted by the semiconductor industry to make sophisticated films [45, 46]. CVD is a vapor-phase process which relies on the chemical reaction of a vapor near the surface of a substrate to form solid deposits and gaseous by-products. The CVD of carbon material started with the pioneer work of Powel, Oxley, and Blocher [47, 48].

Theory: The optimization of the CVD process depends on the carbon source (hydrocarbon gases), the method of activating the decomposition reaction (thermal, plasma, laser, etc.), and the deposition variables (like temperature, pressure, gas flow, etc.). The studies of all of these will give valuable data to optimize the process and the reaction mechanisms. The data on the calculation of the change in the free energy will lead to the information on the feasibility of the reaction. Experimentation shows that Gibb's free energy ($\Delta G$) should be minimized in order to get a dense defect-free deposition. For very small value of ($\Delta G$), the reaction will be very slow. On the other hand, higher value of ($\Delta G$) can lead to the formation of soot. Computer programs are now available to set the optimum value before doing many experiments.

## 6.6 Experimental

Pyrolytic graphite is deposited by CVD method by thermal decomposition (pyrolysis) of hydrocarbons. The most common precursor is methane (CH4). It is generally pyrolyzed at a temperature of 1100 °C over a wide range of pressure (from 100 to 105 Pa). The chemical reaction is as follows:

$$CH_4 \rightarrow C + 2H_2 \tag{6.1}$$

Figure 6.10 shows the schematic of a CVD process for the production of pyrolytic graphite. The other common precursors are ethylene ($C_2H_6$) and acetylene ($C_2H_2$). The activation energies for these precursors are not accurately known, and the reported values scatter considerably. For methane ($CH_4$), it varies from 80 to 108 kcal/g.mole; for ethane ($C_2H_6$), it varies between 60 and 86 kcal/g.mole; and for acetylene ($C_2H_2$), it varies between 30 and 50 kcal/g mole. Figure 6.10 shows the schematic of a CVD process for the production of pyrolytic graphite.

Methane gas is introduced into the system, which might be diluted if necessary with hydrogen. The flow of the gas is regulated by a flow meter attached to the inlet system. The initial gas pressure can be monitored and regulated and finally be recorded. The gas then entered into the reaction chamber where a substrate is placed to deposit the pyrolytic graphite. The chamber is heated to a suitable temperature, and the reacted gases that escape are monitored and analyzed with a gas chromatograph (spectroscopic analysis has shown that the escaped gases contain acetylene, ethylene, benzene, and a variety of compounds consisting mostly of the polyaromatic hydrocarbons (PAH)). The gases once detected are released and filtered. Finally, they are condensed through a liquid nitrogen trap, and the remaining is released in the atmosphere. The CVD system can be of different types, namely cold wall reactor CVD, chemical vapor infiltration (CVI) CVD, fluidized bed CVD, or plasma CVD.

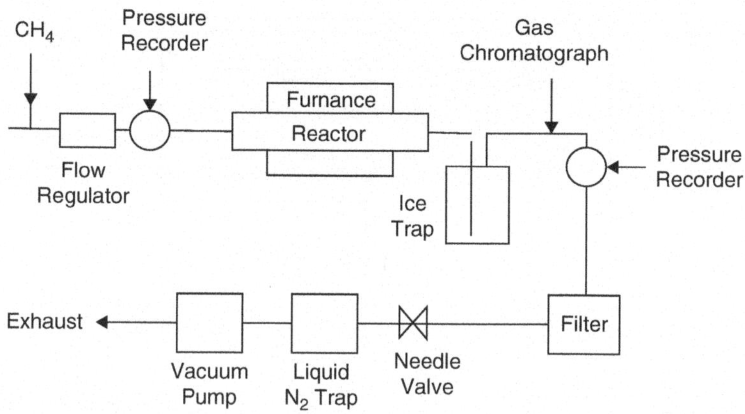


**Fig. 6.10** Schematic of a CVD process for the production of pyrolytic graphite

The cold wall reactor CVD does not require a furnace; rather, the substrate is heated by induction heater (which has a high frequency (450 kHz)). The decomposition reaction for the deposition of pyrolytic graphite is endothermic. The substrate is a molded graphite disk which is rotated during deposition to improve deposition uniformity. The pressure inside the chamber is kept around 500 Pa [49].

In CVI (chemical vapor infiltration) process, the same chemistry and thermodynamics are followed but the kinetics is different. In this process, gas is infiltrated through a porous material like foam, or fibrous mat or weaves. This process is extensively used for the production of C–C materials.

In fluidized CVD, a special technique is adopted as the coated material is nuclear fuel. The fluidizing gas is usually methane, helium, or another nonreactive gas. The density and the size of the particles to be coated determine proper fluidization of the material. The major application of the method is the production of biomedical components such as heart valves. It is also used to coat uranium carbide and thorium carbide nuclear fuel particles.

The deposition of graphite can also be done by plasma CVD. In this process, plasma of the gas is produced at low temperature in the vicinity of the substrate. Like sputtering, here the energetic ions impart their energy and momentum on the reactant gas molecules and atoms. The energy transfer breaks up the molecules and aids the chemical reactions. The deposition temperature is very low compared to thermal CVD; yet, the quality of the film is superior. Figure 6.11 presents a schematic of a plasma CVD.

The process is used extensively to deposit polycrystalline diamond and diamond-like carbon (DLC). The plasma CVD is more expensive than thermally aided CVD.

The pyrolytic graphite is an aggregate or graphite crystallites and have various degrees of orientation. It can have columnar, laminar, or isotropic structure depending on the deposition conditions, such as pressure, temperature, and composition of the input gases [50]. In columnar structure, the crystallites are deposited

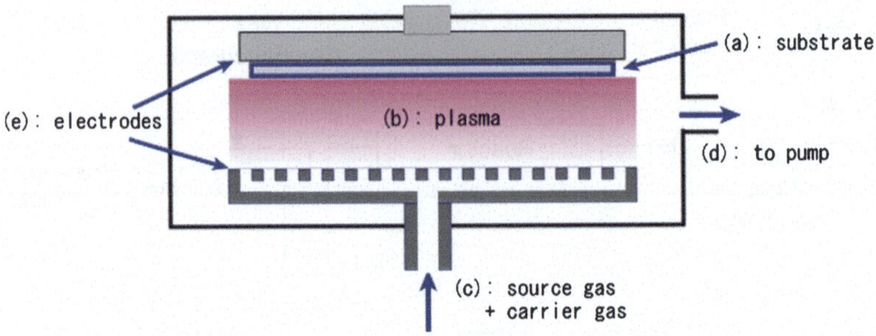


**Fig. 6.11** A schematic of a plasma CVD system (Courtesy: S. Kei, WIKI)

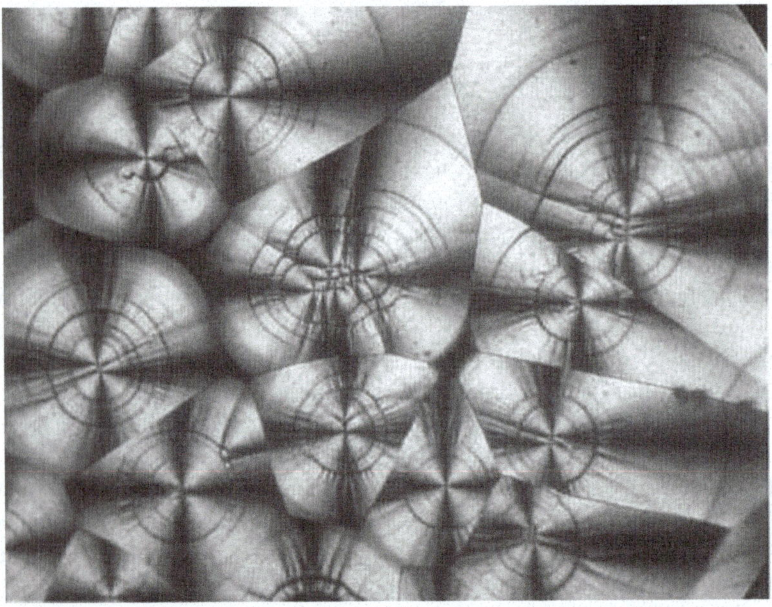

**Fig. 6.12** Laminar structure of pyrolytic graphite (Courtesy: Jack Chin, La Costa, CA)

with the basal planes (ab direction) essentially parallel to the deposition surface. The laminar structure, on the other hand, consists of essentially parallel layers (or concentric shells when deposited on a fiber). Both columnar and laminar structures are optically active to polarized light and possess similar physical properties. Figure 6.12 shows the laminar structure of pyrolytic graphite.

The isotropic structure, however, has little graphite characteristic and the material does not have any optical activity. Therefore, the isotropic variety is known as isotropic carbon rather than isotropic graphite. It is obtained during fluidized bed deposition. Figure 6.13 shows a picture of the surface of an isotropic variety of carbon when observed through transmission electron microscope (TEM) (viewing

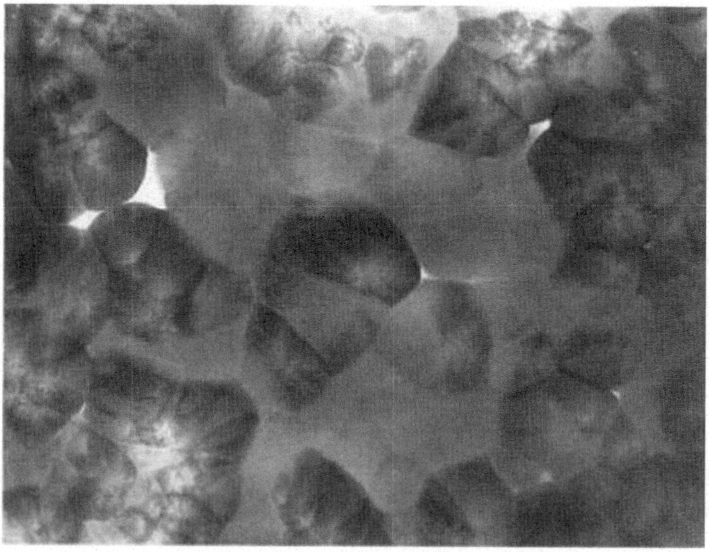

**Fig. 6.13** TEM image of a highly dense isotropic structure of pyrolytic carbon (Courtesy: J.L. Kaae, General Atomics, San Diego, CA)

plane is parallel to deposition plane). The density of the isotropic structure of the film had been detected to be approximately 2.0 g/cm$^3$.

The deposition of the film over the substrate depends on the ratio of methane: hydrogen ($CH_4$:H). It also depends on the purity of the carrier gas, the angle between the receiving gas molecules, the substrate, the substrate temperature, and chamber pressure. Pressure controls the thickness of the surface boundary layer and consequently the degree of diffusion. By operating at a low pressure, surface kinetics becomes rate-controlling because of the minimized diffusion process [51].

Postheating to 2500 °C minimizes imperfections like vacancies, stacking faults, dislocations, and rotational disorders, and it also increases the structure of the crystallites. Heating at ~2700 °C under pressure, both the columnar and the laminar structures relieved their stresses between the layers, and the result is a highly oriented pyrolytic graphite (HOPG) film. The mechanical properties of pyrolytic varieties (both columnar and laminar) are similar to the molded variety.

Table 6.2 shows the physical properties of graphite. Pyrolytic graphite boats and crucibles are used in semiconductor processing especially in molecular beam epitaxy (MBE), liquid phase epitaxy (LPE), and vapor phase epitaxy (VPE), where high-quality and high-purity boats are needed. Because of its good electrical characteristics and refractories, pyrolytic graphite is extensively used for high-temperature resistance heating elements too.

Silicon carbide and alumina provide the reinforcement for metal or ceramic matrices to form refractory composites, but they react at high temperatures and form intermetallics or other compounds that degrade the properties of the composite. A pyrolytic coating of graphite is used as a diffusion barrier. It is used to densify

Table 6.2 Physical properties of graphite

| Properties | Molded graphite | Pyrolytic graphite density |
|---|---|---|
| Density (g/cm$^3$) | 2.20–2.23, depending on the porosity; graphite crystal: 2.25 | 2.10–2.24 |
| Thermal conductivity W/m.k | Petroleum coke: 159<br>Lampblack:31.4 | ab direction: 190–390<br>c direction: 1–3 |
| Thermal expansion 10–6/m.k | With grain: 1.2–2.2<br>Across grain: 2.9–5.1 | ab direction: −1-1<br>c direction: 15–25 |
| Electrical resistivity μΩ.m | ~50 | ab direction: 2.5–5.0<br>c direction: 1 k–3 k |

carbon–carbon structures by infiltration, and they are used in rocket nozzles, aircraft disk brakes, and other aerospace components [52].

## 6.7 Nuclear Graphite

It is a high-grade, special type of graphite used as a moderator or reflector within a nuclear reactor. In the year 1942, Enrico Fermi decided to use graphite as a moderator in the first attempt to create a self-sustaining nuclear chain reaction.

Graphite is used because of its purity. Care must be taken to see that the graphite bar used in moderator should not have any impurity such as boron (B) (Boron is extensively used as a seed electrode in commercial graphite deposition system, and it can withstand very high temperatures).

Figure 6.14 shows a nuclear reactor with graphite core. In the advanced pebble bed nuclear reactor (PBR), radioactive uranium dioxide fuel is used as tiny flakes. These flakes are embedded in round graphite tennis-ball sized (pebble) shell coated in a number of chemical layers. The graphite in the pebbles is a mixture of 75% natural graphite and 25% pyrolytic graphite. The small uranium dioxide spheres are coated with a layer of porous carbon, high-density pyrolytic carbon, silicon carbide, and then another layer of pyrolytic graphite (carbon). Figure 6.15 shows the section of the pebble with a layer of pyrolytic graphite, silicon carbide, and porous pyrolytic graphite.

Pebble bed reactors are supposed to withstand meltdown because the reaction is slowed down when the temperature rises inside the core (Fig. 6.16).

The graphite used in high-temperature nuclear reactors are pure (99.5% C), with ash content less than 0.5%, and the density is better than 1.5 g/cm$^3$. The pebble nuclear reactor contains about 360,000 pebbles. Three-fourths of the total spheres are fuel spheres (Fig. 6.16), and the rest of the spheres are combined natural and pyrolytic graphite. Typically, each pebble is cycled 10–15 times, making one total

**Fig. 6.14** A nuclear reactor with graphite core (Courtesy: ORNL (WIKI)

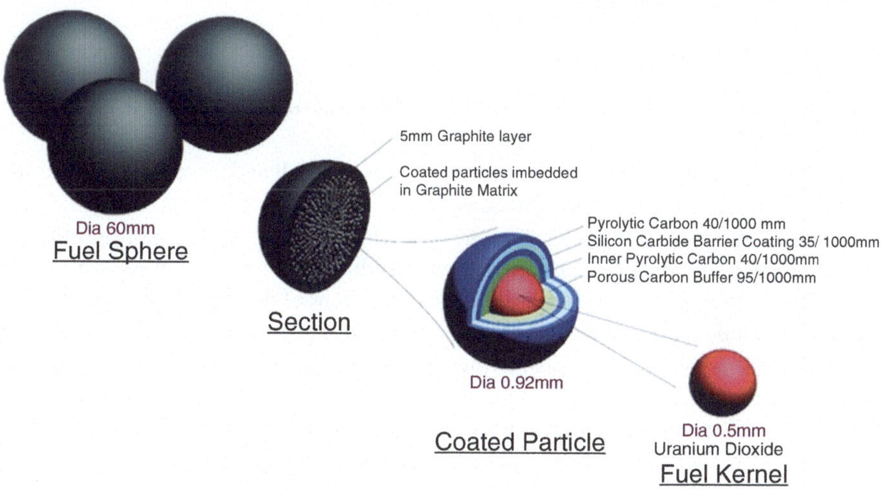


**Fig. 6.15** Section of the pebble with a layer of pyrolytic graphite, silicon carbide, and porous pyrolytic graphite (Courtesy: MEGA graphite, Ontario, Canada)

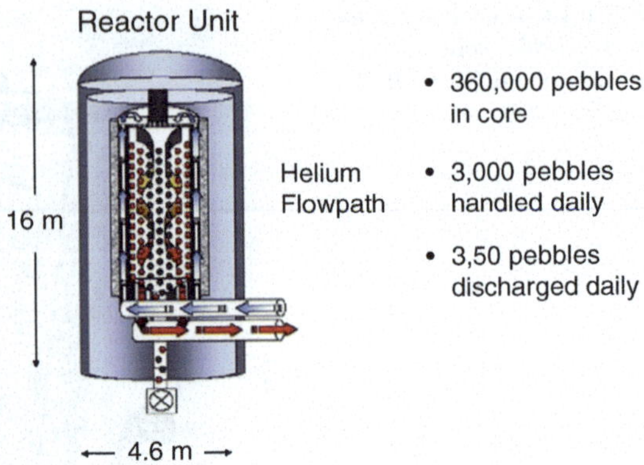


**Fig. 6.16** A sectional diagram of a pebble nuclear reactor (Courtesy: MEGA Graphite Inc., Ontario, Canada)

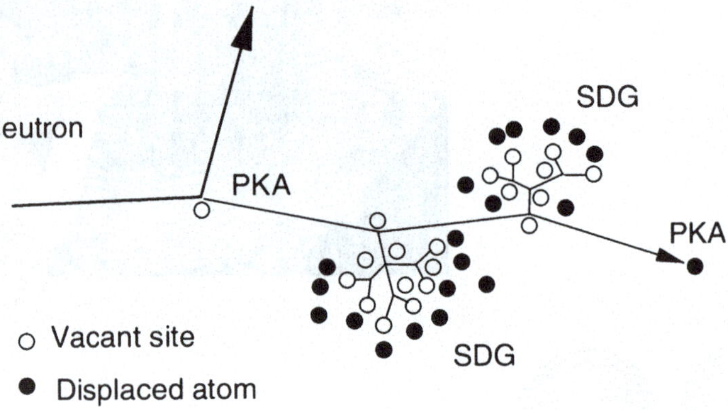


**Fig. 6.17** A schematic showing damages caused by irradiation of neutrons (Ref. [54])

fuel load last just over 42 years (Excerpts from: Nuclear Reactor Applications, MEGA Graphite Inc., Ontario, Canada).

As these graphite fuel cells are used in nuclear reactor, it will be important to calculate their life span because of neutron irradiation. The neutron irradiation on graphite can change various properties like coefficient of thermal expansion (CTE), Young's modulus, and thermal resistivity [53, 54], and structural changes that are mainly interstitials and vacancies. Dimensional change due to neutron irradiation is substantial (Fig. 6.17).

Figure 6.17 shows defects caused by the irradiation of neutrons on graphite. The cascades of defects are mainly primary knock-on atom (PKA) and secondary displacement group (SDG). It was further noticed that basal dislocations lie at the heart of nearly all these effects, which was originally thought exclusively due to Frenkel pairs formed from irradiation. The ground state for a single interstitial atom between two graphite layers has spiro-interstitial (Fig. 6.18a) structure with carbon K-edges for graphite system, as predicted from theoretical calculations (Fig. 6.18b) [55, 56].

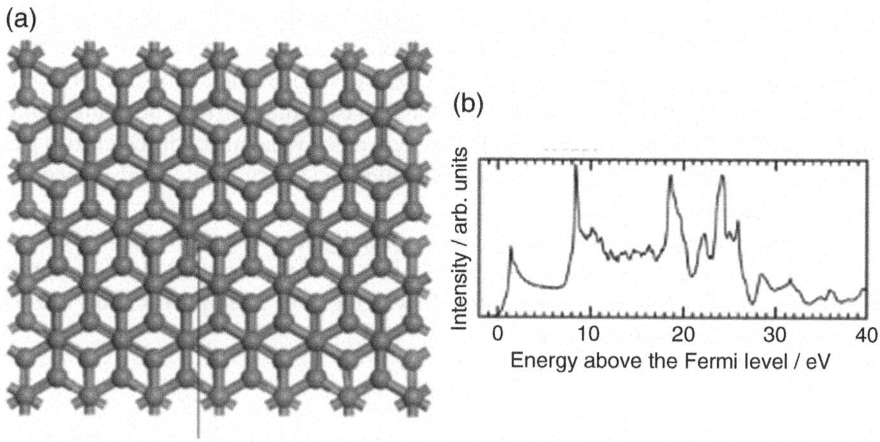


**Fig. 6.18** (**a**) Spiro-interstitial atom between two graphite layers and (**b**) predicted carbon K-edges for graphite system

There have been two major accidents in graphite moderator nuclear reactors, one in Windscale fire in October 1957, which was the worst nuclear accident before Chernobyl disaster near Ukraine, on April 26, 1986 (which was at that time under USSR). In Windscale fire, it was detected that during annealing process the untested graphite was responsible for overheating in the unmonitored areas of the core, leading directly to the ignition of the fire [57, 58]. However, in Chernobyl, the graphite moderator was not primarily responsible for the primary event. A massive power excursion during a mishandled test caused the catastrophic failure of the reactor vessel [59].

## 6.8 Summary

Graphite is considered to be the highest grade of black coal, above anthracite, and alternatively called meta-anthracite, although it cannot be used as a fuel like coal. It is the most stable state carbon under standard conditions. It has a dihexagonal crystal structure, with dipyramidal symmetry. It appears iron black to steel gray in daylight and deep blue in transmitted light. The Mohs scale hardness is between 1 and 2 depending on the porosity. Density is in between 2.09 and 2.23 g/cm$^3$, and it is soluble in molten nickel (Ni).

Graphite reacts with metals and forms carbides of the metals. As for example, it can react with the metals of group IV, V, and VI, such as molybdenum (Mo), titanium (Ti), vanadium (V), and tantalum (Ta), and forms hard carbides of the respective metals. *Graphitization* is defined as the transformation of a turbostratic graphite material (i.e., *carbon*) into a well-ordered graphite structure, when carbon is heated at high temperatures in excess of 2500° C.

Pyrolytic graphite is deposited by CVD method by thermal decomposition (pyrolysis) of hydrocarbons. The most common precursor is methane ($CH_4$). It is generally pyrolyzed at a temperature of 1100° C over a wide range of pressure (from 100 Pa to $10^5$ Pa). The deposition of the film over the substrate depends on the ratio of methane:hydrogen ($CH_4$: H). It also depends on the purity of the carrier gas and the angle between the receiving gas molecules and the substrate. It also depends on the substrate temperature.

A high-grade (without any impurities, especially boron) graphite is used as a moderator or reflector within a nuclear reactor. In the year 1942, Enrico Fermi decided to use graphite as a moderator. It was the first attempt to create a self-sustaining nuclear chain reaction.

## References

1. P.R. Wallace, The band theory of graphite. Phys. Rev. **71**(9), 622 (1947)
2. D.W. Mckee, Carbon and graphite science. Ann. Rev. Mater. Sci. **3**, 195 (1973)
3. S.E. Boulfelfel, A.R. Oganov, S. Leoni, P.R. Wallace, Understanding the nature of super hard graphite. Nat. Scientif. Repts. **2**, 471 (2012)
4. K.-H. Kochling, B. McEnaney, S. Muller, E. Fitzer, International committee for characterization and terminology of carbon "first publication of 14 further tentative definitions". Carbon **23** (5), 601 (1985)
5. D.D.L. Chung, Review graphite. J. Mater. Sci. **37**, 1475 (2001)
6. Y. Wang, J.E. Panzik, B. Kiefer, K.K.M. Lee, Crystal structure of graphite under room temperature compression decompression, science. Sci. Rept. **2**, Article # 520 (2012)
7. L.J. Brillson, E. Burnstein, A.A. Maradudin, T. Stark, in *The Physics of Semimetals and Narrow gap Semiconductors*, ed. by D. L. Carter, R. T. Bate (Pergamon Press, London, 1981), p. 187
8. C.C. Barry, M.G. Norton, Beneficiation, in *Ceramic Materials: Science and Engineering* (Springer, Berlin, 2007)
9. J.W. Anthony, A.R. Bideaux, W.K. Bladh, W. Kenneth, C.M. Nicjols, Graphite, in *Handbook of Mineralogy* (US Metallurgical Soc. Am., Virginia, 1990)
10. P. Delhaes, *Graphite and Precursors* (CRC Press, Boca Raton, 2001)
11. H. Lipson, A.R. Stokes, A new structure of carbon. Nature **149**(3777), 328 (1942)
12. A.H. Taylor, *Graphite Industrials and Rocks*, 7th edn. (Littleton Co. AIME –Soc. Of Mining Engineers, 2005)
13. D.M. Sutphin, D.B. James, Disseminated flake graphite and amorphous graphite deposit types; an anlysis using grade and tonnage models. CIM Bull. **83**(40), 85 (1990)
14. W.G. Wuckoff, *Crystal Structures* (John Wiley, New York, 1942)
15. C. Lorraine, *Graphite a Refractory Material* (Technical Vrochure, Gennevilliers, 1990)
16. G.M. Jenkins, K. Kawamura, *Polymer Carbons* (Cambridge University Press, Cambridge, 1976)
17. E. Fitzer, The Comosite of carbon-carbon composites. Carbon **25**(2), 163 (1987)
18. K.L. Smith, S.D. Smoot, T.H. Fletcher, R.J. Pugmire, *The Structure and Reaction Processes of Coal* (Plenum, New York, 1994)
19. P.R. Solomon et al., Analysis of coal by TG-FTIR and pyrolysis modeling. J. Anal. Appl. Pyrolysis **19**, 1 (1991)
20. D.F. Eggers Jr., N.W. Gregory, J.D. Halsey Jr., B.S. Rabinovitch, *Physical Chemistry* (Wiley, New York, 1964)

21. R.E. Franklin, Crystallite growth in graphitizing and non-graphitizing carbons. Proc. Royal Soc. London **209**, 196 (1951)
22. R.E. Franklin, The structure of graphitic carbons. Acta Cryst **4**, 253 (1951)
23. C.R. Hoska, B.E. Warren, X-ray study of the graphitization of carbon black. J. Appl. Phys. **25**, 1503 (1954)
24. J. Maire, J. Maring, Graphitization of soft carbons, in *Chemistry and Physics of Carbon*, vol. 6, ed. by P. L. Walker (Dekker, New York/Basel, 1970), pp. 125–189
25. P. von Rague Schleyer, Introduction: aromaticity. Chem. Rev. **101**(5), 1115 (2001)
26. P. von Rague Schleyer, Introduction: destabilization pi and sigma. Chem. Rev. **105**(10), 3433 (2005)
27. E. Huckel, Quantentheoretische Beitrage zum Benzolproblem. Z. Phys. **70**(3–4), 204 (1931)
28. H.C. Longuet, Some studies in molecular orbital theory I., Resonance structures and molecular orbitals in unsaturated hydrocarbons. The J. Chem. Phys. **18**(3), 265 (1950)
29. J.C. Fitzer, The chemistry and analysis of the large polycyclic aromatic hydrocarbon, in *Polycyclic Aromatic Compounds* (Wiley, New York, 2000)
30. G.M. Jenkins, K. Kawamura, Structure of glassy carbon. Nature **321**, 175 (1971)
31. T.K. Gupta, Effect of temperature on the photosensitivity of a photoresist (AZ 1350 J). Euro. Poly. J **17**(11), 1127 (1981)
32. H.O. Pierson, *Graphite, Diamonds and Fullerene* (Elsevier, Amsterdam, 2015)
33. P.R.A. Andrews, The benefication of Canadian graphite ores: a review of presenting study at CANMET. CIM Bull. **85**, 76 (1992)
34. P. Ramachandran, C. Eswaraiah, P. Manisankar, Benefication of low grade graphite ore deposits of Tamilnadu (India). Ultra Chem **8**(2), 159 (2012)
35. P.J. Pugh, Non-ionic polyethylene oxide Frothers in graphite floatation. Min. Eng. **13**, 151 (2000)
36. I.W. Gazda, *Twentieth Biennial Cof. On Carbon* (University California Santa Barbara, Calfornia, 1991)
37. A.J. Massaro, *Primary and Mating Ring Materials* (Pure Indust. Pub, St. Marys, 1987)
38. F. Fitzer, K.H. Kochling, H.P. Boehm, H. Marsh, Pure Appl. Chem. **67**(3), 473 (1995)
39. J.L. White, P.M. Sheaffer, Pitch based processing of carbon-carbon composite. Carbon **27**, 697 (1989)
40. S. P. Appleyard, B. Rand, Carbon-Carbon composites, in *Design and Control Structure of Advanced Carbon Materials for Enhanced Performance* (Kluwer, New York, 2001)
41. B.T. Kelly, *Physics of Graphite* (Applied Science, London, 1981)
42. F. Banhart, Irradiation effects in carbon nanostructures. Rept. on Prog. In Phys. **62**, 1181 (1999)
43. F. Banhart, P.M. Ajayan, Carbon onions as nanoscopic pressure cells for diamond formation. Nature **382**, 433 (1996)
44. Report from High Beam Business, Chicago, Il. 60602, USA
45. T.K. Gupta, *Handbook on Thick and Thin Film Hybrid Microelectronics* (Wiley, Hoboken, 2003)
46. T.K. Gupta, *Copper Interconnect Technology* (Springer, New York, 2009)
47. C.F. Powel, J.H. Oxley, J.M. Blocher Jr., *Vapor Deposition* (Wiley, New York, 1966)
48. W.E. Sawer, A. Man, US Patent, 229335, June 29, (1880)
49. H.O. Pierson, *Handbook of Chemical Vapor Deposition* (Noyes, Park Ridge, 1992)
50. J.C. Bokros, in *Chemistry and Physics of Carbon*, vol. 5, ed. by P. L. Walker (Marcel Dekker Inc., New York, 1969)
51. H.O. Pierson, *Handbook of Chemical Vapor Deposition* (Noyes, Westwood, 1992)
52. J.D. Buckley, Carbon-carbon, an overview. Am.Ceramic Soc. Bull **67**(2), 364 (1988)
53. B.T. Kelly, J.E. Brocklehurt, High dose fast neutron irradiation of highly oriented Pyrolytic graphite. Carbon **9**, 783 (1971)
54. V.M. Agranovich, L.P. Semenov, A contribution to the theory of the effect of irradiation on certain properties of graphite. J. Nuclr. Energy, Parts A/B, Reactor Sci. and Tech. **18**,

141 (1964) and also R.H. Telling, M.I. Heggie, Radiation damage in graphite. Phil. Mag. **87** (31), 4797 (2007)
55. P. Moreau et al., Electron energy loss spectra calculations and experiment as a tool for identification of lamellar $C_3N_4$ compound. Phys. Rev. B **73**, 195111 (2006)
56. R.H. Telling et al., Wigner defects bridge the graphite gap. Nat. Mater. **2**, 333 (2003)
57. D. McGeoghegan et al., Mortality and cancer registration experience of the sellafield workers known to have been Involved in the 1957 Windscale accident: 50 years floowup. J. Radiol. Protec **30**, 407 (2010)
58. R. Wakeford, Editor in chief: editorial, the wind scale reactor accident – 50 years. J. Radiol. Protec. **27**, 211 (2007)
59. IAEA Rept. On atomic energy, The information on Accident at the Chernobyl NPP and its Consequences, **61**, 308 (1986)
60. G. Galli, Structure, stability and electronic properties of nanodiamonds, in *Computer aided Modeling of Novel Carbon System and Their Properties* ed. by L. Colombo, A. Fasolino (Springer-Nature, NY, 2010)

# Chapter 7
# Graphene

## 7.1 Introduction

*Graphene* The nineteenth century saw many wondrous discoveries, elemental discoveries, reports of new materials, and electromagnetic phenomena. In the year 1859, Benjamin Collins Brodie recognized the highly layered structure of thermally reduced graphite oxide, and he reported the atomic weight of the material in the Philosophical Transaction of the Royal Society of London. Later on, in the year 1916, the layered structure of graphite was established by X-ray diffraction [1–5]. Figure 7.1 shows an atomic scale honeycomb lattice of graphene made of carbon atoms.

In the year 1918, V. Kohlschutter and P. Haenni studied the first thermally reduced graphite oxide. In 1947, while attempting to gain insight regarding the electronic properties of graphite, P.R. Wallace first explored the electronic structure of graphene, the starting point for the electronic structure of graphite [6]. It is well known that the history of graphene spans more than a century and there is a vast chemical literature of literally hundreds of papers from the early 1960. Studies of graphite oxide membranes were performed by Hanns-Peter Boehm, the German scientist who invented the term *graph* (graphite)*ene* (the suffix -ene is used for fused polycylic aromatic hydrocarbons) in the year 1962, a single carbon layer of the graphite structure (Fig. 7.1), which should be used to designate the individual carbon layers in graphite intercalation compounds [7, 8]. *By graphene, one understands single-carbon hexagonal networks within the structure of the graphite. However, the term graphene is today frequently applied to stacks of a few graphene layers, which often adhere to one another and are partially overlapping* [9].

The rising star graphene on the horizon of materials science and condensed matter physics became very popular, and the research on this particular field became fascinating. The strictly two-dimensional material exhibits exceptionally high crystal and electronic quality (ballistic conveyance of charges, large quantum oscillations are a few of its intrinsic characteristics) [10]. Epitaxial graphene has


T. Gupta, *Carbon*, DOI 10.1007/978-3-319-66405-7_7

**Fig. 7.1** An atomic scale honeycomb lattice of graphene made of carbon atoms (Courtesy: AlexanderAlus, Creative Comm., WIKI)

been produced since 1970s [11–13]. Often, this type of graphene contains variations, which alter the electronic structure. In 1987, graphene was listed as a *constituent of graphite intercalation* compounds containing the *same old prehistoric carbon.*

A team from Manchester, England, Columbia University, NY, Georgia Institute of Technology, Georgia, Atlanta continued their endeavors to characterize the amazing properties of graphene. Through the combined efforts of Geim's team at Manchester University, UK, and Philip Kim's group at Columbia University, NY in 2005, a demonstration proved that the quasiparticles of graphene were Dirac fermions without mass. Since then, the interest in graphene had intensified [14–16]. Hundreds of researchers had turned their attention in finding a way of getting atomic single-layer graphite. In 2010, A.K. Geim and K.S. Novoselov, from the University of Manchester, UK, were awarded the Nobel Prize in Physics for groundbreaking experiments of separating a single atomic layer of graphene with a roll of transparent adhesive tape and a chunk of graphite [17]. According to Connor, *the creation of graphene – a wonder material, and a piece of brilliant serendipity stands alongside the accidental discovery of penicillin by Alexander Fleming* [2].

Although graphene is expected to find use in a whole range of applications, such as low cost, foldable electronic screens, the next generation of food packaging, tough plastics, and in protective coatings, it is currently produced by chemical vapor deposition (CVD) which can only be able to produce meter-scale sheets [18–20]. It is expected that the study which was led by Jonathan Coleman, Professor of Chemical Physics, at Trinity College, Dublin, Ireland, and his published papers in the journal *Nature Materials* [21], can be a potential avenue for a commercial product. Thomas Swan Ltd., a UK-based company, claimed that their process could yield hundreds of liters of high-quality graphene.

Indeed, graphene represents a conceptually new class of material only one atom thick, and it is expected that it will offer new inroads into two-dimensional physics and will continue to provide a fertile ground for applications. The *transparent graphene* (which absorbs only 10% of incident light) can *stretch* up to 20% of its length, which shows amazing strength, with *no bandgap*, and having *ballistic conduction* with *highest current density* (a million times that of copper) and having *thermal conductivity* that can beat diamond [22–26]. According to Lucy, *when first discovered, graphene was odd. Now odd is too small a word for a material seemingly set on winning all the records a material can win* [27].

## 7.2 Optical Phenomena

### 7.2.1 Dynamical Conductance

Reflection and transmission of light waves occur because, the light waves do not match the natural frequencies of vibration of the object. When light waves of these frequencies strike an object, the electrons in the atoms of the object begin vibrating. But instead of vibrating in resonance at large amplitude, if the electrons vibrate for brief periods of time with small amplitudes of vibration; then the energy is re-emitted as a light wave. If the object is transparent, then the vibrations of the electrons are passed to the neighboring atoms through the bulk of the material and transmitted on the opposite side of the object [28, 29].

The electronic band structure of graphene is simple, and near energy $\varepsilon = 0$ the energy bands are conical with no gap in between valence band and conduction band, with $\varepsilon = 1, 2\,(p) = vp$, $v$ being the velocity of electrons at $K$ points in the 2D Brillouin zone. It has also been established by theoretical and experimental observations that the main mechanism of carrier relaxation in a single atomic layer of graphene is provided by the long-range scatterers and gives a collision rate $\tau \sim (2\pi I^2 e^4\, n_{sct})/\,(\hbar\, \varepsilon \epsilon_g^{\,2})$, where $\varepsilon$ is the characteristic electron energy, $\epsilon_g$ is the dielectric constant of graphene, and $n_{sct}$ is the density of scatterers per unit surface [30–32].

According to the relativistic theory, the optical transmittance $T_{opt}$ and reflectance $R$ for a suspended graphene when interacted by an incident light can be expressed as $T_{opt} = (1 + 2\pi G/c)^{-2} = (1 + ½\ \pi\alpha)^{-2} \approx (1 - \pi\alpha) \approx 0.977$(optical transmittance of a freestanding monolayer of graphene is frequency independent) and $R = (¼\ \pi^2\alpha^2 T)$, where $G$ is the high-frequency (dynamic) conductivity for Dirac fermions (I) in graphene, c is the velocity of light, and $\alpha$ is the fine structure constant, and is equal

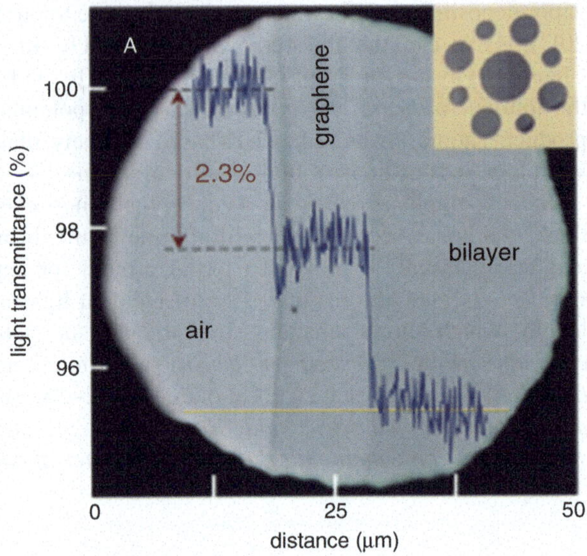


**Fig. 7.2** Photograph of graphene in transmitted light (Reproduced with kind permission of A.K. Geim)

to $(e^2\hbar^{-1}c^{-1}) \sim 1/137$, where $\hbar$ is Planck's constant. This yields the opacity of graphene $(1-T) \sim \pi\alpha$ by calculating the absorption of light by two-dimensional Dirac fermions with Fermi's Golden rule. Measurements of a single atomic layer graphene's opacity has been reported to be $\approx 2.3 \pm 0.1\%$ with negligible reflectance ($<0.1\%$), whereas optical spectroscopy shows that the opacity is practically independent of wavelength $\lambda$. The value of $G = (1.01 \pm 0.04) \equiv e^2(4\hbar)^{-1}$ $\approx 6.08 \times 10 - 5\ \Omega^{-1}$ [33, 34].

The right-hand corner of Fig. 7.2 shows the metal holes used to put the graphene sheet. To compare the light transmittance through the single-layer and bilayer graphene, light normal to the plate is passed through a hole without any graphene sheet (air), and then passed through different holes with single atomic layer; bilayer or triple layer of graphene and the corresponding light transmittance through them were also measured. To avoid any error due to the contamination of any carbonaceous material in graphene, the wavelength ($\lambda$) of the incident light ($<800$ nm) normal to the graphene plate was chosen. Transmittance ($T_{\text{opt}}$) vs wavelength of light used (over a range of photon energies) from near-infrared to violet is presented in Fig. 7.3. The spectra for single atomic layer (open blue circles) and for the bilayer (open red circles) show deviation from the ideal Dirac fermions (dotted straight lines). Due to graphene's unique electronic structure, one atom thick graphene layer is found to absorb almost 2.3% of the normal incident light. The ideal Dirac fermions' values of $T_{\text{opt}} \approx 1-\pi\alpha$ for single layer and $T_{\text{opt}} \approx 1-2\pi\alpha$ for bilayer are also shown in Fig. 7.3.

The opacity of graphene is $\approx 2.3\%$ which equates to universal dynamic conductivity value of $G = e^2(4\hbar)^{-1}(\pm 2.3\%)$. It has been observed that once optical intensity reaches a certain threshold (known as the saturation fluence), saturable absorption takes place (very high intensity of light causes a reduction in

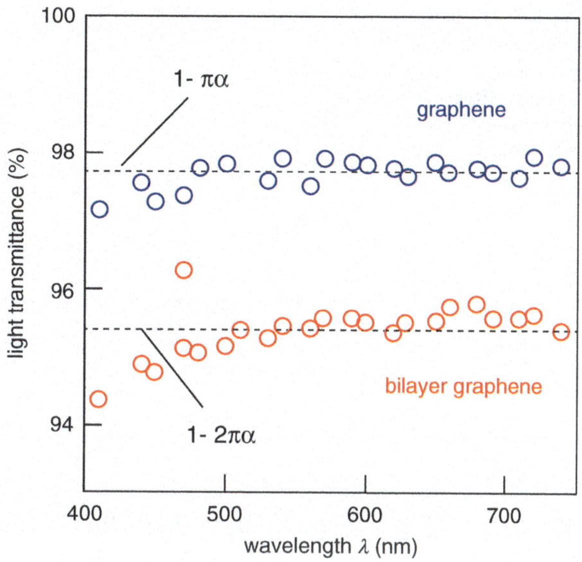


**Fig. 7.3** Transmittance spectrum of graphene (Reproduced with kind permission of A.K. Geim)

absorption). This is an important characteristic with regard to the mode-locking of fiber lasers. The thickness of a single atomic layer of graphene has been reported to vary from 0.3 to 1.16 nm [35, 36].

At energies considerably smaller than the bandwidth ($\leq$2 eV), the dispersion of a monolayer graphene features massless electron and hole conical bands $\varepsilon_{e,h}(k) = \pm\hbar v_F|k - k_D|$ formed by $p_z$ orbitals, as shown in Fig. 7.4, where $k_D$ is the momentum of Dirac point, and $v_F \approx 10^6$ m/s is the Fermi velocity. This type of dispersion is qualitatively different from more common quadratic massive bands, and has been demonstrated by the equation {En (B) $\alpha$ sinh ($n$) √(|n|B)} which relates to field dependence of Landau levels [16, 37].

In graphene, the two Fermi points, each with a twofold band degeneracy, can be described by a low-energy continuum approximation. The interaction between light and Dirac fermions can be presented by the Hamiltonian ($\hat{\mathrm{H}}$) as

$$\hat{\mathrm{H}} = \hat{\mathrm{H}}_0 + \hat{\mathrm{H}}_{\mathrm{int}} = v_F\sigma^{\rightarrow}.\pi^{\rightarrow} = v_F\sigma^{\rightarrow}.\{(p - (e/c\mathrm{A}^{\rightarrow})\} \tag{7.1}$$

where $v_F$ is the Fermi velocity, $\sigma^{\rightarrow}$ is Pauli's spin matrices, $\mathrm{A}^{\rightarrow} = (ic/\omega\ \varepsilon)$is the vector potential. $\hat{\mathrm{H}}_{\mathrm{int}} = -v_F\sigma^{\rightarrow}.(e/c\mathrm{A}^{\rightarrow}) = v_F\sigma^{\rightarrow}.(c/i\omega\ \varepsilon)$ describes the interaction with the electromagnetic field.

The absorption, € = ($W_i/W_a$), where $W_i$ is the incident energy{($c/4\pi$) $|\varepsilon|^2$}, $W_a$ is the absorbed energy ($\eta\hbar\omega$), $\varepsilon$ is the electric field, $\omega$ is the angular frequency, $\eta$ is the absorption events per unit time, $\hbar$ is Planck's constant, and $c$ is the velocity of light (Figure 7.4).

One way to incorporate absorption or attenuation into the mathematical description of electromagnetic wave is via absorption coefficient or opacity:

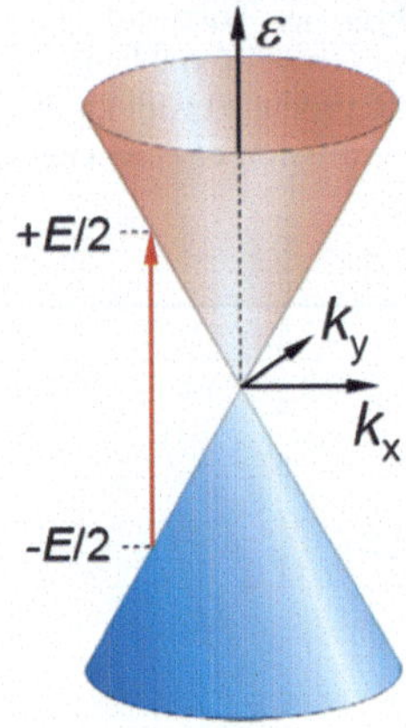


**Fig. 7.4** Excitation processes responsible for absorption of light in graphene (Reproduced with kind permission of A.K. Geim)

$$E\,(z,t) = \exp\,(-\alpha_{\rm abs})\,(z/2))\;{\rm Re}\,(E_0)\;\exp\,\{{\rm i}(kz-\omega t)\} \tag{7.2}$$

where $E$ is a vector in $X$–$Y$ plane with units of electric field (the vector is in general a complex vector, which allows all possible polarization and phases), $\alpha_{\rm abs}$ is the absorption coefficient, Re indicates the real part, $k$ is the angular wave number of the electromagnetic wave (EMW), $\omega$ is the angular frequency of the EMW propagating in the $z$-direction, $\hbar$ is Planck's constant, $c$ is the velocity of light. The intensity of the wave in this case satisfies [33]

$$\begin{aligned}&I\,(z)\;\alpha\;|\;\exp\,\{(-\alpha_{\rm abs})\,z/2\}\;(E_0)\;\exp\,({\rm i}(kz-\omega t))\;|^2 = |\;E|^2\;\exp\,((-\alpha_{\rm abs})\,z)\\ &{\rm Or,}\,I(z)\equiv I_0 \exp\,((-\alpha_{\rm abs})\,z)\\ &{\rm Or}\;I\,(z)/I_0 = \exp\,((-\alpha_{\rm abs})\,z)\sim(1-\pi\alpha),\;{\rm where,}\;\alpha = e^2(4\hbar)^{-1}\end{aligned} \tag{7.3}$$

The optical constants (refractive index, $n$; extinction coefficient, $k$,) of a two-dimensional (e.g., an atomically thin graphene layer) material can be used to describe the optical behavior of the material in the visible range, where the intraband contribution to absorption is negligible and the material's opacity (sometimes called as absorption coefficient) is only determined by the universal optical conductivity [38]. We can set up a mathematical general equation of absorption coefficient ($\alpha_{\rm abs}$) for the propagation of light through an absorbing material as

$$k = (\lambda/nt)\;\ln\;[(I/I_0)\;(1/(1-R)] \tag{7.4}$$

where $k$ is the extinction coefficient, $n$ is the refractive index ($RI$), $t$ is the thickness of the material, $I_0$ is the incident light, $I$ is the transmitted light, $R$ is the fraction of the reflected light, and $\lambda$ is the wavelength of the light used for the experiment. It has been shown that in the case of graphene in the visible range, if $(I/I_0)\sim(1-\pi\alpha)$, and $(1-R)\sim 1$, then the above equation will transform to

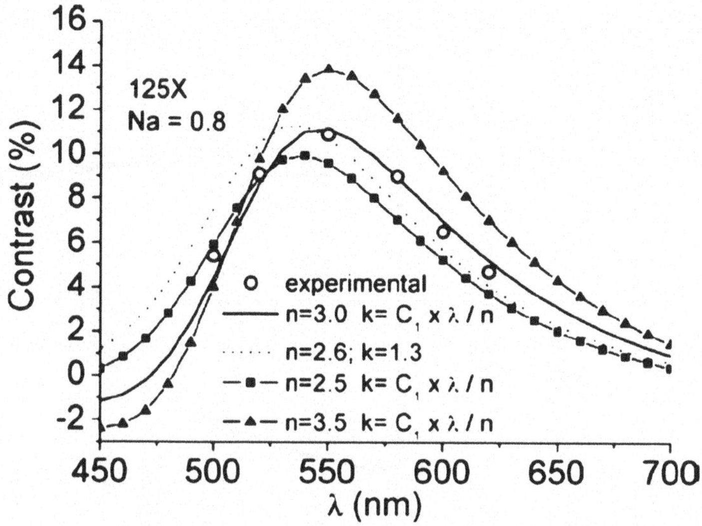


**Fig. 7.5** Contrast of a graphene monolayer on a 291 nm thick $SiO_2$ film over Si (Reproduced with permission, from S. Borini, Ref. [39])

$$k = (\lambda/4\pi nt)\ \ln\ [(1\text{-}\pi\alpha)] \equiv C_1/(\lambda/n) \tag{7.5}$$

In our calculations, we have considered the thickness ($t$) of a single atomic layer of graphene which is ~0.345 nm. The spectra calculated with $k$ given by Eq. (7.5) clearly depend on the value of $n$ and shows a good agreement at $n \sim 3.0$ [39]. Using $n = 3.0$ in Eq. (7.4), the experimental contrast is very well reproducible than by using the constant refractive index of graphene. Fig. 7.5 shows the contrast of a graphene monolayer on a 291 nm thick silicon dioxide ($SiO_2$) film, when the numerical aperture of the objective is 0.8 and the amplification is 125×.

It is interesting to note that the optical constants of graphene in the visible range can be estimated in a simple way by imposing their consistence with experimental spectra in the framework of Fresnel coefficients calculation.

In thin films following Fresnel's equation, the optical transmittance of a freestanding monolayer graphene is frequency-dependent and can be represented by [34]:

$$T_{\text{opt}} = \{1 + (\pi\alpha/2)\}^{-2} \approx (1 - \pi\alpha), \text{where}, \alpha = \left(e^2/\hbar c\right) \tag{7.6}$$

where $\alpha \approx 1.42$, e is the electronic charge, and $\hbar$ is Planck's constant. Equation (7.6) gives us the value of $T_{\text{opt}} \approx 97.7\%$, when all the values of the parameters are inserted into the equation [33]. We can establish a relation between transmittance ($T$) and the sheet resistance ($R_s$) of the atomic layer graphene as [40]

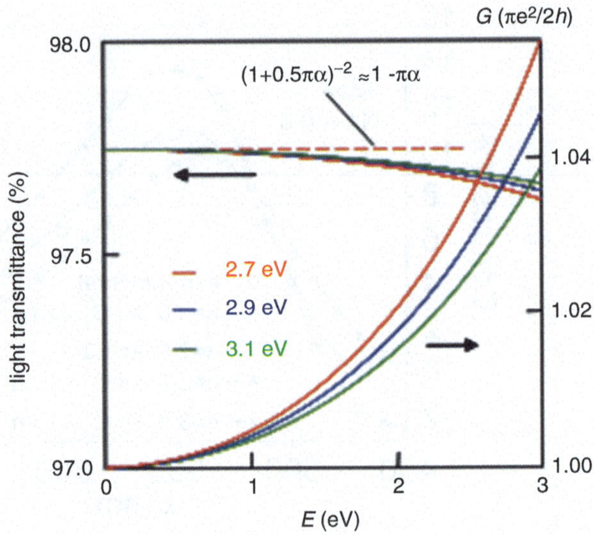


**Fig. 7.6** Light transmittance (%) as a function of photon energy, *E* (Reproduced with permission from A.K. Geim)

$$T_{\mathrm{opt}} = \{1 + (Z_0/R_s)\,(G_0/\sigma_{2D})\} \tag{7.7}$$

where $Z_0 = \{1/(\varepsilon_0 c)\}, = 377, R_s \approx 6\mathrm{k}\Omega, G_0 = (e^2/4\hbar) = 6.08 \times 10^{-5}\Omega^{-1}$, $\sigma_{2D} = (e^2/\hbar)$. Inserting these values in Eq. (7.7), we can get the value of $T_{\mathrm{opt}} \approx 96.5\%$, which is close to the value of $T_{\mathrm{opt}}$ as obtained from the article of R.R. Nair et al. [33].

Fig. 7.6 shows the light transmittance (%) as a function of photon energy, *E* (eV). Optical properties of thin films can be described in terms of dynamic or optical conductivity, *G*. For a two-dimensional Dirac spectrum with a conical dispersion relation, G is predicted to exhibit a universal value $G_0 \equiv (e^2/4\hbar)$. This approximation is only valid near the Dirac point, which is represented by the red dotted line in Fig. 7.6. For visible higher energies (2–3 eV), one has to take account of the band structure as triangular warping and nonlinear. Therefore, for visible energies of 2–3 eV, which are comparable with the nearest neighbor hopping energy ≈3 eV, we can expect the breakdown of the Dirac Fermion approximation used in the calculation of $G_0$, as evident from the calculated dynamic conductivity *G* and light transmittance *T* (%) when the finite *E* effects are included (*E* is the photon energy of graphene).

### 7.2.2 Raman Spectroscopy

Graphene is a bipartite lattice of carbon atoms arranged in a honeycomb lattice as shown in Fig. 7.7. Ever since the discovery of graphene, micro-Raman

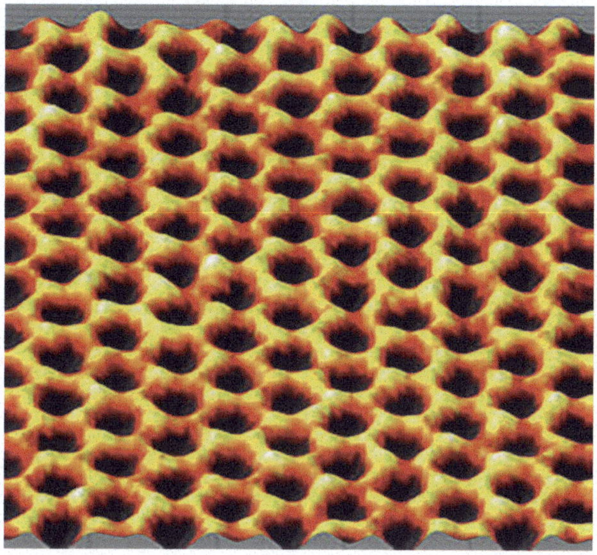

**Fig. 7.7** Honeycomb structure of graphene (Photo courtesy: US Army Mat. Commd. Creative Comm., WIKI)

spectroscopy and imaging have emerged as a powerful tool to characterize the quality of a graphene layer. As for example, the number of layers present, their stacking order, quantitative measurements of electron–phonon coupling, doping levels, strain, and phonon dispersion near the *k*-point of Brillouin zone can also be obtained from Raman spectroscopy [41–44]. The phonons have wave vector $q \approx 0$ around Γ point. They also have wave vector $q \neq 0$ on higher order combination modes with hexagonal Brillouin zone [45–47].

One monolayer of carbon atoms comprises a two-dimensional honeycomb lattice structure shown in Fig. 7.7. It is the building block for other carbon nanomaterials, such as 0-dimensional fullerenes, one-dimensional carbon nanotubes, and two-dimensional nanographite sheets. Studies of Raman spectroscopy of single atomic layer of graphene have revealed many interesting phenomena such as the single and sharp second order Raman band (2D) which have been widely used to identify the single monolayer graphene (shown in Fig. 7.8) [48]. The G band (~1580 $cm^{-1}$) and the 2D (also called the G') (~2670 $cm^{-1}$) bands are the two most prominent bands that can be seen in Fig. 7.8. The G band originates from in-plane vibration of $sp^2$ carbon atoms and is a doubly degenerate (transverse optical phonon, TO and longitudinal optical phonon, LO) phonon mode ($E_{2g}$ symmetry) at the Brillouin zone center [49]. The 2D (or G') band of a single-layer graphene is very sharp, and the intensity is higher compared to G band. The 2D band originates from two phonon double resonance Raman sources [50]. However, in graphite, the 2D band can be resolved into two different peaks due to coupling between layers in a three-dimensional crystal (Fig. 7.9).

In highly oriented pyrolytic graphite (HOPG), the intensity of the 2D band is shorter than the G-band peak which is contrary to the observed phenomenon in a single-layer graphene. In addition to the shortening of the peak, the 2D band in

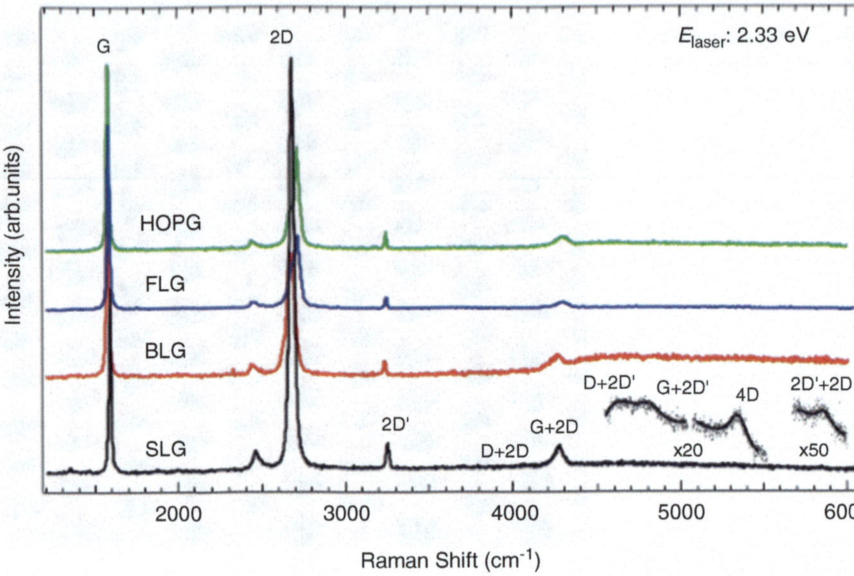


**Fig. 7.8** Raman spectra for single-layer graphene (SLG), bilayer graphene (BLG), few-layer graphene (FLG), and the highly oriented pyrolytic graphite (HOPG) when irradiated with laser beam having energy ~2.33 eV (Reproduced with permission from R. Rao, Ref. [53])

**Fig. 7.9** Raman spectra for atomic layer graphene and graphite (Reproduced with permission: Springer, Nano Res. Ref. [51])

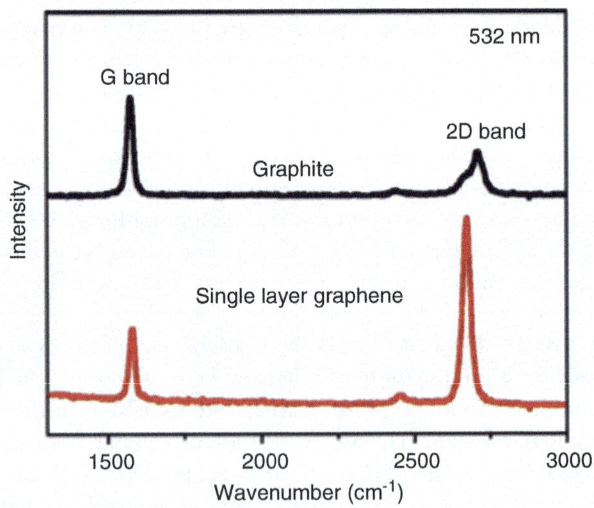


HOPG broadens due to the splitting of the electronic band. Cancado et al. also gave a detailed description on how the multi-peak 2D structure in graphite is observed for a stacking of more graphene layers due to dispersion of $\pi$ electrons [51, 52]. The 2D peaks in single-layer (SLG), bilayer (BLG), and few-layer (FLG) graphene can be fitted with one, four, and two Lorentzian peaks, respectively (Fig. 7.8) [53].

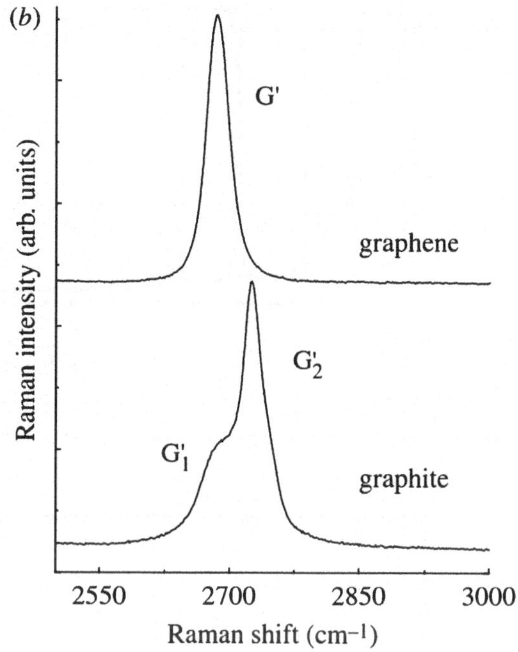


**Fig. 7.10** Broadening and splitting of G' (2D) band in graphite

The split of G' peak has also been observed in a strained carbon nanotube. Both peaks shift toward red with increasing strain. It has been concluded by Mohiuddin et al. that Gruneisen parameters for phonons are responsible for G, D, and D' peaks [43] (Fig. 7.10).

From the Raman spectra (Fig. 7.10) of single-layer graphene and graphite, we could distinguish a single-layer graphene from multilayer graphene (graphite). Raman spectroscopy has become a powerful tool for detecting the artificial disorder introduced by irradiations, such as electron beam exposure and oxygen plasma, focusing on deactivated Raman D peak. The first order D peak is not visible in pristine graphene because of the crystal symmetry [54, 55]. Figure 7.11 shows the Raman spectrum of a single-layer graphene irradiated by electron beam.

The intrinsic strength of a monolayer pristine (defect-free) graphene, i.e., the maximum stress the atomic layer can withstand before all the atoms in a given cross section are pulled apart at the same time is found to be 42 $Nm^{-1}$ [56]. Enhancement of lattice defect signatures in graphene and ultrathin graphite has been measured by using tip-enhanced Raman spectroscopy (TERS) [57]. Lattice defects in a material are important because they affect the electronic properties of a device while also offering a means by which to functionalize the local properties.

Figure 7.12 illustrates an example of the tip-enhanced Raman spectroscopy (TERS) response from the edge of an ultrathin highly oriented pyrolytic graphite (HOPG) flake. The enhanced spectral response is shown (in Fig. 7.12) (tip-approached) alongside the far-field Raman response (tip-retracted). The Raman spectra of graphite and graphene can share four main peaks, D, G, D',

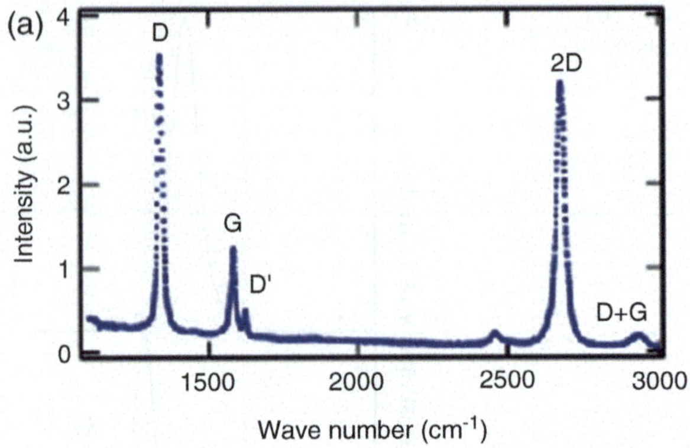


**Fig. 7.11** Raman spectrum of a single-layer graphene irradiated by electron beam (Reproduced with permission from Y.P. Chen, Purdue Univ., IN)

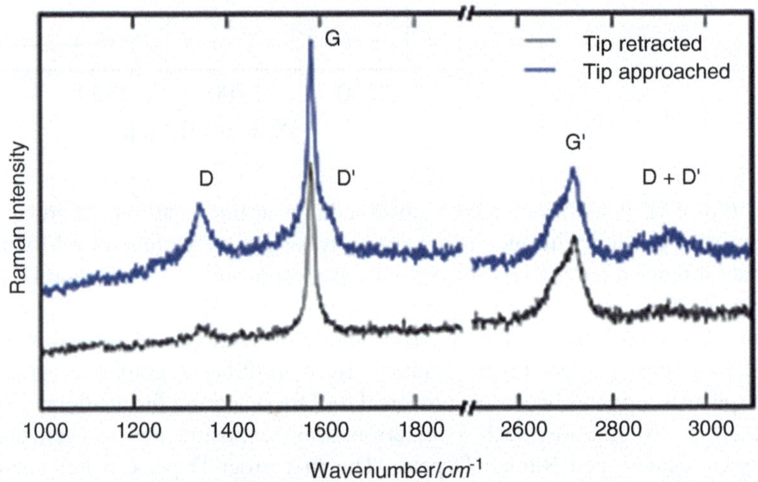


**Fig. 7.12** A tip-enhanced Raman spectroscopy measurement with the tip located at the edge boundary of an ultrathin highly oriented pyrolytic graphite (HOPG) flake (~25 nm thick) (Reproduced with permission: John Wiley & Sons, Ltd., 2013, Ref. [57])

and G' at 1350, 1600, 1620, and 2700 $cm^{-1}$, respectively, and these can be clearly seen in HOPG flake, in particular the enhanced peaks related to defective lattice sites [57].

Both the G + 2D and D + 2D peaks are shown in Fig. 7.13. The intensity of the G + 2D peak is the highest for single-layer graphene (SLG), which decreases in intensity as the number of layers increases. We have also seen that the intensity of the 2D peak is the most intense for SLG compared to multilayered graphene peak (Fig. 7.8) [58]. G + 2D peak in SLG can be fitted with a single Lorentzian peak. On the other hand, the G + 2D peaks in bilayer, few-layer, and high-oriented pyrolytic

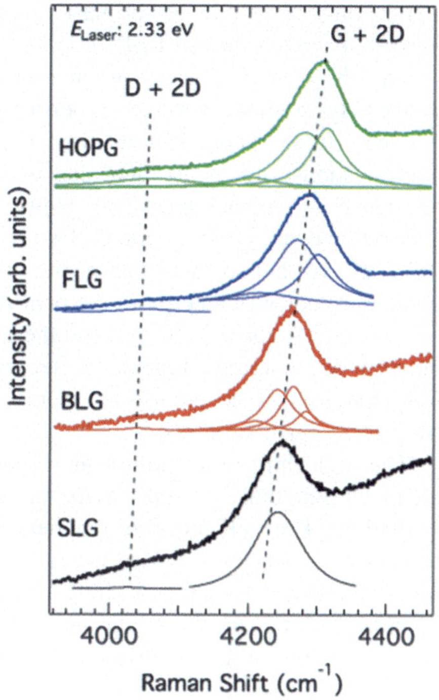


**Fig. 7.13** Higher order combination modes between 4000 and 4400 cm$^{-1}$ in graphene samples with $E_{Laser} = 2.33$ eV. All spectra have been normalized with respect to the G + 2D peak intensity (Reproduced with permission from R. Rao)

graphite can be deconvoluted into multiple peaks, indicating the changes in electronic and band structure brought about by coupling between graphene layers. Figure 7.13 presents the higher order combination modes between 4000 and 4400 cm$^{-1}$ in graphene samples, with $E_{Laser} = 2.33$ eV.

## 7.3 Photonics and Optoelectronics Applications

Graphene has been used in many *optoelectronics* applications, e.g., as *photoconductive material*, *channel* for the *transparent conductor window*; *photoactive material*, *channel* for *charge transport*; and *catalyst*. Graphene has also been used as window electrodes in inorganic and organic and dye-synthesized solar cell devices [59]. The energy conversion efficiency ($\eta$) has been reported $\approx 0.3\%$ [60]. Yang et al. have reported the use of graphene in nanostructured titanium dioxide ($TiO_2$) (for electron transport) using a graphene transparent conductive film [61, 62].

In *light-emitting devices* (LEDs), graphene layer has been successfully implemented as a flexible and cheap transparent conducting film [63]. Matyba et al. demonstrated a light-emitting electrochemical cell based solely on dispersion-processable carbon-based materials. It is a totally organic low-voltage, inexpensive and efficient LED [63].

The theoretical models on photoelectrical response of graphene have been thoroughly studied, verified experimentally, and reported from time to time [64, 65]. It has been demonstrated that graphene photodetectors (GPDs) show photo-response up to 40 GHz due to the limitations resulting from the device resistance and capacitance [66]. However, the internal photocurrent efficiencies and external responsivity for GPDs are low compared to current photodetectors [67]. It is hoped that graphene might be a promising candidate for photothermoelectric energy in the near future.

Radiation in the 0.3–10 THz (terahertz) region of graphene may be attractive for biomedical imaging, homeland security, remote sensing, and spectroscopy [68]. Devices made out of graphene can be used for terahertz detection and frequency conversion [40].

The high third order nonlinear response of graphene with a wavelength-scale localized photonic crystal cavity is used in ultralow power optical bistable switching. The self-induced regenerative oscillations and coherent four-wave mixing at femto-joule cavity energies on graphene–silicon platform have been reported recently by Gu et al. [69]. The research group has fabricated a large-area monolayer graphene sheet onto an air-bridged silicon photonic crystal to demonstrate graphene–silicon hybrid optoelectronics behaviors.

## 7.4 Electrical Phenomena

Electrons propagating through the two-dimensional (2D) structure of a monolayer graphene exhibit a linear relationship between energy and momentum, and behave as massless Dirac fermions [6, 45, 70]. As a result, graphene behaves like a 2D gas of charged particles following the relativistic Dirac equation rather than nonrelativistic Schrodinger equation, which is proven from its other behaviors such as specific integer and fractional quantum Hall effects [71], a minimum conductivity ($\sigma$) of ~ ($4e^2/\hbar$) even when the carrier concentration tends to zero [45], and the Shubnikov-de Haas oscillations with a $\pi$-phase shift due to Berry's phase. Mobilities ($\mu$) of up to $10^6$ cm$^2$ V$^{-1}$ s$^{-1}$ are observed in suspended atomic layered graphene samples [40].

The electronic structure of one atomic layer of graphene can be described using a tight-binding Hamiltonian. Due to bonding and antibonding, $\sigma$-bands are well separated. Figure 7.14 (a) represents the Brillouin zone with Dirac cones located at the $K$ and $K'$ points and (b) represents the honeycomb lattice structure of graphene, where $a_1$ and $a_2$ are lattice unit vectors, $\delta_i$, $i = 1,2,3$ are nearest neighbor vectors. Brillouin zone center can be neglected in semiempirical calculations, retaining only two remaining $\pi$-bands. The electronic wave functions of graphene from different atoms on the honeycomb lattice overlap. However, any such overlap between $p_z$ ($\pi$) and the s or $p_x$ and $p_y$ orbitals is strictly zero by symmetry.

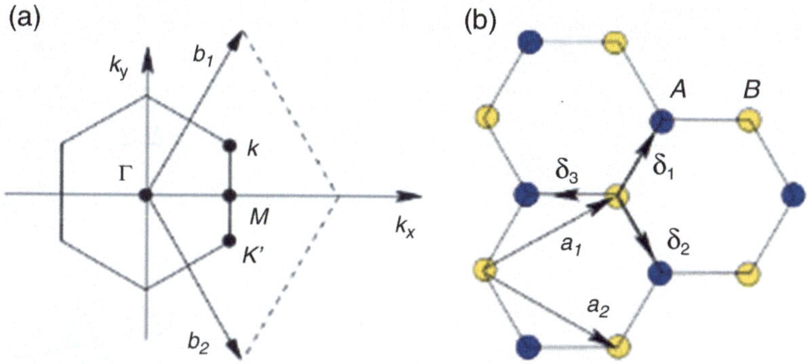


**Fig. 7.14** (**a**) The Brillouin zone with Dirac cones located at the $K$ and $K'$ points; (**b**) The honeycomb lattice structure of graphene; $a_1$ and $a_2$ are lattice unit vectors; $\delta_i$, $i = 1,2,3$ are nearest neighbor vectors

The low-energy band structure of graphene is cone-shaped and located at two inequivalent Brillouin zone corners:

$$\pm K = \left(\pm 4\pi/\surd 3a, 0\right), \tag{7.8}$$

where $a$ is the lattice constant. In these cones, the two-dimensional energy dispersion relation $\{\pm E\ (k_x, k_y)\}$ can be established as

$$\{\pm E\left(k_x, k_y\right)\} = \pm y_0 \surd\{\left(1 + 4\cos\left(\left(\surd 3k_x a\right)/2\right)\right)\cos\left(\left(k_y a\right)/2\right) + 4\cos\left(\left(k_y a\right)/2\right)\} \tag{7.9}$$

The + band (positive energy band) is totally empty, whereas the negative energy band (−) is fully occupied. These unoccupied and occupied bands touch each other at $K$ ($k_x$, $k_y$) points (Fig. 7.15).

Figure 7.15 represents Dirac cone in graphene. In the cones, the two-dimensional (2D) energy dispersion relation is linear $\{\pm E\ (k_x, k_y)\} = \hbar v_F |\ k \pm K|$, for $k \approx \pm K$, where, $v_F \approx 10^6$m/sis the Fermi velocity, spinor wave functions are parameterized by Berry's phase $\theta$. The wave vector $K$ is a Dirac point on Brillouin zone boundary. The electrons near the $K$, $K'$ points are described by Dirac Hamiltonian in Weyl's representation and can be expressed mathematically as

$$H_D^{\pm} = v_F\left(\pm \alpha_x p_x + \alpha_y p_y\right) = v_F(\pm \sigma_x p_x)\gamma_5, \boldsymbol{\alpha} = \begin{pmatrix} 0 & \sigma \\ \sigma & 0 \end{pmatrix}, \gamma_5 = \begin{pmatrix} 0 & 1 \\ 1 & 0 \end{pmatrix} \tag{7.10}$$

This is a (2 + 1)-dimensional system, and the electrons are massless. The physics of relativistic electronics can be studied in controllable nanoelectronic circuits with a graphene device. It also allowed to establish a relationship between massless Dirac fermions in two dimensions in an electromagnetic field and atoms in quantum optics.

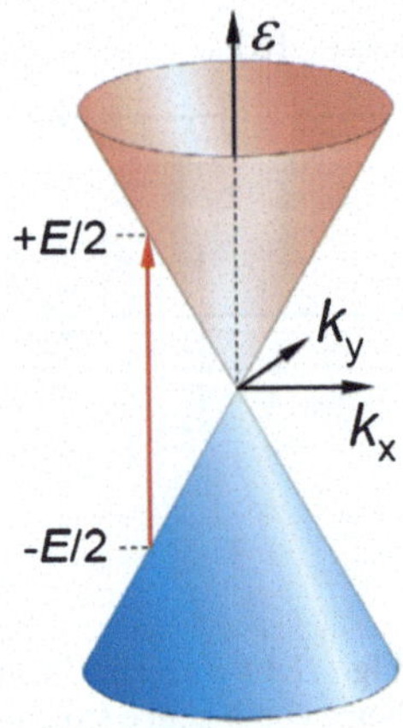


**Fig. 7.15** Dirac cone in graphene (Reproduced with permission from A.K. Geim)

A tight-binding calculation for the $\pi$-bands reveals that graphene is a gapless semimetal with linear dispersion around two valleys in Brillouin zone, where the conduction and the valence bands meet at the Dirac point. At zero temperature, the $\pi$ electrons from each carbon atom fill the states up to the Dirac point, where the density of states vanishes [72].

Patterning graphene into narrow channels has been the primary route to opening up the bandgap. It was indeed found that the conductance of narrow ribbons can be made vanishingly small within a width-dependent confinement gap [73]. But the size of the transport gap exceeds the estimates based on lateral confinement effects alone, and at low temperature the gapped region also contains many sharp conductance resonance [74, 75].

Indeed, in 2006, some researchers from the University of California, Berkeley, tried to do the bandgap engineering to fabricate field effect transistors with bilayer graphene. As in monolayer graphene, the carbon atoms are arranged in chicken wire configuration with no bandgap between the Dirac cones (Fig. 7.15), the bilayer graphene was tried for the experiment. The doping of the layers was unsuccessful in the sense that the charge carriers were found to be uncontrollable, and tiny amounts of impurity or defect doping created a big change in the resistance of the graphene. Instead, they applied a variable voltage to vary the bandgap and measure the optical transmission by using an infrared beam line 1.4 (synchrotron beam line). The results were astonishing, in the sense that manipulating the voltage of the two gates, the researchers could control two important parameters, the size of the bandgap and the degree of doping in the bilayer graphene. In essence, they created a virtual semiconductor from a material that was not inherently a semiconductor [76].

The screening of the surface potential is important from the point of view that the fluctuations in local electrostatic potential can contribute to elastic scattering in graphene [77]. The surface potential on a graphene film grown on silicon-terminated 6H-SiC (0001) surface was measured by using kelvin probe microscopy (KPM). The study shows the ability of graphene to screen the electric field from the substrate. The voltage required to null the change in electrostatic field microscopy (EFM) phase was found to decrease with increasing thickness of the graphene layers, implying that multilayer graphene is more efficient in screening the surface potential [78].

When a magnetic field $\{(\mathbf{B} = \nabla \times \mathbf{A} = (0,0,-B)\}$ is applied along the $z$-axis with the graphene sheet taken along the xy plane, then with the minimal substitution,

$$p \rightarrow P = (\hbar/i)\nabla + e\mathbf{A} \tag{7.11}$$

An electron in magnetic field undergoes cyclotron motion with angular frequency $\omega_c$. This circular motion can be decomposed into harmonic motion in the $x,y$ directions, because the electrons have spin magnetic moment, and the levels are further split by the Zeeman effect. As the magnetic field is varied, the system exhibits the Shubnikov-de Haas oscillations observed by Novoselov et al. [79–82]. The oscillations can also be studied by varying the gate voltage across the graphene sheet in the $x$-direction, keeping the magnetic field constant.

Bilayer graphene represents an interesting material to study the fractional quantum Hall effects. It has three distinct two-valued quantum numbers such as layer index, electron spin, and the sublattice index of graphene honeycomb. Researchers observe a number of interesting effects, most notably a series of transitions in the fractional quantum Hall states, as the electric displacement field is varied [83]. A. M. Fischer et al. studied magnetoplasmons or neutral collective excitations of graphene in a strong perpendicular magnetic field. The SU (special unitary) (4) symmetry of graphene arising from spin and valley pseudo-spin degrees of freedom has been explored using Young's diagrams to correctly predict the degeneracies of these excitations [84]. Further studies by the researchers at the University of Columbia also found that the tilted magnetic fields were able to break the SU (4) symmetry states in graphene by their spin polarization. It was found that graphene turns into either a spin ferromagnet or some variety of density waves [85].

## 7.5 Physical Characteristics

Figure 7.16 shows that there is a shift of the 2D + G band to a lower wave number and that the rate of band shift per unit strain tends to decrease as the number of graphene layers increases [86]. The band broadening takes place (Fig. 7.16) during tensile deformation for monolayer and bilayer graphene, but band narrowing occurs when the number of graphene layers increases. It is also found that the characteristic asymmetric shape of 2D Raman band for graphene with three or more layers changes to symmetrical shape. The change in Raman band shape and width has been interpreted as being due to a reversible loss of Bernal stacking in the few-layer graphene during deformation.

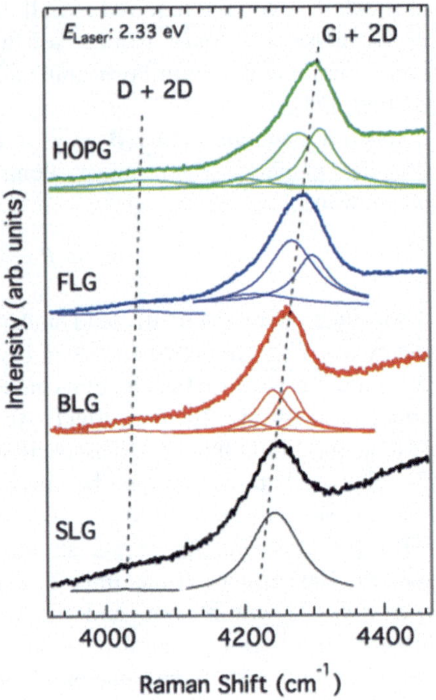


**Fig. 7.16** Upshift of the peaks in frequency with increasing layers (Reproduced with permission from R. Rao)

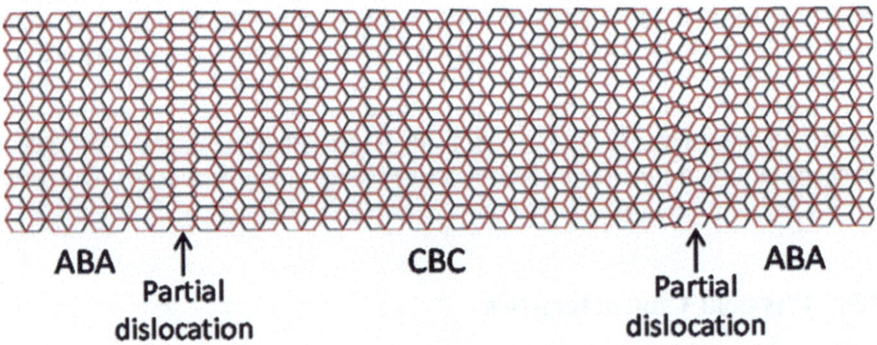


**Fig. 7.17** Bernal-stacked trilayer graphene lattice structure (Reproduced with permission from Robert Young)

According to Gong et al., the loss of Bernal stacking might not be observed or distinguishable with two or three atomic layers of graphene. But, with a stack of 20–50 atomic layers of graphene, formation of arrays of partial dislocations and stacking faults have been observed [86–88] (Fig. 7.17). The fault will be of the CBC type (and hence still Bernal-stacked).

The upshift of peaks in frequency in D + 2D is very small compared to G + 2D. The peaks in G + 2D are dependent on the number of layers and can be explained on the basis of linear chain model. The ($N-1$) shear modes of a number of layers ($N$) in $N$-layer graphene (NLG) can be computed by diagonalizing the corresponding $N \times N$ (trigonal) dynamical matrix. The peak position ($P$) of G + 2D (in $cm^{-1}$) for an NLG can be given by [89]:

$$P(G+2D)_N = [\{(1/\sqrt{2}\pi c)\}\ \{\sqrt{(\alpha/\mu)}\}\ \{\sqrt{\{1+\cos\ (\pi/N)\}}] \tag{7.12}$$

where $N$ is the number of layers, $\alpha$ is the interlayer (or shear force per unit area) force constant, $\mu$ is mass per unit area, and $c$ is the speed of light. The value of $\alpha$ has been given in the literature as ~$12.8 \times 10^{18}$ $Nm^{-3}$, and the value of α for a single-layer graphene ($N = 1$) is $7.6 \times 10^{-27}$ kg $Å^{-1}$ [90].

Tan et al. have introduced density functional theory (DFT) and density functional perturbation theory to explain in-plane shear modes for 2LG (two-layer graphene, $N = 2$) to 5LG (five-layer graphene, $N = 5$) and bulk graphite [89]. For a given $N$, there are $N-1$ shear modes, which can be either Raman-active or infrared-active, or both, but for $N > 2$, $N-2$ of those have a different displacement pattern.

Figure 7.18 shows the shear modulus when a force $F$ is applied. The shear modulus or modulus of rigidity is defined by the ratio of shear stress ($F/A$; $F$ is the shear force and $A$ is the area, and $F = \alpha Ax$, where $A$ is the graphene surface area and $x$ is the shear displacement) to the shear strain ($\Delta x / l = \tan\theta$). Also, by definition, the shear modulus of the layer-to-layer interface is $C_{44} = \alpha l$, where $l$ is the equilibrium distance between two adjacent graphene layers. The measured value of $C_{44}$ as reported by Tan et al. [89] is 4.3 GPa for graphite. By using triple-grating micro Raman spectrometer and effective filters, Zhao et al. have reported the existence of two shear modes and two breathing modes in ultralow frequency (<55 $cm^{-1}$) region for few-layer $MoS_2$ and $WSe_2$. The simple linear chain model fits into 2D atomic layer graphene and 2D-layered transitional metal dichalcogenides (TMDs). It predicts that the in-plane (shear) and out-of-plane (breathing) force constants will remain the same for 2D atomic layer graphene and the bulk graphite, and will be independent of the number of layers [89, 90].

The atomic (mono) layer of graphene can be modeled into two dimensions (2D). The studies on the stress–strain relation and the shear modulus on an atomic (mono) layer graphene performed by James Hone et al. [91] show that the nonlinear elastic

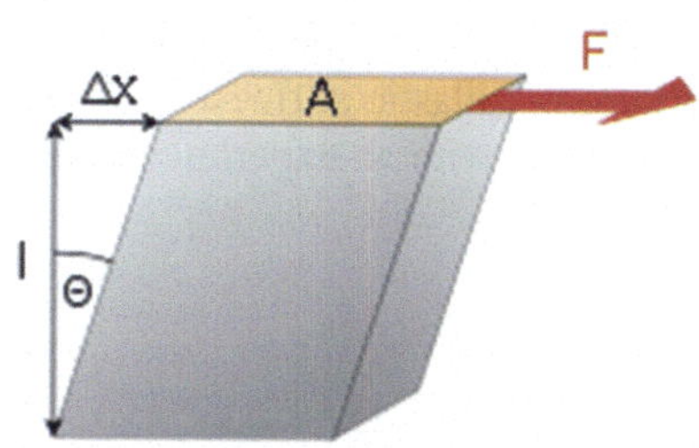


**Fig. 7.18** Shear modulus when a force F is applied (Courtesy: WIKI)

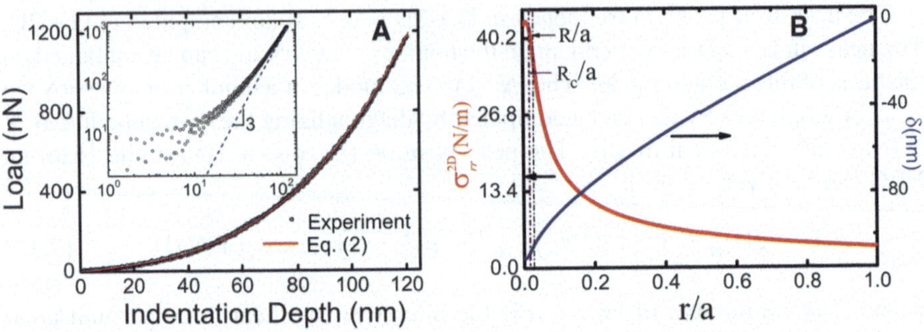


**Fig. 7.19** (**a**) Loading/unloading curve, and curve fitting to Eq. (7.13); (**b**) Stress and the maximum deflection of graphene membrane versus normalized radial distance at maximum loading (Reproduced with permission from J. Hones)

model (Fig. 7.19b) can be ignored in 2D graphene. Further studies on the simulations performed by James et al. show that the force–displacement curve is insensitive to the indenter tip radius when $R<<a$, where $R$ is the tip radius and represents the graphene membrane diameter. It is also verified that if the indenter tip is located within (a/10) of the center of the graphene film, force–displacement response due to the applied force becomes insensitive. Taking into consideration all of the tests, the force–displacement behavior can be approximated [91–93] as

$$F = {\sigma_0}^{2D}(\pi a)\,(\delta/a) + E^{2D}(q^3 a)\,(\delta/a) \tag{7.13}$$

where $F$ is the applied force, ${\sigma_0}^{2D}$ is the pretension in the graphene film, $E^{2D}$ is the elastic constant, and $\delta$ is the deflection at the center.

Figure 7.19a shows the least squares curve fit of one set of experimental data based on Eq. (7.13), considering ${\sigma_0}^{2D}$ and $E^{2D}$ as free parameters; Figure 7.19b shows stress and the maximum deflection of graphene membrane versus normalized radial distance at maximum loading. Figure 7.20a shows the distribution of the derived values of $E^{2D}$. The mean value of $E^{2D}$ is 342 $\mathrm{Nm}^{-1}$, with a standard deviation of 30 $\mathrm{Nm}^{-1}$. Figure 7.20b shows the derived values for the film pretention, which range from 0.07 to 0.74 $\mathrm{Nm}^{-1}$ [91].

Figure 7.21a shows the plot of indentation depth vs. load for both tip radii as well as diameters. The films break at large deflections (above 100 nm) and forces of about 1.8μN and 2.9μN for the smaller and the larger indenter tips, respectively. In Fig. 7.21b, we can see that the breaking force for both the tips follows Gaussian distribution. From the experimental results, the values of the second order and the third order elastic stiffness ($E^{2D}$) for monolayer graphene are found to be around 340 $\mathrm{Nm}^{-1}$ with an experimental error of ±50.

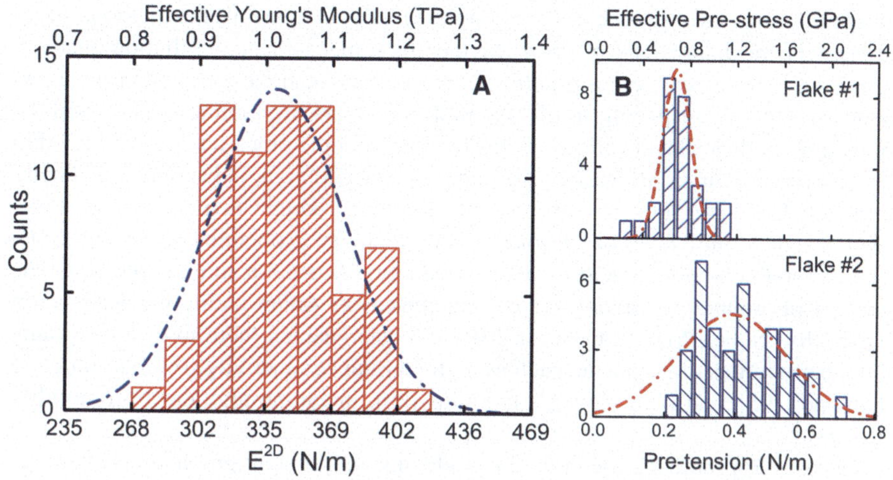


**Fig. 7.20** (**a**) Histogram of elastic stiffness; (**b**) histogram of film pretensions (Reproduced with permission from J. Hone)

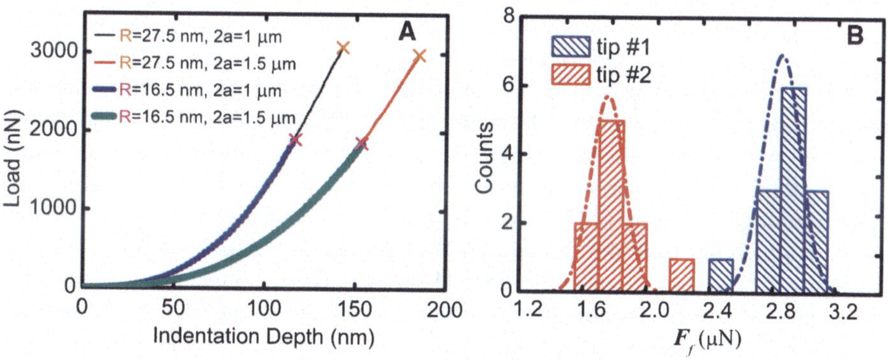


**Fig. 7.21** Fracture test results. (**a**) Four typical tests with different tip radii and film diameters; fracture loads are indicated by X marks. (**b**) Histogram and Gaussian distribution of breaking force for both tips (Reproduced with permission from J. Hone)

## 7.6 Methods toward the Preparation of Graphene and Its Unique Properties

Graphene is a very special material (two-dimensional carbon), since it has the advantage of being both conducting and transparent. The mechanical and the electrical properties of a monolayer graphene are so unique that it has surpassed many materials that have been known for their mechanical strength, electrical behavior, and their broad-range applications in electronic devices (by bandgap engineering). As a result, it has been a primary task for the engineers and the scientists to synthesize monolayer graphene. There are several methods to produce single (atomic) layer graphene. The simplest method among all the methods so far applied is mechanical exfoliation. In the micromechanical exfoliation method, graphene is detached from a graphite (three-dimensional carbon) crystal using an adhesive tape. After peeling it off, the multiple layers of graphite remain, and the remaining multiple layers are separated by repeating the procedure until a very thin monolayer remains. Afterward, the tape is attached to a substrate, and finally removed from the substrate by acetone. Finally, a single layer of the graphite (which is an atomic layer of carbon), called graphene, is obtained on the substrate. The method is laborious, and for mass production it is not a suitable method. The success of isolating a monolayer of graphene won Andre Geim and Konstantin Novoselov Nobel Prize in the year 2010 and also enabled them to discover many unique properties of graphene, such as high thermal conductivity (twice as much as diamond) [95], and high Young's modulus [96], including high electron mobility [81, 94].

One of the promising methods for producing single-layer graphene is chemical vapor deposition (CVD). The method is applied to deposit a few layers of graphite on a suitable substrate and exfoliate a single layer by wet transfer method [97]. It involves a wet chemical etching of metals on which graphene grows with a layer of polymethyl methacrylate (PMMA) coated on top of graphene. Finally, a single layer of graphene can be exfoliated from PMMA surface by using acetone. However, contaminations are induced mostly during the performance of the wet process.

A different method to reduce the contamination has been tried by using thermal release tape instead (developed by Samsung) of PMMA. Some researchers have also used polydimethylsiloxane (PDMS) and claimed to have little or no contamination [98–101]. Another successful method to detach a single layer of graphene from ultrahigh vacuum (UHV) grown graphite is called dry contact transfer method, which has been developed by J.W. Lyding and his students [102, 103].

Figure 7.22 shows the inert Si (100): H surface. The nanometer-sized graphene sheet (pieces) is formed spontaneously during mechanical exfoliation of graphite. The silicon (Si) substrate is hydrogen-passivated by leaking $H_2$ into the ultrahigh vacuum (UHV) chamber where H atom is created by cracking $H_2$ molecules on hot (1500 °C) tungsten filament [102]. Graphene is deposited onto the Si (1000)-2x1: H surface using in-situ dry contact transfer (DCT).

Figure 7.23 shows the scanning tunneling microscopy (STM) of the graphene layer over Si (100): H surface. The STM topography of the graphene layer shows several parallel lines spaced approximately 2.5 Å, consistent with 2.5 Å spacing between nearest neighbor carbon hexagons [104].

M.J. McAllister et al. have produced individual functionalized graphene sheets (PGSs) through the oxidation and thermal exfoliation of natural graphite [105]. Exfoliation takes place when the decomposition rate of the epoxy and hydroxyl sites of graphite exceeds the diffusion rate of the evolved gases, thus yielding pressures that exceed the van der Waals forces holding the graphene sheet together.

There are some reports [106] on the wet ball milling process to exfoliate graphite platelets into graphene in a liquid medium. Multilayered graphite nano sheets with a

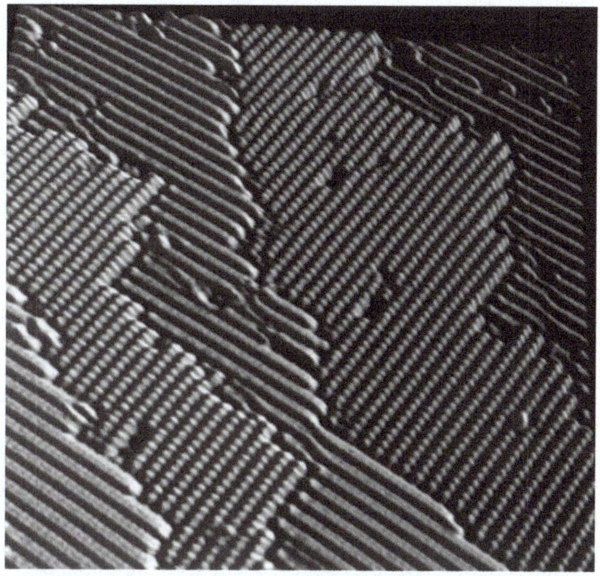

**Fig. 7.22** Inert Si (100): H surface (Photo courtesy: J.W. Lyding, reproduced with permission)

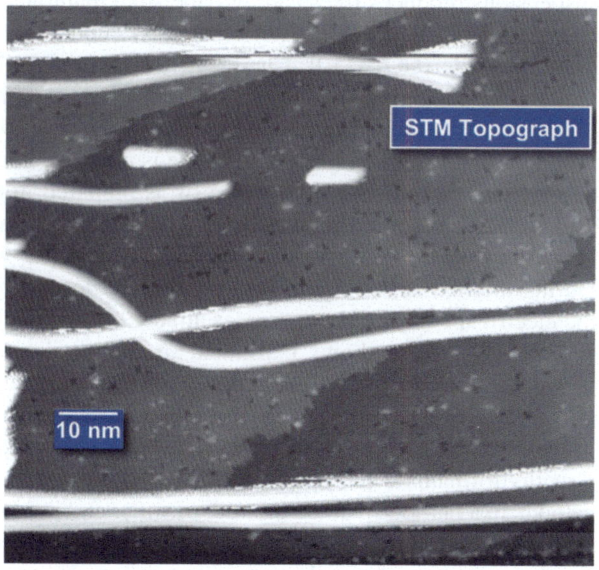


**Fig. 7.23** Scanning tunneling microscopy (STM) of the graphene layer over Si (100): H surface (Photo courtesy: J.W. Lyding reproduced with permission)

thickness of 30–80 nm are dispersed into N, N-dimethylformamide (DMF) and exfoliated by shear force dominated ball milling carried out in a planetary mill. After high-speed configuration, irregular-shaped single-layer and few-layer graphene sheets with thickness between 0.8 and 1.8 nm are found to form.

Chemical tuning of chemically modified graphene (CMG) produced by simple sonication of water-dispersed graphite oxide is a promising method of producing

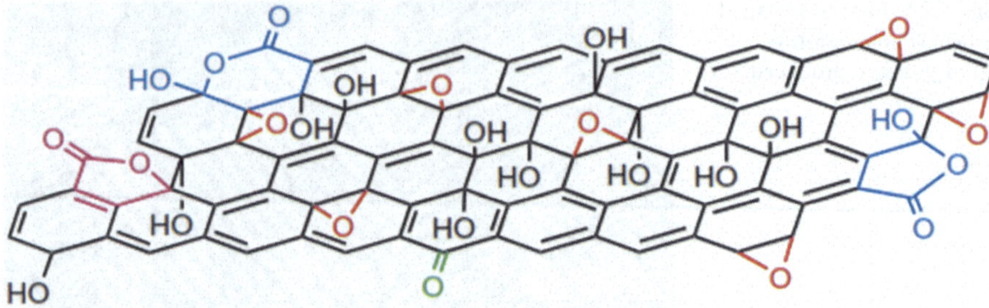


**Fig. 7.24** Chemical structure of graphite oxide following the Lerf-Klinowski model

single-layer graphene. The dispersion of graphite oxide has been done directly in several polar solvents such as ethylene glycol, DMF (N-N-Dimethylformamide), NMP (N-methyl-2-pyrrolidone), and THF (tetrahydrofuran). Graphite oxide is a layered structure of graphene oxide and is hydrophilic. It can be completely exfoliated to produce aqueous colloidal suspensions of graphite oxide sheets by sonication and by stirring the mixture for a long time [107]. The dispersed particles are negatively charged, and filtration of CMG (chemically modified graphene) suspensions produces freestanding paper-like materials [108, 109].

Figure 7.24 shows the Lerf-Klinowski model of the chemical structure of graphite oxide (GO) [110]. Chemical reduction of GO is one of the established methods of making large volume of graphene. As a matter of fact, liquid phase exfoliation of graphite turns out to be a facile way to make graphene, although production of a single layer of graphene is not yet achieved [111].

Figure 7.25 shows the two faces of as-prepared graphene paper (top) and the scanning electron microscopy (SEM) analysis of the graphene paper (bottom). The side-view looks exactly like some stacked up paper with smooth shiny surface. During thermal annealing, the graphene paper loses oxygen from the surface, and the electrical conductivity increases almost linearly with temperature [109].

Figure 7.26 (1) shows exfoliated graphite reintercalated with sulfuric acid between layers; (2) shows insertion of tetrabutylammonium (TBA) and sonication with polyethylene glycol, and (3) shows graphene chemically dried and modified graphite sheet coated with polymer molecule and a photo of the suspension in dimethylformamide (DMF). Figure 7.26b shows graphite rods used as anode and cathode in a phase-separated mixture of water and imidazolium-based ionic liquids. As voltage is applied between the anode and the cathode, ionic-liquid functionalized graphene sheets precipitated at the anode could be redispersed in polar aprotic solvents [112]. At the middle of Fig. 7.16 (3), the exfoliation of a chemically dried and modified graphite (CMG) sheet is shown.

Graphene can be viewed as an individual atomic plane extracted from graphite, and as an unrolled single-walled carbon nanotube (SWCNT) or as an extended flat fullerene molecule. Recently, the synthesis of high-quality graphene nanosheets in large scale through electrochemical reduction of exfoliated graphite oxide precursor has been reported by a research group [113].

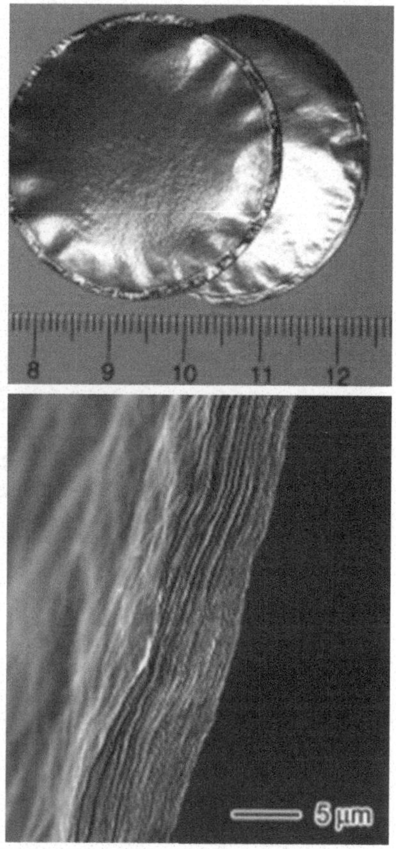


**Fig. 7.25** Top: two pieces of freestanding graphene paper; bottom: side-view SEM images of the graphene paper (Produced with permission, Wiley Interscience, Ref. H. Chen et al.)

The electrochemically reduced graphene nanosheets (GNSs) have layered structure and have a 2D system like monolayer graphene with very strong $sp^2$ bonds, causing a threefold coordinated planar structure. The sheet has a large $\pi$-electronic surface, and much cheaper than single-wall nanotubes (SWNTs) [114, 115]. High-quality graphene accelerates the electron transfer rate in dopamine electrochemistry. This approach opens up the possibility for assembling graphene biocomposites for electrolysis and the construction of biosensors [113]. Graphene nanosheet/polystyrene composites synthesized by a liquid-phase blend route exhibit a percolation threshold of 0.1 vol. % for room temperature electrical conductivity, and at only 4.19 vol. %, this composite has a conductivity of 13.84 $Sm^{-1}$, which is 3–15 times that of polystyrene composites filled with single-walled carbon nanotubes (SWCNTs) [116].

Recently, under the supervision of Professor James Hone, the research group at Columbia University, NY, has created the world's thinnest light bulb out of graphene (Fig. 7.27). The temperature of the light bulb when it emits light reaches above 2500 °C. The visible light is very bright and is visible even to the naked eye. At this temperature, the electron temperature is much higher than that of acoustic

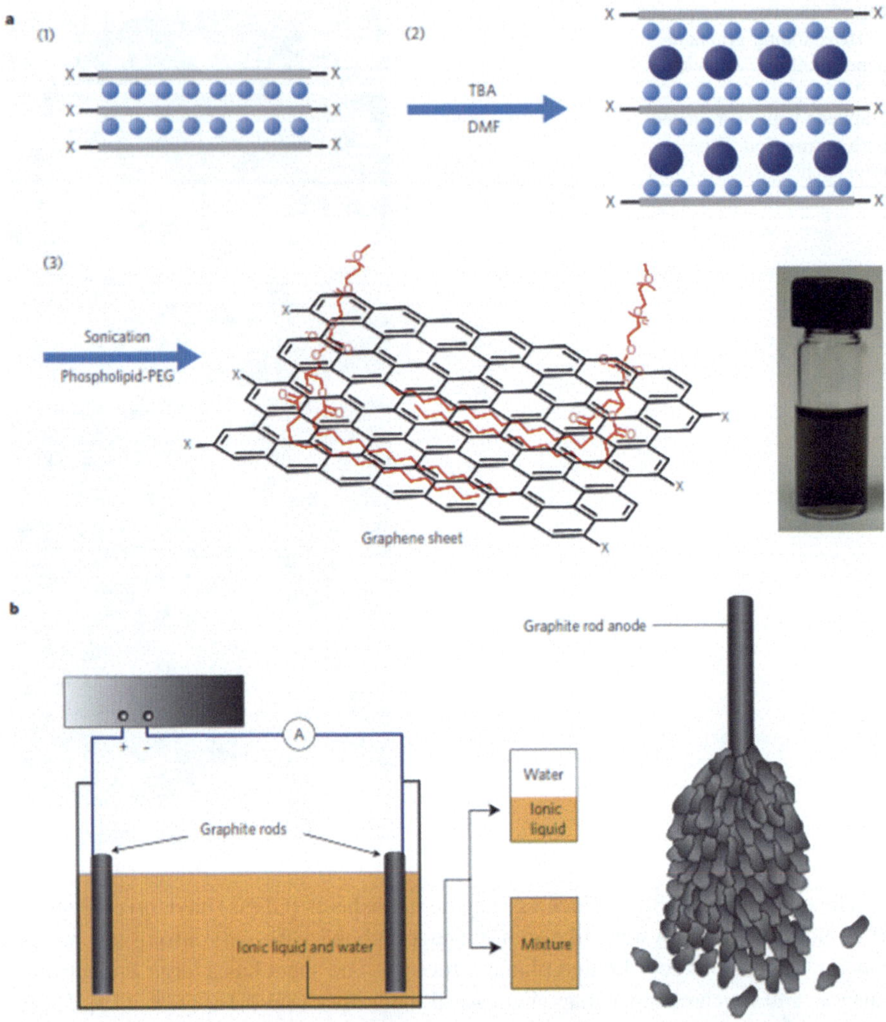


**Fig. 7.26** Experimental setup to fabricate ionic-liquid functionalized graphene nanosheets (Reproduced with permission from Wiley, Ref. Liu et al., [116])

vibrational modes of the graphene lattice; so, less energy is needed to attain temperatures needed for visible light emission. Another interesting thing is that graphene is a poor conductor of heat; so, the high temperature is confined to a small hot spot in the center. Professor Hone is excited about his dream to construct a micro-hot plate that could be heated to thousands of degrees in a fraction of a second to study high-temperature chemical reactions or catalysis [117]. Figure 7.27 shows the broadband light emitter device fabricated out of graphene.

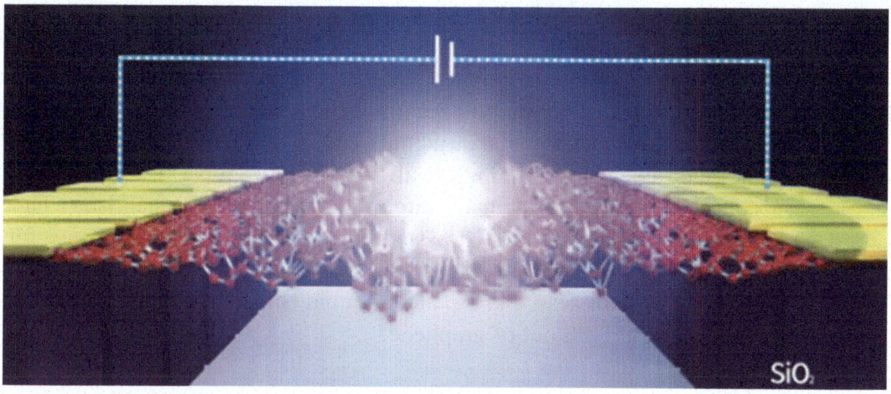


**Fig. 7.27** Broadband light emitter device fabricated out of graphene (Reproduced with permission from J. Hone et al. Photo credit: M-Ho Bae)

## 7.7 Summary

The rising star graphene on the horizon of materials science and condensed matter physics became popular, and the research on this particular field became fascinating. Indeed, graphene represents a conceptually new class of material that is only one atom thick. It offers new inroads into two-dimensional physics that has never ceased to surprise. The *transparent graphene* (which absorbs only 10% of incident light) can *stretch* up to 20% of its length, showing amazing strength, with *no bandgap*, and having *ballistic conduction* with *highest current density* (a million times that of copper) and having *thermal conductivity* that can beat diamond.

As a result of these characteristics, it has been a primary task for the engineers and the scientists to produce a monolayer of graphene. There are several methods to produce single (atomic) layer graphene. The simplest method among all the methods so far applied is mechanical exfoliation. The other promising method of producing single-layer graphene is chemical vapor deposition (CVD). The process can deposit a few layers of graphite on a suitable substrate, and later can be exfoliated by wet transfer method, which involves a wet chemical etching of metals.

The opacity of graphene is ≈2.3% (which equates to universal dynamic conductivity value of $G = (e^2/4\hbar)$ (±2.3%)). It has been observed that once optical intensity reaches a certain threshold (known as the saturation fluence), saturable absorption takes place (very high intensity light causes a reduction in absorption). This is an important characteristic with regard to the mode-locking of fiber lasers.

The atomic (mono) layer of graphene can be represented as a two-dimensional (2D) model. The studies on the stress–strain relation and the shear modulus on an atomic (mono) layer graphene show that the nonlinear elastic model cannot be fitted in 2D graphene. The loss of Bernal stacking might not be observed or distinguishable with two or three atomic layers of graphene, but with a stack of 20–50 atomic layers of graphene, formation of arrays of partial dislocations and stacking faults have been observed.

The electrochemically reduced graphene nanosheets (GNSs) are layered structures and have a 2D system like monolayer graphene with very strong $sp^2$ bonds, causing a threefold coordinated planar structure. The sheet has a large $\pi$-electronic surface and is much cheaper than single-wall nanotubes (SWNTs). High-quality graphene accelerates the electron transfer rate in dopamine electrochemistry. This approach opens up the possibility for assembling graphene biocomposites for electrolysis and the construction of biosensors. Recently, under the supervision of Professor James Hone, the research group at Columbia University, NY, has created the world's thinnest light bulb out of graphene.

## References


1. D. Bradley, A Chemical History of Graphene, Materials Today, June 10th, (2014)
2. S. Connor, The Graphene Story, The Independent, Science, July 1st, (2015)
3. A. Codos (ed.), The month in physics history, Discovery of Graphene, Am. Phys. Soc. (2004)
4. J.C. Slonczewski, P.R. Weiss, Band structure of graphite. Phys. Rev. **109**, 271 (1958.) and also J.W. McClure, Diamagnetism of Graphite. Phys. Rev. **104**, 666 (1956)
5. A.K. Geim, K.S. Novoselov, The rise of graphene. Science **306**, 666 (2004)
6. P.R. Wallace, The band theory of graphite. Phys. Rev. **71**, 622 (1947)
7. H.-P. Boehm, R. Setton, E. Stumpp, Nomenclature and terminology of graphite intercalation compounds. Pure Appl. Chem. **66**(9), 1893 (1994)
8. E. McCann, Asymetry gap in electronic band structure of bilayer graphene. Phys. Rev. **B74**, 161403 (2006)
9. H.P. Boem, Graphene – how a laboratory curiosity suddenly became extremely interesting. Angew. Chem. Int. Ed. **49**, 9332 (2010)
10. A.K. Geim, K.S. Novoselov, The rise of grephene. Nat. Mater. **6**, 183 (2007)
11. H. Tetlow et al., Epitaxial growth of graphene: theory and experiment. Phys. Rep. **542**(3), 195 (2014)
12. W. Norimatsu, M. Kusunoki, Epitaxial graphene on SiC (0001): advances and perspective. Phys. Chem. Chem. Phys. **16**, 3501 (2014)
13. P. Sutter, Epitaxial graphene: how silicon leaves the scene. Nat. Mater. **8**, 171 (2009)
14. S.Y. Zhou et al., First direct observation of Dirac fermions in graphite. Nat. Phys. **2**, 595 (2006)
15. V.M. Apalkov, T. Chakraborty, The fractional quantum hall states of Dirac electrons in graphene. Phys. Rev. Lett. **97**, 126801 (2006)
16. K.S. Novoselov et al., Two dimensional gas of Massless Dirac Fermions in graphene. Nature **438**, 197 (2005)
17. M.S. Dresselhaus, P.T. Araujo, Perspective on the 2010 Nobel prize for graphene. ACS Nano **4**(11), 6297 (2010)
18. A.J. Strudwick et al., Chemical vapor deposition of high quality grephene films from carbon dioxide atmosphere. ACS Nano **9**(1), 31 (2015)
19. Y. Zhang, L. Zhang, C. Zhou, Review of chemical vapor deposition of graphene and related applications. Acc. Chem. Res. **46**(10), 2329–2013
20. A. Guermoune et al., Chemical vapor deposition of graphene on copper with methanol, ethanol and propanol, precursors. Carbon **46**(3), 4204 (2011)

21. Paton et al., Scalable production of large quantities of defect free few-layer graphene by shear exfoliation in liquids. Nat. Mater. **13**, 624 (2014)
22. K.S. Novoselov et al., Electronic field effect in atomically thin carbon films. Science **306**, 666 (2004)
23. Y. Zhang, J.W. Tan, H.L. Stomer, P. Kim, Experimental observation of quantum hall effect and Berry's phase in graphite. Nature **438**, 201 (2005)
24. B. Partoens, F.M. Peeters, From graphene to graphite: electronic structure around the K-point. Phys. Rev. **B74**, 075404 (2006)
25. M. Frreitag, Graphene: nanoelectronics goes flat out. Nat. Nanotechnol. **3**(8), 455 (2008)
26. Y. Zhang et al., Direc observation of a widely tunable bandgap bilayer graphene. Nature **459**, 820 (2009)
27. L. Sherriff, ZDNet Microsoft newsletter now, London, UK. June 10, (2011)
28. A.R. Wright, C. Zhang, Dynamic conductivity of graphene with electron-LO-phonon interaction. Phys. Rev. B **81**, 165413 (2010)
29. P. Carbotte, E.J. Nicol, S.G. Sharapov, Effect of electron-phonon interaction on spectroscopies in graphene. Phys. Rev. **B81**, 045419 (2010)
30. L.A. Falkovsky, Optical properties of graphene. J. Phys. Conf. Ser. **129**, 012004 (2008)
31. D. Fathi, A review of electronic band structures of graphene and carbon nanotube using tight binding. J. Nanotechnol. **2011** (2011.) Article ID:471241
32. C.R. Poole, J. Owens, *Introduction to Nanotechnology* (Wiley, 2003)
33. R.R. Nair, P. Blake, A.N. Grigorenko, K.S. Novoselov, T.J. Booth, T. Stauber, N.M.R. Peres, A.K. Geim, Fine structure constant defines visual transparency of graphene. Science **320**, 1308 (2008)
34. A.B. Kuzmenko, E. van Heumen, F. Carbone, D. van der Marel, Universal dynamical conductance of graphite. Phys. Rev. Lett. **100**, 117401 (2008)
35. K.S. Novoselov, A.K. Geim, et al., Electric field effect in atomically thin carbon films. Science **306**, 666 (2004)
36. A. Gupta et al., Raman scattering from high frequency phonons in supported n-graphene layer films. Nano Lett. **6**, 2667 (2006)
37. M.L. Sadowski et al., Landau level spectroscopy of ultrathin graphite layers. Phy. Rev. Letts **97**, 266405 (2006.) and also Z. Jiang et al., Landau level spectroscopy of graphene, Phys. Rev. Lettrs. **98**, 197403 (2007)
38. T. Stauber, N.M.R. Press, A.K. Geim, Optical conductivity of graphene in the visible region. Phys. Rev. B **78**, 085432 (2008)
39. M. Bruna, S. Borini, Optical constants of graphene layers in the visible range. Appl. Phys. Lett. **94**, 031901 (2009)
40. F. Bonaccorso, Z. Shun, T. Hassan, A.C. Ferrari, Graphene photonics and optoelectronics. Nat. Photonics **4**, 611 (2010)
41. L.M. Malard, M.A. Pimenta, M.S. Dresselhaus, Raman spectroscopy in graphene. Phys. Rep. **473**, 51 (2009)
42. M.S. Dresselhaus, A. Jorio, R. Saito, Characterizing, graphene, graphite, and carbon nanotube by Raman spectroscopy. Annu. Rev. Condens. Matter Phys. **1**, 89 (2010)
43. T.M.G. Mohiuddin et al., Uniaxial strain in graphene by Raman spectroscopy. Phys. Rev. **B79**, 205433 (2009)
44. A.C. Ferrari, D.M. Basko, Raman spectroscopy as a versatile tool for studying the properties of graphene. Nat. Nanotechnol. **8**, 235 (2013)
45. A.S. Geim, K.S. Novoselov, The rise in graphene. Nat. Mater. **6**, 183 (2007)
46. A.H. Castro Neto, F. Guines, N.M.R. Peres, K.S. NOvoselov, A.K. Geim, The electronic properties of graphene. Rev. Mod. Phys. **81**, 109 (2009)
47. N.M.R. Peres, Transport properties of graphene: an introduction. Rev. Mod. Phys. **82**, 63 (2010)

48. A.C. Ferrari, J.C. Meyer, V. Scardaci, C. Casiraghi, M. Lazzeri, F. Piscanec, D. Jiang, K.S. Novoselov, S. Roth, A.K. Geim, Raman structure of graphene and graphene layers. Phys. Rev. Lett. **97**, 187401 (2006)
49. M.A. Pimenta, G. Dresselhaus, M. Dresselhaus, L.G. Jorio, A. Saito, Studying disorder in graphite based systems by Raman spectroscopy. Phys. Chem. Chem. Phys. **9**, 1276 (2007)
50. L.G. Cancado, A. Reina, J. Kong, M. Dresselhaus, Geometrical approach for the study of G' band in Raman Spectrum of monolayer graphene and bulk graphite. Phys. Rev. B **77**, 245408 (2008)
51. Z. Ni, Y. Wang, T. Yu, Z. Shen, Raman spectroscopy and imaging of graphene. Nano Res. **1**, 273 (2008)
52. R. Narula, S. Reich, Double resonant Raman spectra in graphene and graphite: a two dimensional explanation of the Raman amplitude. Phys. Rev. B **78**, 165422 (2008)
53. R. Rao, D. Tishler, J. Katoch, M. Ishigami, Raman scattering in graphene. Phys. Rev. **B84**, 113406 (2011)
54. F. Tuinstra, L.J. Koenig, Raman spectrum of graphite. J. Chem. Phys. **53**, 1126 (1970)
55. I. Childres et al., Raman spectroscopy of graphene and related materials, Chapt-19, in *New Developments in Photons and Materils Research*, ed. by J. J. Jang (Ed), (Nova, 2013)
56. Belle Dumé, Graphene has Record Breaking Strength, Phys. World, IOP pub. UK, July 17, (2008)
57. R.H. Rickman, P.R. Dunstan, Enhancement of lattice defect signatures in graphene and ultrathin graphite by using tip-enhanced Raman spectroscopy. J. Raman Spectrosc. **45**, 15 (2014)
58. R. Rao, D. Tishler, J. Katoch, M. Ishigami, Multiphonon Raman scattering in graphene. Phys. Rev. **B84**, 113406 (2011)
59. F. Wang et al., Gate variable optical transitions in graphene. Science **320**, 206 (2008)
60. X. Wang et al., Transparent carbon films as electrodes in organic solar Cella. Angew. Chem. **47**, 2990 (2008)
61. N. Yang, J. Zhai, D. Wang, Y. Chen, L. Jiang, Two dimensional graphene bridges enhanced Photoinduce charge transport in dye-sensitized solar cell. ACS Nano **4**, 887 (2010)
62. T.K. Gupta et al., *Screen Printed Dye-Sensitized Large Area Nano-Crytalline Solar Cell, Nanophase and Nanocomposite Materials III, 581, 653* (MRS, Warrendale, 2000)
63. T. Matyba et al., Graphene and mobile ions: the key to all plastic , solution-processed light emitting devices. ACS Nano. **4**, 637 (2010.) and also J. Wu et al., Organic light emitting diodes on Solution Processed graphene Transparent Electrodes. ACS Nano. **4**, 43 (2009)
64. T. Muller, F. Xia, P. Avouris, Graphene Photodetectors for high speed communication. Nat. Photonics **4**, 297 (2010)
65. F.T. Vasko, V. Ryzhii, Photoconductivity of intrinsic graphene. Phys. Rev. B **77**, 195433 (2008)
66. F. Xia, T. Mueller, Y.M. Lin, A. Valdes-Garcia, P. Avouris, Ultrafast graphene photodetector. Nat. Nanotechnol. **4**, 839 (2009)
67. A.B.E. Saleh, M.C. Teich, Chapt-18, in *Fundamentals of Photonics*, (Wiley, Hoboken, 2009), p. 784
68. X.-C. Zhang, J. Xu, *Introduction to THz Wave Photonics* (Springer, New York, 2010)
69. T. Gu et al., Regenerative oscillation and four-wave mixing in graphene optoelectronics. Nat. Photonics **6**, 553 (2012)
70. J.C. Charlier, P.C. Eklund, J. Zhu, A.C. Ferrari, Electron and phonon properties of graphene: their relationship with carbon nanotubes. Top. Appl. Phys. **111**, 673 (2008)
71. X.I. Du et al., Fractional quantum hall effect and insulating phase of Dirac electrons in graphene. Nature **462**, 192 (2009)
72. M.R. Connolly, C.G. Smith, Nano analysis of graphene layers using scanning probe techniques. Phil. Trans. R. Soc. A **368**, 5379 (2010)
73. M.Y. Han, B. Ozyilmaz, Y. Zhang, P. Kim, Energy bandgap engineering of graphene nanoribons. Phys. Rev. Lett. **98**, 206805 (2007)

74. F. Molitor, A. Jacobsen, C. Stampfer, J. Guttinger, T. Ihn, K. Ensslin, Transport gap in side-gated graphene constrictions. Phys. Rev. **B79**(075426) (2009)
75. K. Todd, H.-T. Chou, S. Amasha, D. Goldhaber-Gordon, Quantum dot behavior in graphene nanoconstrictions. Nano Lett. **9**(416) (2009)
76. Y. Zhang et al., Direct observation of a widely tunable bandgap in bilayer graphene. Nat. Lett **459**, 820 (2009.) and also N. Savage, Graphene Makes Transistor Tunable. IEEE Spectrum, August 31, (2009)
77. J. Martin et al., Observation of electron-hole puddles in graphene using a scanning single electron transistor. Nat. Phys. **4**, 144 (2008)
78. S.S. Datta, D.R. Strachan, E.J. Mele, A.T. Johnson Charlie, Surface potentials and layer charge distributions in few layer graphene films. Nano Lett. **9**, 7 (2009)
79. K.S. Novoselov et al., Electric field effect in atomically thin carbon films. Science **306**, 5696 (2004)
80. K.S. Novoselov et al., Unconventional quantum hall effect & Berry's phase of $2\pi$in bilayer graphene. Nat. Phys. **2**, 177 (2006)
81. K.S. Novoselov, A.K. Geim, et al., Two dimensional gas of Massless Dirac Fermions in graphene. Nature **438**, 197 (2005)
82. K.S. Novoselov, A.K. Geim, et al., Unconventional quantum hall effect in graphene. Phys. Rev. Lett. **95**, 146801 (2005)
83. P. Maher et al., Tunable Fractional Quantum Hall Phases in Bilayer Graphene. Science **345** (6192), 61 (2014)
84. A.M. Fischer, R.A. Romer, A.B. Dzyubenko, Magnetoplasmons and SU(4) symmetry in graphene. J. Phys. **286**, 012054 (2011)
85. A.F. Young et al., Tunable symmetry breaking and helical edge transport in graphene quantum spin hall state. Nature **505**, 528532 (2014)
86. L. Gong et al., Reversible loss of Bernal stacking during deformation of few layer graphene in Nanocomposites. Nano **7**(8), 7287 (2013)
87. S. Amelinckx, P. Delavignette, Electron optical study of basal dislocations in graphite. J. Appl. Phys. **31**, 687 (1960)
88. C. Baker, A. Kelly, The effect of neutron irradiation on elastic moduli of graphite single crystal. Philos. Mag. **9**, 927 (1964)
89. P.H. Tan et al., The shear mode of multilayer graphene. Nat. Mater. **11**, 294 (2012)
90. Y. Zhao et al., Inter layer breathing and shear modes in few-tri-layer $MoS_2$ and $WSe_2$. Nano Lett. **13**, 1007 (2013)
91. C. Lee, X. Wei, J.W. Kysar, J. Hone, Measurement of elastic properties and intrinsic strength of monolayer graphene. Science **321**, 385 (2008)
92. K.T. Wan, S. Guo, D.A. Dillard, A theoretical and numerical study of a thin clamped circular film under an external load in presence of a tensile residual stress. Thin Solid Films **425**, 150 (2003)
93. U. Komaragiri, M.R. Begley, The mechanical response of freestanding circular elastic films under point and pressure load. J. Appl. Mech. **72**(2), 203 (2005)
94. K.I. Bolotin, K.J. Sikes, Z. Jiang, M. Klima, G. Fudenberg, J. Hone, P. Kim, H.L. Stormer, Ultrahigh electron mobility in suspended graphene. Solid State Commun. **146**(9–10), 355 (2008)
95. A. Balandin, S. Ghosh, W. Bao, I. Calizo, D. Teweldebrhan, F. Miao, C.N. Lau, Superior thermal conductivity of single-layer graphene. Nano Lett. **8**(3), 902 (2008)
96. J.U. Lee, D. Yoon, H. Cheong, Estimation of Young's modulus of graphene by Raman spectroscopy. Nano Lett. **12**(9), 4444 (2012)
97. X. Li, Y. Zhu, W. Cai, M. Borysiak, B. Han, D. Chen, R.D. Piner, L. Colombo, R.S. Ruoff, Transfer of large-area graphene films for high-performance transparent conductive electrodes. Nano Lett. **9**(12), 4359 (2009)

98. W. Xin, Z.B. Liu, Q.W. Sheng, M. Feng, L.G. Huang, P. Wang, W.S. Jiang, F. Xing, Y.G. Liu, J.G. Tian, Flexible graphene saturable absorber on two-layer structure for tunable mode-locked soliton fiber laser. Opt. Express **22**(9), –10239 (2014)
99. M. Liu, Characterization study of bonded and unbonded polydimethylsiloxane aimed for bio-micro-electromechanical systems-related applications. J. Micro/Nanolithogr. MEMS MOEMS. **6** (2), 023008 (2007)
100. Y.Y. Wang, P.J. Burke, A large-area and contamination-free graphene transistor for liquid-gated sensing applications. Appl. Phys. Lett. **103**(5), 052103 (2013)
101. R.W.R.L. Gajasinghe, S.U. Senveli, S. Rawal, A. Williams, A. Zheng, R.H. Datar, R.J. Cote, O. Tigli, Experimental study of PDMS bonding to various substrates for monolithic microfluidic applications. J. Micromech. Microeng. **24**(7), 075010 (2014)
102. K.A. Ritter, J.W. Lyding, Characterization of nanometer-sized, mechanically exfoliated graphene on the H-passivated Si (100) surface using scanning tunneling microscopy. Nanotechnology **19**(1), 015704 (2008)
103. P.M. Albrecht, J.W. Lyding, Ultrahigh-vacuum scanning tunneling microscopy and spectroscopy of single-walled carbon nanotubes on hydrogen-passivated Si(100) surfaces. Appl. Phys. Lett. **83**(24), 5029 (2003)
104. M.S. Dresselhaus, G. Dresselhaus, P.C. Eklind, *Science of Fullerenes and Carbon Nanotubes* (Academic, San Diego, 1996)
105. M.J. McAllister et al., Single Sheet functionalized graphene by oxidation and thermal expansion of graphite. Chem. Mater. **19**(18), 4396 (2007)
106. W. Zhao et al., Preparation of graphene by Exfoliationof graphite by using ball milling. J. Mater. Chem. **20**, 5817 (2010)
107. I. Jung et al., Simple approach for high contrast optical imaging and characterization of graphene based sheets. Nano Lett. **7**, 3569 (2007)
108. D.A. Dikin et al., Preparation and characterization of graphene oxide paper. Nature **448**, 457 (2007)
109. H. Chen, M.B. Muller, K.J. Gilmore, G.G. Walace, D. Li, Mechanically strong, electrically conductive, and biocompatible graphene paper. Adv. Mater. **20**, 3557 (2008)
110. A. Lerf, H. He, M. Forster, J. Klinowski, Structure of graphite oxide revisited. J. Phys. Chem. **B102**, 4477 (1998)
111. W. Gao, L.B. Alemany, L. Ci, P.M. Ajayan, New insights into the structure and reduction of graphite oxide. Nat. Chem. **1**, 403 (2009)
112. S. Park, R.S. Ruoff, Chemical methods for the production of graphenes. Nat. Technol. **5**, 574 (2010)
113. H.L. Guo et al., A green approach to the synthesis of graphene nano-sheets. ACS Nano **3**(9), 2653 (2009)
114. W.G. Zheng, S.C. Wong, Electrical conductivity and dielectric propertirs of PMMA/ expanded graphite composite. Compos. Sci. Technol. **63**, 225–2003
115. Y.B. Zhang, J.P. Small, M.E.S. Amori, P. Kim, Electric field modulation of Galvanomagnetic properties of Mesoscopic graphite. Phys. Rev. Lett. **94**, 176803 (2005)
116. N. Liu et al., One step ionic liquid assisted electrochemical synthesis of ionic liquid functionalized graphene sheets directly from graphite. Adv. Funct. Mater. **18**, 1518 (2008)
117. Y.D. Kim et al., Bright visible light emission from graphene. Nat. Nanotechnol. Lett. **10**, 676 (2015)

# Chapter 8
# Carbon Nanotube (CNT)

## 8.1 Introduction

Carbon Nanotubes: *"Everything when miniaturized to the sub-100 nanometer scale, has new properties, regardless of what it is"* said Chad Mirkin, Professor of Chemistry (and materials science, engineering, medicine, biomedical engineering, and chemical and biological engineering) at Northwestern University, Chicago, IL. Indeed, nanoscale materials are used from sunscreen to chemical catalysis to antibacterial agents from the mundane to life-saving. Researchers are developing nanoscale assays to screen cancer, and detect infections and genes [1–8].
Gold nanoparticles doped with DNA (*deoxyribonucleic acid*) are being used to detect bacteria in the blood stream and to detect a person's immune system that reflects the presence of any cancerous cell. The smaller you go, the atoms in nanomaterials begin to contribute significantly to the overall properties of the material. Nanoscale materials also play an important role in modern electronics for the development of tiny computers and communication devices [9–11].

Nanomaterials have relatively larger surface area than the same mass of material produced in a larger form. As a result, materials are more chemically active, and are having higher strength (including tensile strength), resilience, and different electrical properties. More importantly, particles below 50 nm do not obey the laws of classical physics, provoking optical, electrical, and magnetic behaviors different from those of the same materials at a larger scale [12–14]. Carbon nanotubes (CNTs), the long thin cylinders of atomic layer of graphite, may be the most significant new materials since plastics and are the most significant of today's nanomaterials [15, 16]. Figure 8.1 shows the projected contribution of nanotechnology to the US economy in 2015.



T. Gupta, *Carbon*, DOI 10.1007/978-3-319-66405-7_8

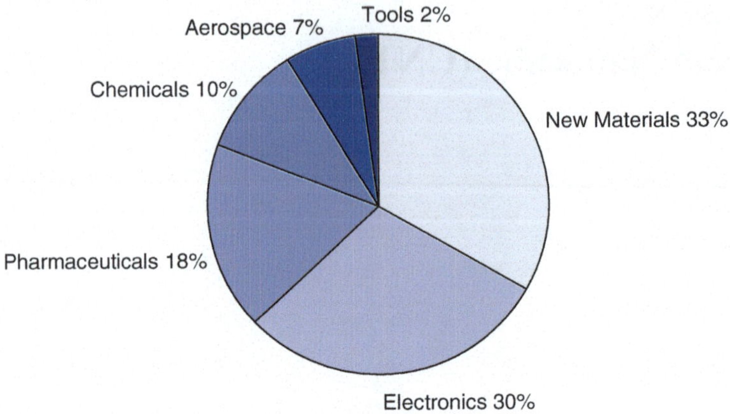


**Fig. 8.1** The projected contribution of nanotechnology to the US economy, 2015 (Source: National Science Foundation, 2003)

## 8.2 CNTs: The Allotropes of Carbon

Carbon nanotubes (CNTs) are the allotropes of carbon and one of the most interesting family members of carbon owing to their physical, chemical, mechanical, optical, and electronic characters. As a result, CNTs have attracted a good deal of attention from scientists and engineers of different disciplines [17–19]. The chemical bonding of carbon nanotubes (CNTs) is composed entirely of $sp^2$ bonds similar to the planar structure of graphite. Fullerenes and CNTs, on the other hand, have hybrid bonding orbitals between $sp^2$ and $sp^3$. However, CNTs having low percentage of $sp^3$ bonds are in this regard nearer to graphite. The difference in chemical activity between CNT caps and sidewalls as well as between straight and curved CNTs is determined by the $sp^2$–$sp^3$ bonds ratio [20]. CNT name is derived from their long, hollow structure with the walls formed by one atomic layer sheets of carbon called graphene. These sheets are rolled at specific and discrete (chiral) angles, and the combination of rolling angle and radius decides the carbon nanotube (CNT) properties. CNTs are categorized as *single-walled* (SWCNT) or *multiple-walled* nanotubes (MWCNTs) [21].

Most SWCNTs have a diameter of approximately 1 nm with a tube length that can be many millions of times longer. The structure of SWCNT can be conceptualized by wrapping a one atom thick layer of graphene into a seamless cylinder. The way the graphene sheet is wrapped is represented by a pair of indices ($n$, $m$). The integer n and m denote the number of unit vectors along two directions in honeycomb crystal lattice of graphene. If $m = 0$, the nanotube is called zigzag nanotube, and if $n = m$, it is called armchair; otherwise, the nanotube is called chiral (Fig. 8.2) [22–26].

The diameter ($d$) of an individual nanotube can be calculated from its indices as follows:

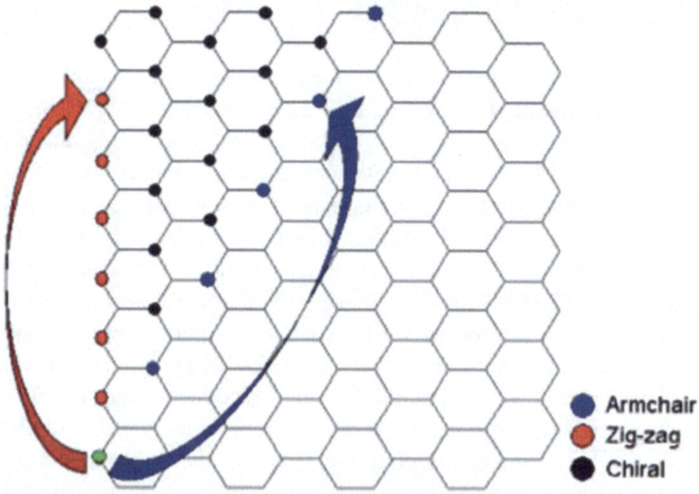


**Fig. 8.2** A zigzag, an armchair, and a chiral carbon nanotube (CNT) geometry in the honeycomb crystal lattice (Courtesy: H. Dai, Stanford Univ., CA)

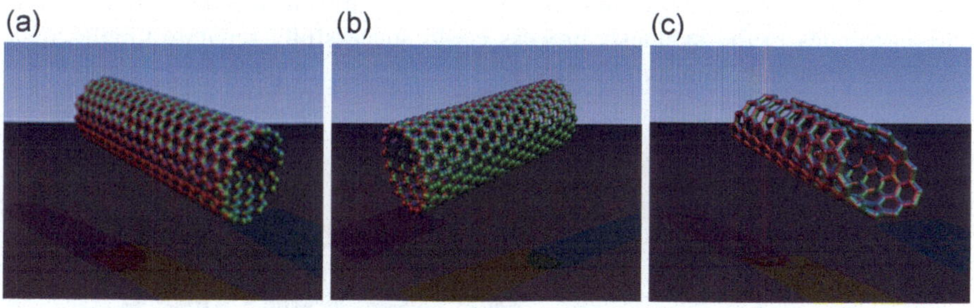


**Fig. 8.3** (**a**) Armchair ($m = n$), (**b**) zigzag ($n$, 0), and chiral ($n$, $m$) structures of SWCNTs (Courtesy: WIKI)

$$d = (a/\pi)\sqrt{(n^2 + nm + m^2)} \tag{8.1}$$

where $a = 0.246$ nm.

The properties of SWCNTs change significantly with the change of ($n$, $m$) values. In particular, their bandgap can vary from 0 to about 2 eV, and their electrical conductivity can show metallic to semiconducting behavior.

Figure 8.3 shows the single atomic layer of graphene wrapped in different configurations of SWCNTs, armchair ($m = n$), zigzag ($n$,0), and chiral ($n$, $m$), according to the arrangements of the unit vectors ($n$, $m$) along the honeycomb lattice of graphene. The integers ($n$, $m$) uniquely determine the tubule diameter and the chiral angle $\theta$. The chirality predestinates the electrical, mechanical, optical, and other properties of CNTs [27–29].

## 8.3 Construction of the Carbon Nanotubes

CNTs can be constructed in two basic forms, single-walled carbon nanotubes (SWCNTs) and multiwalled carbon nanotubes (MWCNTs). MWCNTs are composed of several concentric tubes of graphene fitted one inside the other. The simplest example of MWCNTs is double-walled carbon nanotubes

(DWCNTs), but the number of walls can be varied. Multiwalled CNTs have several advantages over SWCNTs in that they have higher stability and stiffness, and the individual CNTs can be doped independently. The scanning tunneling electron microscopy (STEM) image of a triple-walled CNT is shown in Fig. 8.4.

The first calculations for double-walled carbon nanotubes [30] were done using tight-binding technique, which sensitively includes all symmetry constraints in a simplified Hamiltonian. More specifically, the calculated energy band structure showed that two coaxial zigzag single-walled nanotubes yield a double-walled nanotube when a weak interlayer coupling between the concentric nanotubes is introduced. Similarly, two coaxial semiconducting tubules remain semiconducting when a weak interlayer is introduced. Despite the prediction that 1D electronic system cannot support superconductivity, superconductivity has been discovered in MWCNTs or nanotube arrays [31, 32]. The ability of experimentalists to study 1D quantum behavior in carbon nanotubes would be greatly enhanced if purification of carbon tubules in the synthesis process could successfully separate tubules of a given diameter and chirality [33, 34].

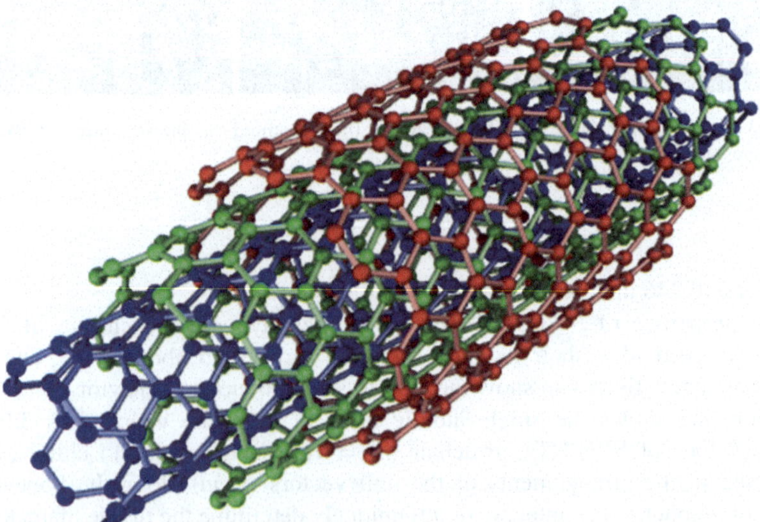

**Fig. 8.4** The structure model of a triple-walled armchair CNT derived from STEM image (Courtesy: Eric Weiser, WIKI)

## 8.4 Processing of CNTs

The current research and excitement in carbon nanotubes (CNTs) were triggered by the work of Sumio Iijima and his subsequent publication of the work in *Nature* [35]. Initial synthesis of $C_{60}$ in macroscopic amount was mostly confined to *arc evaporation* method discovered by Huffman and Kratschmer [36]. However, the material deposited on the cathode by Iijima was analyzed and found to be macroscopic amounts of multiwalled carbon nanotubes (MWCNTs) and faceted graphite particles. Further work of Iijima and Ichihaschi confirmed the synthesis of SWCNTs [37]. At the same year, Bethune and the coworkers from IBM Research Laboratory reported their discovery of SWCNTs [38]. Recently, CNTs synthesis is not confined to arc evaporation, but to a variety of other cost-effective methods. Figure 8.5 shows some of the most recently adopted methods to produce CNTs.

There are two main avenues to synthesize CNTs. The first one consists of sublimation of graphite and subsequently desublimation of the product. The second one is the decomposition of carbon-containing compounds. The first one requires

**Fig. 8.5** Some of the most recently adopted methods to produce CNTs

very high temperatures (~4000 °C) which can be achieved by electric arc furnace [39–41], laser ablation [42–44], focused solar radiation [45], or by resistive heating of graphite [46].

### *8.4.1 Arc Evaporation or Arc Discharge Method*

The arc discharge method which can be operated either by alternating or direct current (AC/DC) is the one by which CNTs were produced by Iijima [35]. Figure 8.6a shows DC arc discharge and Fig. 8.6b shows the AC arc discharge systems.

The methods require very high temperatures (above 1700 °C). The graphite rods used in the arc discharge systems are usually between 6 and 12 mm in diameter and are water-cooled. The chamber is kept at a subatmospheric pressure. Figure 8.6a shows the DC arc system to produce CNTs. DC provides higher yields of CNTs over AC arc discharge (Fig. 8.6b), and the CNT is deposited on the cathode.

To stabilize the current during the operation, a constant distance of ~1 mm between the graphite electrodes is maintained [47]. Gram-scale synthesis of MWNTs by arc discharge has been achieved in the presence of helium (He) and $CH_4$ gases at a pressure of 50 Torr. An arc current of 20 A is applied to the anode with a diameter of 6 mm [48, 49]. A graphite rod containing a metal catalyst (Fe, Co, etc.) is used as the anode with a pure graphite cathode to produce SWNTs in the form of soot [50, 51]. It was found that the presence of hydrogen ($H_2$) gas in the growth region gives the optimum synthesis of MWNTs with high crystallinity (having regular graphene sheets at an interlayer spacing of 0.34 nm) and few coexisting carbon nanoparticles [52–59]. It was also found that the presence of $H_2$ gas inside the chamber produces little or no carbon smoke compared to the presence of helium (He) gas inside the chamber.

Jiang et al. [60] studied the influence of $NH_3$ atmosphere on the arc discharge growth of CNTs and demonstrated that the arc discharge method in $NH_3$ atmosphere is highly efficient for CNTs preparation. Besides arc discharge, single-pulse arc discharge has been successfully applied in the production of near-vertically oriented MWCNTs on Ni/glass using a graphite counter-electrode in an ambient air [61, 62].

The growth of SWCNTs can be done by arc discharge method with or without the use of catalyst precursors. On the other hand, the MWCNTs are produced without a catalyst precursor. In SWCNTs, the anode is made of graphite and a metal catalyst such as nickel (Ni), cobalt (Co), palladium (Pd), titanium (Ti), or platinum (Pt), or the mixtures of Co–Ni, Fe–Ni, Ni–Ti, etc. is used. The metal catalyst plays an important role in the process yield. The gap distance between the electrodes is held constant to ensure high efficiency. The production of CNTs is mostly affected by the presence of impurities. Calcination in air at 400 °C for 2 h is

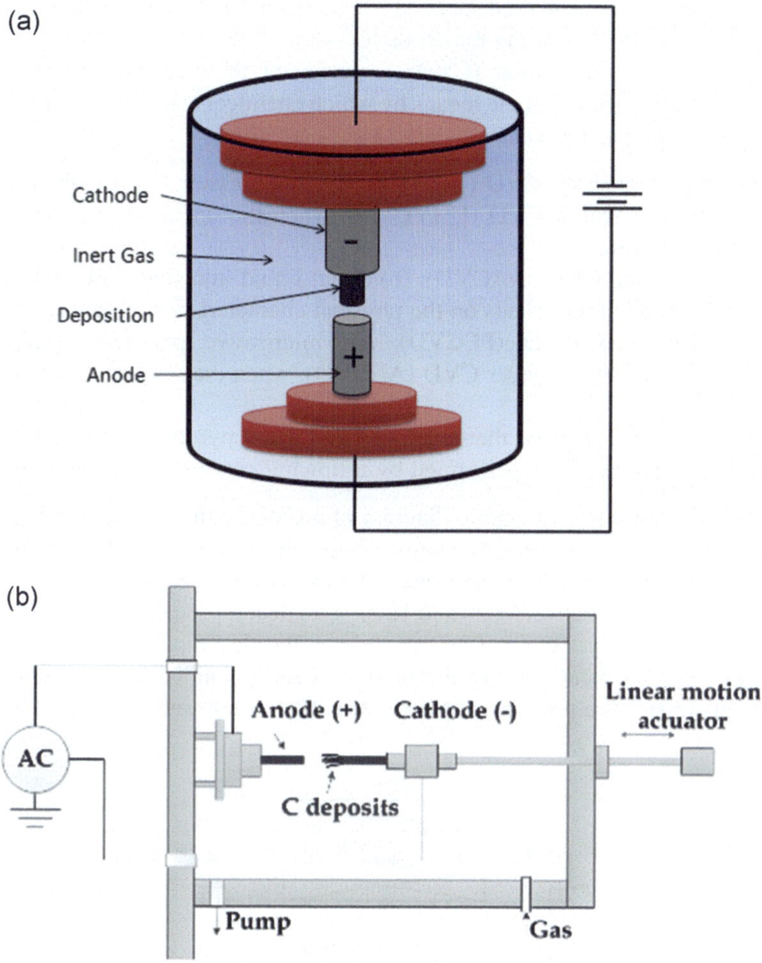


**Fig. 8.6** (**a**) DC arc discharge apparatus for CNT growth (Courtesy: WIKI); (**b**) AC arc discharge systems (Reproduced with permission; J.P. Gore)

very effective in the removal of carbon nanoparticles deposited during DC arc discharge. Heating by IR irradiation in air at 500 °C is also being used to remove the unwanted carbon nanoparticles.

### *8.4.2 Chemical Vapor Deposition of CNTs*

Chemical vapor deposition (CVD) process is a versatile process in which gas-phase molecules are decomposed to reactive species, leading to film or particle growth

[63]. This process has been used successfully to deposit a wide range of conducting, semiconducting, and insulating materials [64–66].

Chemical vapor deposition (CVD) is practiced in different formats. These formats generally differed in the means by which chemical reactions are performed. They are classified as follows:

(a) Atmospheric pressure CVD (APCVD), (b) low-pressure CVD (LPCVD), and (c) ultrahigh vacuum CVD (UHVCVD), when the operation depends on the chamber pressure.
(b) Aerosol-assisted CVD (AACVD), (b) direct liquid injection CVD (DLICVD), when the operation depends on the physical characteristics of vapor.
(c) Plasma-enhanced CVD (PECVD), (b) microwave plasma-assisted CVD (MPCVD), (c) atomic layer CVD (ALCVD), when the operation is controlled by plasmas.
(d) Hot filament CVD, rapid thermal CVD, and flame pyrolysis CVD in open air. All these operations are performed by following combustion technology.

Figure 8.7 shows the schematic diagram of a CVD setup used for CNT growth. The method involves passing a hydrocarbon gas ($CH_n$) (typically 15–60 min) through a tubular reactor in the presence of a catalyst material. The tube is heated to a high temperature between 600 and 1200 °C to decompose the hydrocarbon. The function of the catalyst in the CVD process is the decomposition of the carbon source by applying heat. During the process, heat is applied to the walls of the reaction chamber. The process is sometimes called as hot wall CVD (Fig. 8.8).

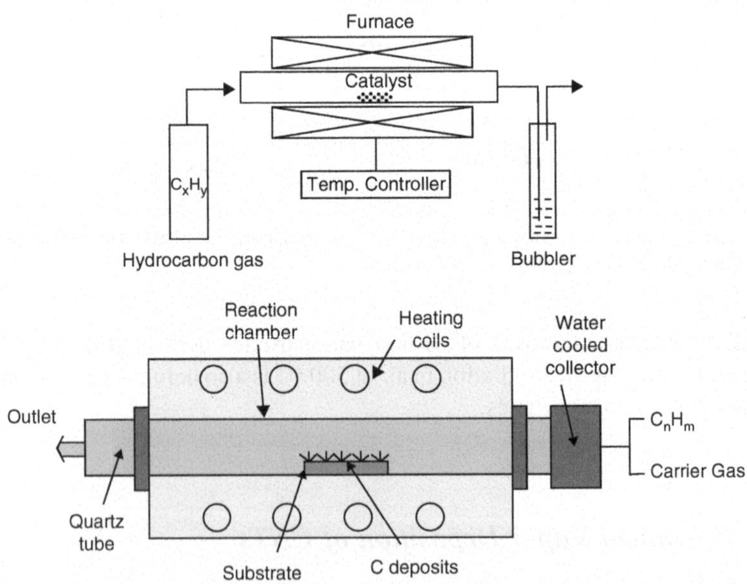


**Fig. 8.7** Schematic diagram of a CVD setup used for CNT growth (Reproduced with permission J.P. Gore)

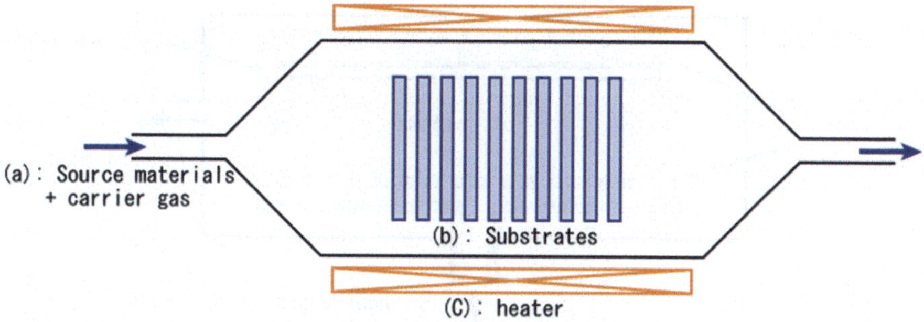


**Fig. 8.8** Schematic of a hot wall CVD (Courtesy: S. Kei, WIKI)

The commonly used substrates are nickel (Ni), silicon (Si), silicon dioxide ($SiO_2$), copper (Cu), or an alloy of copper (Cu/Ti/Si). The choice of the catalyst is one of the most important parameters affecting the growth of CNTs. In catalytic chemical vapor deposition (CCVD) method, which may be thermal or plasma-assisted, the catalysts are prepared by the combustion route using either urea or citric acid as fuel [67]. Acetylene and benzene have also been used by some researchers as catalysts [68, 69].

Most important catalysts are transition metal alloys. Water-stimulated catalytic activity had been applied to grow massive superdense and vertically aligned nanotube forests with height up to 2.5 mm. The water-assisted synthesis method had been reported to address many critical problems that plague carbon nanotubes [68].

The plasma-enhanced CVD (PECVD) is a low-temperature operation and can be also used in several different modes (radio frequency (RF), direct current (DC), diffusion (D), or microwave (MW)). The major difference between RF and DC mode PECVD is the higher concentration of reactive radicles in the former [70, 71].

Figure 8.9 shows the schematic of a plasma-enhanced CVD deposition system. Chemical reactions are involved in the process, which occur after the creation of plasma of the reacting gas either by radio frequency (RF) or DC discharge between two electrodes. Processing plasmas are typically operated at pressure of a few millimeters to a few torr, although arc discharges and inductive plasmas can be ignited at atmospheric pressure. Figure 8.10 shows DC plasma (violet) enhanced growth of carbon nanotubes (CNTs).

A high electric field in the sheath generated in plasma enables self-standing and vertically aligned growth of carbon nanotubes (CNTs). Power sources of direct current (DC) with or without hot filaments [72, 73], microwave [74], and radio frequency (RF) [75, 76] have been used for generating plasma in vertically aligned growth of CNTs by PECVD.

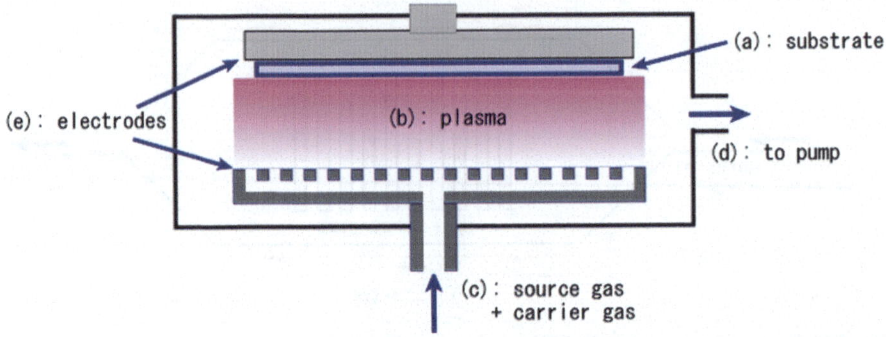


**Fig. 8.9** Schematic of a plasma-enhanced CVD deposition system (Courtesy: S. Kei, WIKI)

**Fig. 8.10** DC plasma (*violet*) enhanced growth of carbon nanotubes (Courtesy: Polyparadigm, WIKI)

### *8.4.3 Laser Ablation*

CNTs have also been deposited by laser ablation (Fig. 8.11). The pulsed laser deposition process is strongly dependent on many parameters such as energy fluence, peak power, continuous wave (cw) versus pulse, repetition rate and oscillation wavelength, the chemical composition of the target, the chamber pressure

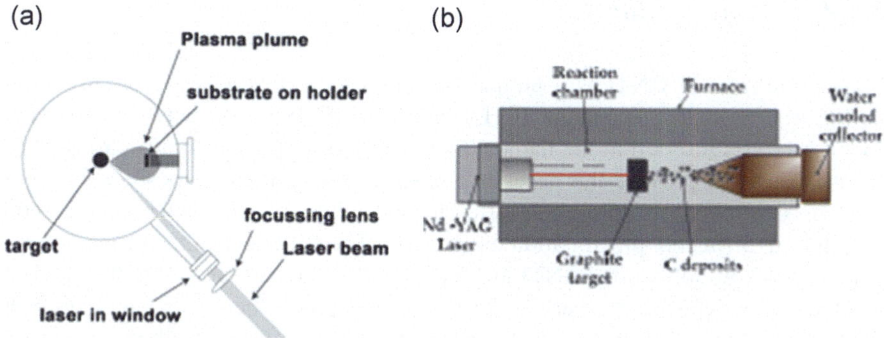


**Fig. 8.11** Schematic of a laser deposition method for CNT (Courtesy: WIKI); (**b**) Nd-YAG laser used to produce a CNT (Reproduced with permission: J. P. Gore)

and flow pressure of the buffer gas, substrate, the ambient temperature, and the distance between the target and the substrate.

When energy from the laser pulse is absorbed by the target, it is first converted to electronic excitation and then into thermal, chemical, and mechanical energy resulting in evaporation, ablation, plasma formation, and even exfoliation. Though the basic setup is simple relative to many other vapor deposition techniques, the physical phenomena of laser target interaction and film growth are quite complex. As for example, the removal of atoms from the bulk material is done by vaporization of the bulk surface region in a state of equilibrium. In this the incident laser beam penetrates into the surface of the material. The penetration depth depends upon several factors as described earlier. The strong electric field generated by the laser pulse is sufficiently strong enough to remove electrons within 10 ps (picosecond) of an ns (nanosecond) laser pulse. The free electrons oscillate within the electromagnetic field (EMF) and can collide with the atoms of the bulk material, thus transferring some of their energy to the lattice of the target material within the surface region. This generates heat within the surface region of the target, and the material is vaporized [77, 78].

Laser ablation, as a crucial step of pulse laser deposition (PLD), is one of the superior methods to grow SWCNTs [79]. Almost all the lasers used for the ablation have been Nd-YAG (Fig. 8.11) [80] and $CO_2$, without applying additional heat to the target. In the year 2010, Lebel et al. synthesized SWCNTs using ultraviolet (UV) laser ablation of graphite target appropriately doped with Co/Ni metal catalyst [81]. It has been argued that UV laser is superior in photochemical ablation to the infrared (IR) laser.

### 8.4.4 Flame Synthesis

Flame synthesis of CNTs is a unique process which uses a fuel generally from hydrocarbons such as methane ($CH_4$), ethylene ($C_2H_4$), and acetylene ($C_2H_2$). Metal catalysts inserted in the flame either in the form of a substrate coating or as aerosol particles provide the necessary reaction sites for the deposition of solid carbon. Nanotubes are believed to form on the catalyst particle via the same carbon dissolution–diffusion–precipitation mechanism. The structures of the grown single-walled (SW) or multiwalled CNTs are dependent on the particle size of the catalyst and the rate of deposition of carbon. Post-flame gas-phase chemistry, temperature at the surface of the catalyst particle, the structure, and the type of catalyst particle are the key controlling factors for the growth of CNTs [82].

Vander Wall et al. carried out a comprehensive study of multiwalled (MW) CNT synthesis in premixed flames. They also studied the effect of catalysts in aerosol form on the growth of CNTs [83]. Primarily, SWCNTs are grown on aerosol catalyst particles using acetylene air flame. From the experimental studies, it appears that there is a need for appropriate gas-phase precursor and catalyst particle size for SWCNT growth. It also appeared that there should be a minimum limit for the presence of CO and $H_2$ and maximum limit for $H_2O$ presence for the production of SWCNTs.

Figure 8.12 shows the schematic of flame synthesis of CNTs. The figure shows premixed flat-flame McKenna burner which is used with stainless steel (SS) chimney for cooling. At the top of the chimney, a circular molybdenum ring held the mesh (SS)-supported catalyst within the post-flame gases. The post-flame temperature is recorded around 1100 K. SS meshes are kept in the flame gases for 10–15 min. Coated meshes show a uniform and dense growth of CNT formation.

The growth mechanism for SWCNTs has been found to be very similar to the mechanism for other forms of solid carbon soot. The precursors for soot are polyaromatic hydrocarbons (PAH) that are formed through the breakdown of $C_2H_2$. However, presence of high concentration of $C_2H_2$ causes the catalyst particle

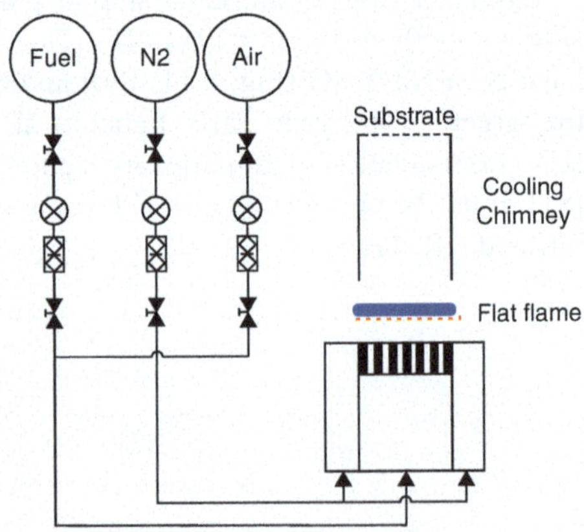


**Fig. 8.12** Schematic of flame synthesis of CNTs (Reproduced with permission: J.P. Gore)

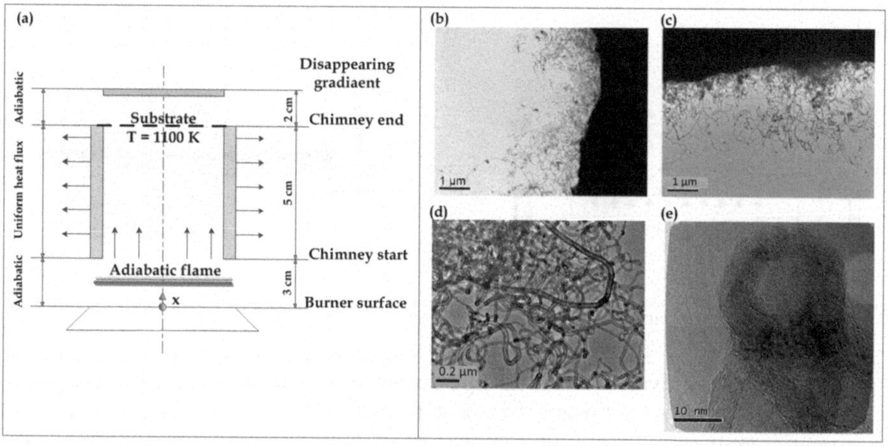


**Fig. 8.13** (**a**) Experimental setup; (**b**) and (**c**) TEM of nanotube growth; (**d**) HRTEM of MWCNTs and HRTEM of closed MWCNT (Reproduced with permission: J.P. Gore)

to be coated with amorphous carbon, inhibiting the growth of SWCNT. Gopinath and Gore observed that CNTs with maximum yield and best morphology are produced with ethylene flame for equivalence ratio range of $1.5 < \varphi < 1.6$ (φ being the equivalence ratio between the air flow and the fuel flow rate) [84]. The experimental setup of the work by Gopinath and Gore is shown in Fig. 8.13a. Figure 8.13b and c show TEM images of nanotube growth on cobalt catalyst at $\varphi = 1.55$. Figure 8.13d shows high-resolution transmission microscopy (HRTEM) image of MWCNTs at $\varphi = 1.55$, and Fig. 8.13e shows the HRTEM image of closed MWCNTs.

The growth process can be broken down into two steps: (i) establishment of the flame that acts as the source for gaseous precursors and heat, and (ii) simultaneous formation of catalyst particles and growth of carbon nanotube occurring at a nanoscale. Each of the process requires mathematical modeling.

### *8.4.5 Pyrolysis*

The process comes under the category of catalytic CVD or (CCVD), which is based on the pyrolysis of mixed liquid aerosols composed of both liquid hydrocarbon and a catalyst precursor. The process is called liquid-phase pyrolysis [85, 86]. On the other hand, when the process contains a solid compound of nitrogen-rich iron salts and a precursor, the process is called solid-state pyrolysis [87].

Figure 8.14 shows a schematic of an ultrasonic spray pyrolysis. Figure 8.14a shows the schematic of the furnace, and Fig. 8.14b shows the ultrasonic device with piezoelectric transducer. Pyrolysis is a thermochemical decomposition of organic materials at elevated temperatures in the absence of oxygen. An aerosol of a base

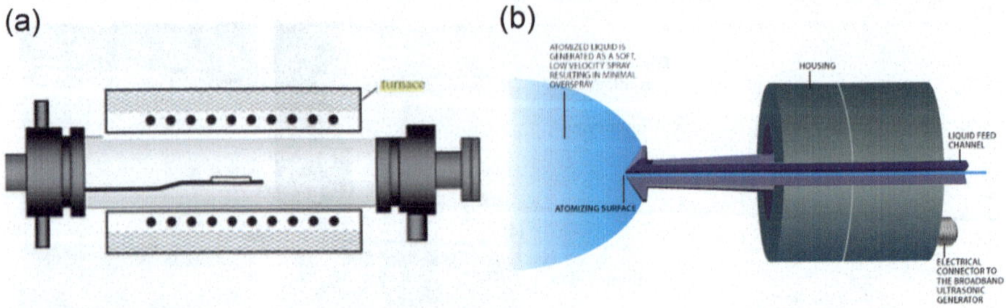


**Fig. 8.14** Schematic of an ultrasonic spray pyrolysis

solution (usually sprayed by an ultrasonic dispenser) is carried with a flow of carrier gas through the hot zone furnace. A special ultrasonic nozzle is used, and a high-frequency sound wave produced by piezoelectric transducers acts upon the nozzle tip that creates capillary waves in the liquid film. Khatri et al. reported that they had used ethanol and bimetallic catalyst of cobalt (Co) and molybdenum (Mo) acetates to produce SWCNTs using ultrasonic spray pyrolysis. They also had used zeolite powder as a catalyst-supporting material for the production of SWCNTs. Camarena and colleagues reported production of multiwalled CNTs (MWCNTs) by spray pyrolysis using toluene as the carbon source and ferrocene as the catalyst [88].

Kacukayan and colleagues synthesized MWCNTs through pyrolysis of sulfuric acid carbonized by product of sucrose. In the year 2009, Kuang et al. reported the synthesis of CNTs in large scale by solid-state pyrolysis of two mixed metal phthalocyanines at 800–950 °C. The as-synthesized CNTs were 15–35 nm in diameter and 200–800 nm in length [62, 89, 90].

Among all the techniques so far discussed, chemical vapor deposition has been the most populous and widely used method for its low-cost high production yield and ease of scale-up. As compared to other methods, CVD requires low temperature and ambient pressure. The CVD-grown CNTs are inferior in crystallinity, compared to arc MWCNTs. CVD process is a well-known method in semiconductor industry, and several research groups have been successful in implementing this process for mass production of CNTs [91]. Kumar and Ando used an ordinary CVD reactor and a desired amount of vaporized camphor ($C_{10}H_{16}O$) at 150–170 °C and Fe–Co impregnated zeolite support at 650 °C for 30 min. at atmospheric pressure to produce CNTs [92].

### *8.4.6 Floating Catalyst CVD*

Large-scale production of CNTs by floating catalyst CVD method (FCCVD) has also been reported by M. Endo et al. and H.M. Cheng et al. [93, 94]. A schematic of the FCCVD method is presented in Fig. 8.15.

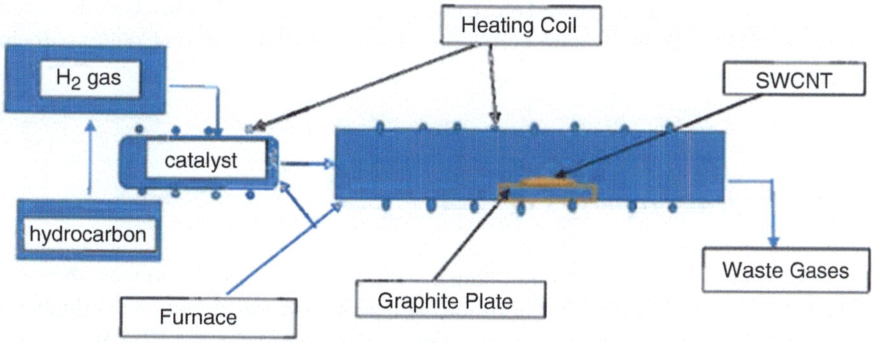


**Fig. 8.15** Schematic of the floating catalyst chemical vapor deposition setup

In this method, benzene or methane has been used as the carbon source (hydrocarbon), hydrogen has been used as the carrier gas, and ferrocene as the catalyst precursor. Ferrocene is vaporized and carried into the reaction chamber along with benzene and thiophene vapor. The ferrocene vapor is reduced by hydrogen into atomic iron, which catalyzes the growth of SWCNTs and carbon nanofibers (CNFs).

## 8.5 Properties of CNTs

### *8.5.1 Superconductor*

Multiwalled carbon nanotubes (MWCNTs) showed superconductivity with a relatively high transition temperature [95, 96]. The ballistic transport of electrons in MWCNTs shows that the mean free path is longer than the nanotube itself [97] and the measured resistance under four-probe is less than an ohm. It exhibits a sharp order of magnitude drop to near-zero resistance at 3.8 K. Moreover, the small resistance means that both the outer and the inner tubes must be metallic. The sharpness of the transition supports the true 1D superconductivity in double-walled CNTs (DWCNTs) [98]. Histogram of the transition temperatures observed in different samples exhibit evidence of superconductivity with a wide temperature range, owing to the different inner tube diameters/chiralities. There are three ways to confirm superconductivity in a particular material such as the observation of zero resistivity, observation of Meissner effect, and the observation of the specific heat anomaly. All of these three effects have been observed in DWCNTs [99, 100].

For SWCNTs, the theoretical limit of conductance is $4e^2\hbar^{-1}$, where $e = 1.602 \times 10^{-19}$ and $\hbar = 6.6 \times 10^{-34}$ Nms [101]. Haruyama reported superconductivity in a highly uniformed thin film of boron-doped MWCNT at the transition temperature of 12 K with entirely end-bonded electrode junction [102, 103]. It has been reported that some MWCNTs include no Fe/Co catalyst, and tend to destroy Cooper pairs in the entire region.

### 8.5.2 Optical Properties of CNTs

The optical absorption spectrum of single-walled CNTS (SWCNTs) was observed at 0.68, 1.2, and 1.7 eV due to π-plasmon. The peaks are dependent on the diameter of the SWCNTs and it can be fitted in an empirical relation $\omega \propto (1/D)$, where $\omega$ is the frequency and $D$ is the diameter of the CNT [104, 105]. It was further investigated by S. M. Bachilo et al. that the detailed composition of SWCNTs can be obtained from the optical spectroscopy [106]. They further predicted that optical transition frequencies deviate from simple diameter dependence more strongly than expected. In the year 2003, Botti et al. demonstrated that SWCNTs synthesized without catalyst with a nonconventional laser-assisted procedure and have their own third order nonlinear optical response at 532 nm. They further concluded that second order hyperpolarizability cannot be fitted with a simple empirical formula because of their different lengths and diameters [107, 108].

Raman spectroscopy has been proved to be a powerful tool to analyze the intrinsic properties of CNTs, such as diameter, chirality, and electronic structure. Individual CNTs prepared by different methods show different Raman shift peak positions. As for example, high-pressure carbon monoxide conversion (HiPCO) in SWNTs is formed by high-pressure decomposition of ferrocene (Fe $(CO)_5$. The SWCNT formed by this process is purified by a two-step process of oxidation and successive washing. The material exhibits strong radial breathing modes (RBMs) and low D-band intensity with respect to G-band intensity, implying a low density of defects in SWNTs [109, 110]. The peak intensity depends on the energy of excitation used. It has been reported that the temperature dependence of Raman shift arises mainly due to anharmonic terms in the lattice potential energy [104].

Optical absorption spectra of four kinds of SWCNTs produced by electric arc method using NiY catalyst with different diameters are shown in Fig. 8.16. Background absorption due to π-plasmon has been subtracted. Inset in Fig. 8.16 shows the Raman spectra of the breathing mode (Fig. 8.16) [111].

### 8.5.3 Electrical Properties and Photoconductivity of Carbon Nanotubes (CNTs)

Plasma-enhanced chemical vapor deposited (PECVD) CNTs have shown resistivity ranging from 3.2 × 10–3 Ω-cm to 4.6 × 10–3 Ω-cm, which is the resistivity of graphite parallel to the basal plane and perpendicular to the basal plane [112]. Klaus

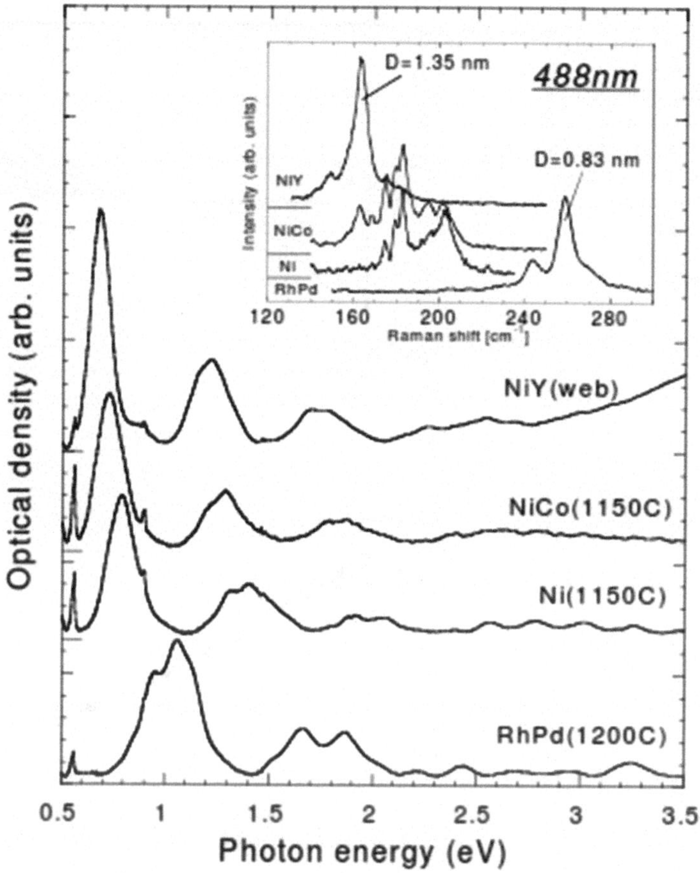


**Fig. 8.16** Optical absorption spectra of four kinds of SWCNTs with different diameters. The lower one presents the spectra for different catalysts used and the sensitivity of the spectra at different temperatures (Reproduced with permission, H. Kataura et al., Elsevier)

Kern's team reported that the PECVD-grown CNTs show I–V characteristics at room temperature as linear when deposited on silicon dioxide ($SiO_2$) grown on a single crystal silicon (Si). The two-probe resistance measurements of the CNT vary between 100 MΩ and 1 GΩ. In general, the resistivity of carbon nanofibers can be fitted into an empirical relation as

$$\rho\,(\theta) = \rho_a \sin^2\theta + \rho_c \cos^2\theta \tag{8.2}$$

where $\rho_a = 4 \times 10^{-5}$ Ω-cm and $\rho_c = 4 \times 10^{-2}$ Ω-cm are the resistivities of graphite parallel and perpendicular to the basal planes, respectively, and $\theta$ is the angle between the graphitic planes with respect to the nanofiber axis [113, 114].

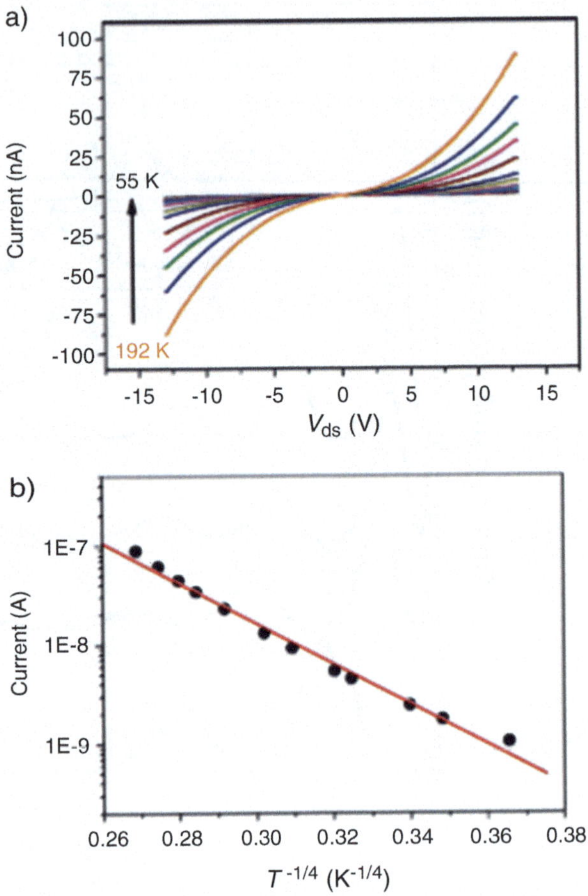


**Fig. 8.17** (**a**) Current–voltage characteristics of a stacked graphene CNT device taken as a function of temperature (between 192 K and 55 K). (**b**) Linear fit of logarithm of the drain current as a function of $T^{-1/4}$ (Reproduced with permission: Wiley-VCH Verlag GmbH)

Temperature-dependent I–V curve fitting suggests that the possible mechanism of charge transport is by variable range hopping (VRH) (drain current as a function of $T^{-1/4}$). Figure 8.17a shows the current–voltage characteristics of a stacked graphene CNT device taken as a function of temperature (between 192 and 55 K). Figure 8.17b is the linear fit of logarithm of the drain current as a function of $T^{-1/4}$ [113, 115, 116]. Figure 8.18a shows the scanning photocurrent microscopy (SPCM) image of a stacked graphene carbon nanotube (SG-CNT) device taken under ambient conditions with $Vds = 5$ V, whereas Fig. 8.17b shows the line profile of the photocurrent response taken from the source to drain contacts.

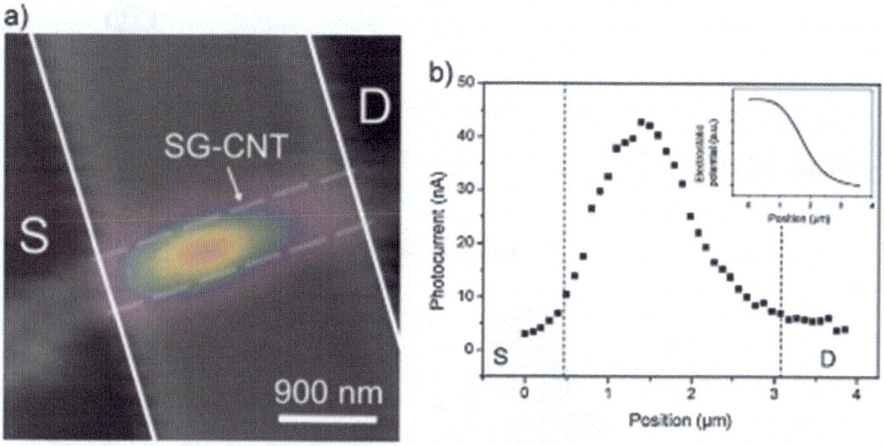


**Fig. 8.18** (**a**) Scanning photocurrent microscopy (SPCM) image of a SG-CNT device taken under ambient conditions with *V*ds = 5 V. (**b**) Line profile of the photocurrent response taken from the source to drain contacts. (Reproduced with permission from Wiley-VCH Verlag GmbH)

## 8.6 Uses of CNTs

### *8.6.1 Medical Science*

R.J. Chen et al. reported the discovery of photo-induced molecular desorption phenomena in SWCNTs. These phenomena were found to be generic to various molecules of the SWCNTs. They further concluded that photodesorption depends strongly on the wavelength of light. The discovery leads to the construction of several miniature electrochemical biosensors and optoelectronic devices [117].

Recent studies demonstrate that CNT can enhance the electrochemical reactivity of important biomolecules [118, 119] and can promote the electron transfer reactions of proteins (including those where the redox center is embedded deep within the glycoprotein shell) [120–122].

The diagnosis and management of diabetes mellitus require a tight monitoring of blood glucose levels. Electrochemical biosensors for glucose play an important role in this direction. The dramatic decrease in the overpotential of hydrogen peroxide ($H_2O_2$) as well as the direct electron transfer of glucose oxidase (GOx) observed at CNT-modified electrodes indicate great promise for the biosensing of glucose. Willner's group demonstrated that aligned GOx on the edge of SWCNT can be linked to an electrode surface [123]. Figure 8.19 represents an assembly of the SWCNT electrically connected glucose oxidase (GOx) electrode [123].

Patlsky et al. first oxidized SWCNTs and covalently coupled with 2-thioethanol/ cystamine mixed monolayer (3:1 ratio) assembled on a gold (Au) electrode. The incorporation of 2-thioethanol in the mixed monolayer was anticipated to prevent nonspecific adsorption of the surfactant-protected SWCNTs onto electrode surface.

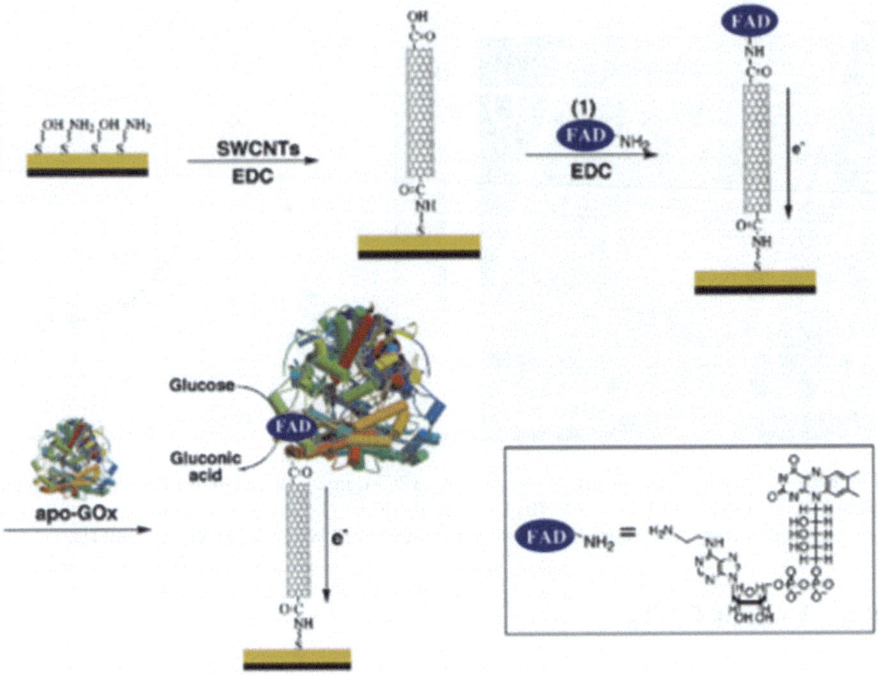


**Fig. 8.19** Representation of an assembly of the SWCNT electrically connected glucose oxidase (GOx) electrode. EDC =1-ethyl-3 (3-dimethyl aminopropyl) carbodiimide hydrochloride, *FAD* Flavin adenine dinucleotide (Reproduced with permission from Wiley-VCH Verlag GmbH & Co.)

The amino-derivative of flavin adenine dinucleotide (FAD) cofactor was then coupled to the carboxylic acid groups at the free top ends of the aligned SWCNTs. Finally, GOx was reconstituted on the FAD unit linked to the ends of the aligned SWCNTs for glucose sensing.

Klaus and his team also reported the photocurrent generation mechanism in the SG-CNTs. Figure 8.18a displays a photocurrent image of an individual contacted SG-CNT acquired with $Vds = 5$ V ($\lambda = 633$ nm and $P = 100$ kW cm$^{-2}$). The image reveals a nonhomogeneous signal distribution along the CNT, with a photocurrent maximum located in the vicinity of the source contact and an asymmetric, exponential decay toward the electrical contacts (Fig. 8.18b) [124].

There are several methods to integrate carbon nanotubes (CNTs) to different gas sensor structures. Li et al. developed a resistive gas sensor by simply casting SWCNTs on interdigitated electrodes (IDEs) [114]. T. Someya et al. measured the conductance of single-walled semiconducting carbon nanotube (CNT) in field effect transistor (FET) geometry in the presence of alcohol vapors and detected significant change in the drain current. This idea led to the construction of a vapor sensor using single-walled carbon nanotubes (SWCNTs) [125, 126]. Figure 8.20 shows the schematic of a SWCNT FET structure that can be used as gas sensor.

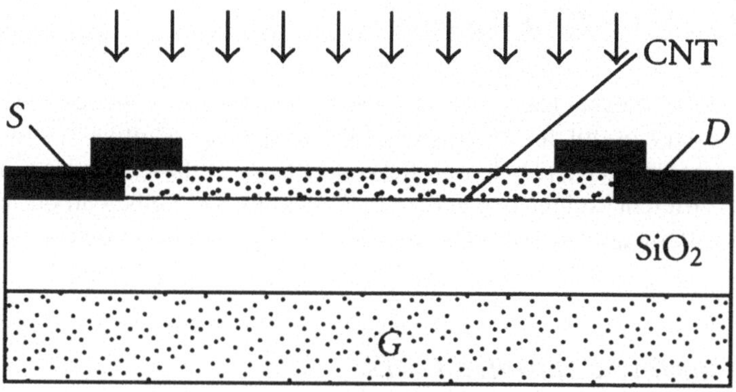


**Fig. 8.20** Schematic of the SWCNT-based FET sensor. $G$ = gate, $D$ = drain, and $S$ = source of the FET

J. Cong et al. found that nanotubes can be used for miniature chemical sensors to detect small concentrations of gas molecules with high sensitivity [127]. Figure 8.20 shows the schematic of the SWCNT-based FET sensor ($G$ = gate, $D$ = drain, and $S$ = source of the FET).

Khamis et al. demonstrated that DNA decorated carbon nanotube (DNA-NT)-based FETs as ultrasensitive chemical sensors. They further demonstrated that DNA-NT vapor sensors readily discriminate between series of chemical homologs that differ by single methyl groups, and the devices also discriminate among structural isomers and optical isomers, a trait common in biological olfactory systems [128].

Nanotube network FETs (NTNFETs) also have been used as selective detectors of DNA immobilization and hybridization. NTNFETs with immobilized synthetic oligonucleotides have been shown to specifically recognize target DNA sequences, including H63D single-nucleotide polymorphism (SNP) discrimination in the *HFE* gene, responsible for hereditary hemochromatosis [129].

Bhirde et al. reported CNT-based drug delivery for cancer therapy and demonstrated efficacy superior to nontargeted bioconjugates. First-line anticancer agent cisplatin and epidermal growth factor (EGF) are attached to SWCNT to specifically target squamous cancer, and the nontargeted control is SWCNT–cisplatin without EGF. Initial in vitro imaging studies with head and neck squamous carcinoma cells (HNSCC) overexpressing EGF receptor (EGFR) using Q-dot luminescence and confocal microscopy show that SWCNT–Q-dot–EGF bioconjugates internalized rapidly into cancer cells [130–132].

### 8.6.2 Use of CNTs in Batteries

The outstanding mechanical properties and high surface to volume ratio of CNTs make them very useful for electrodes [133, 134] or as additives in lithium-ion battery system. In lead acid batteries, the conductivity of the electrodes is increased by adding different weight percentage of CNTs. In fuel cells, carbon nanotubes decorated with metal nanoparticles have doubled the fuel performance [135, 136].

### 8.6.3 Chirality-Controlled CNTs

Chirality is a geometric property of some molecules and ions. However, in SWCNT, the structure of CNT is defined by a pair of integers known as chiral indices (*n*,*m*), in other words, *chirality*. Chirality defines the optical and electrical properties of CNTs, and controlling is a key to exploiting their practical applications [137]. Although significant progress has been made in chirality separation of SWCNTs by a number of post-synthesis approaches, such as density gradient configuration, DNA wrapping chromatography, and multicolumn gel chromatography, direct selective growth of SWCNTs with a narrow chirality distribution is still regarded as an irreplaceable approach to achieve the goal of deteriorating the pristine structure and modifying the electronic properties of SWCNTs [138]. - Chirality-controlled networks of SWCNTs can be processed from solution and have excellent mechanical properties [139, 140]. They are highly flexible and stretchable. Depending on the types of nanotubes (semiconductor or metallic), they can be used as replacements of metal or transparent conductive oxide electrodes or as semiconducting layers for field-effect transistors (FETs) with high carrier mobilities. They are thus competitive alternatives to other solution-possible materials for flexible and printed electronics [141].

## 8.7 Summary

Carbon nanotubes (CNTs) are allotropes of carbon and one of the most interesting family members of carbon owing to their physical, chemical, mechanical, optical, and electronic properties. As a result, CNTs have attracted a good deal of attention from scientists and engineers of different disciplines. CNTs are categorized as single-walled (SWCNTs) or multiple-walled nanotubes (MWCNTs). Most SWCNTs have a diameter of approximately 1 nm, with a tube length that can be many millions of times longer.

Synthesis of CNT can be performed by arc evaporation or arc discharge method, either by alternating or direct current (AC or DC) or by chemical vapor deposition (CVD) process (it can be operated at atmospheric, low pressure, or at ultrahigh vacuum) in which gas-phase molecules are decomposed to reactive species, leading to film or particle growth. It can also be synthesized by pulsed laser deposition process. Flame synthesis of CNTs is also a unique process which uses a fuel, generally hydrocarbons such as methane ($CH_4$), ethylene ($C_2H_4$), or acetylene ($C_2H_2$).

Multiwalled carbon nanotubes (MWCNTs) have showed superconductivity with a relatively high transition temperature. The ballistic transport of electrons in MWCNTs show that the mean free path is longer than the nanotube itself and that the measured resistance under four-probe is less than an ohm. It exhibits a sharp drop to near-zero resistance at 3.8 K.

The optical absorption spectrum of single-walled CNTS (SWCNTs) was observed at 0.68, 1.2, and 1.7 eV due to π-plasmon. The peaks are dependent on the diameter of the SWCNTs, and it can be fitted in an empirical relation $\omega \propto (1/D)$, where ω is the frequency and D is the diameter of the CNT.

Raman spectroscopy has been proved to be a powerful tool to analyze the intrinsic properties of CNTs, such as diameter, chirality, and electronic structure. The reported discovery of photo-induced molecular desorption phenomena in SWCNTs was found to be generic to various molecules of the SWCNTs. Recent studies demonstrate that CNT can enhance the electrochemical reactivity of important biomolecules and can promote the electron transfer reactions of proteins (including those where the redox center is embedded deep within the glycoprotein shell).

The outstanding mechanical properties and high surface to volume ratio of CNTs make them very useful for electrodes or as additives in lithium-ion battery system. This is not all. CNTs have also been used as DNA-decorated carbon nanotube (DNA-NT)-based FETs, as ultrasensitive chemical sensors besides its use as a drug delivery agent for cancer therapy where it has demonstrated efficacy superior to nontargeted bioconjugates. First-line anticancer agent cisplatin and epidermal growth factor (EGF) are attached to SWCNTs to specifically target squamous cancer.

## References


1. C. Gerber, H.P. Lang, How the doors to the nanoworld opened. Nat. Technol. **1**(1), 3–5 (2006)
2. R.A. Vega, C.K.F. Shen, D. Maspoch, J.G. Robach, R.A. Lamb, C.A. Mirkin, Monitoring single cell infectivity from virus-particle nanoarrays fabricated by parallel dip-pen nanolithography. Small **3**, 1482 (2007)
3. I. Bulbaugh, V. Blog, Engineered virus builds carbon nanotubes, Real Clear Sci. (2011)
4. Y.J. Lee et al., Fabricating genetically engineered high Poer lithium ion batteries using mutiple virus genes. Science **324**(5930), 1050 (2009)
5. X. Dang et al., Researchers harness viruses to build solar cells. Nat. Technol. **6**, 377 (2007)
6. J. Xiao, X. Pan, S. Guo, P. Ran, X. Bao, Towards fundamentals of confined catalysis in carbon nanotubes. J. Am. Chem. Soc. **137**(1), 477 (2015)

7. D. K. Yee, C. Georgia (eds.), *Nanobiomaterials: Developments and Applications* (CRC Press, Boca Raton, 2013)
8. S. Kang, M. Pinault, L.D. Pfefferle, M. Elimelech, Single walled VCarbonnanotubes exhibit strong antimicrobial activity. Langmuir **23**, 8670 (2007)
9. M.S. Islam, Nanoscale materials and devices for future communication networks. IEEE Comm. Mag. **48**(96), 112 (2010)
10. J. Liu, Y. Lu, A calorimetric lead biosensor using DNA zyme directed assembly of gold nano particles. J. Am. Chem. Soc. **125**, 6642 (2003)
11. S. Das et al., Designs for ultra-tiny special purpose nanoelectronic circuits. IEEE Trans. On Circuits and Syst. **54**(11), 2528 (2007)
12. R. Boyle, 7 amazing ways nanotechnology is changing the world, Science, (2012)
13. G.L. Hornyak, H.F. Tibals, J. Dutta, J.J. Moore, *Introduction to Nanoscience and Nanotechnology* (CRC Press, Boca Raton, 2008)
14. C. Binns, *Introduction to Nanoscience and Nanotechnology* (Wiley, Oxford, 2010)
15. M.C. Roco, Nanotechnology's future. Sci. Am. **295**(2), 21 (2006.) and also The future of national nanotechnology initiative, NSF, Nov. 7, (2003) and OECD Information Technology Outlook Ch-7, p-264 (2004)
16. M.C. Roco, *National Nanotechnology Initiative – Past, Present, Future, in Handbook on Nanoscience Engineering and Technology*, 2nd edn. (Taylor and Francis, Oxford, 2007), pp. 3.1–3.26
17. J.M. Bonard, H. Kind, T. Stockli, L.O. Nilsson, Field emission from carbon nanotubes: the first five years. Solid State Electron. **45**, 893 (2001)
18. M. Daenen, R.D. de Fouw, B. Hamers, P.G.A. Janssen, K. SSchouteden, M.A.J. Veld, *The Wondrous World of Carbon Nanotubes. A Review of Current Carbon Nanotube Technologies* (Endhoven University of Technology, Eindhoven, 2003), p. 93
19. T.W. Odom, J.L. Huang, C.M. Lieber, Single walled carbon nanotubes: from fundamental studies to new device concepts. Ann. N. Y. Acad. Sci. **960**, 203 (2002)
20. E.G. Rakov, Chemistry of carbon nanotubes, in *Carbon Nanomaterials* ed. by Y. Gogotsi (CRC Press, Boca Raton, 2006), p. 78
21. E.N. Ganesh, Single walled and multiple walled carbon nanotubes structure, synthesis, and applications. Int. J. Innov. Technol. Explorating Eng. (IJITEE) **2**(4), 34 (2013)
22. T.W. Odom, H. Jin-Lin, P. Kim, C.M. Lieber, Atomic structure and electronic properties single walled carbon nanotubes. Nature **391**, 62 (1998)
23. R. Matel et al., Ambipolar electrical transport in semiconducting single wall carbon nanotubes. Phys. Rev. Lett. **87**, 256805 (2001)
24. R.J. Chen et al., Molecular photodesorption from single walled carbon nanotubes. Appl. Phys. Lett. **79**, 2258 (2001)
25. Y. Li, W. Kim, Y. Zhang, M. Rolandi, D. Wang, H. Dai, Growth of single walled carbon nanotubes from discrete catalytic nanoparticles of various sizes. J. Phys. Chem. B **105**, 11424 (2001)
26. J. Kong, H.T. Soh, A. Cassell, C.F. Quate, H. Dai, Synthesis of single walled carbon nanotubes on patterned silicon wafers. Nature **395**, 878 (1998)
27. M.S. Dresselhaus, G. Dresseihaus, R. Saito, Physics of carbon nanotubes. Carbon **33**(7), 883 (1995)
28. M.S. Dresselhaus et al., Nanowires and nanotubes. Mater. Sci. Eng. C **23**, 129 (2003)
29. T. Brown, H. Lemay, B. Bursten, *Chemistry, The Central Science*, 8th edn. (Prantice Hall, Upper Saddle River, 2002)
30. R. Saito, G. Dresselhaus, M. Dresselhaus, Electronic structure of double-layer graphene tubes. J. Appl. Phys. **73**(2), 494 (1993)
31. H. Frohlich, Theory of superconductivity state.1. The ground state at the absolute zero temperature. Phys. Ther. Rev. **79**, 845 (1952)
32. J. Bardeen, N. Cooper, J.R. Schrieffer, Theory of superconductivity. Phys. Ther. Rev. **108**(5), 1175 (1957.) and also W. Shi, et al., superconductivity in bundles of double wall carbon

nanotubes, Nature Sci. Repts., -2 Article No. 625, (2012), and M. Ferrier, Appl. Phys. D, **43**, 374003 (2010)
33. J.P. Issi et al., Electronic properties of carbon nanotubes: experimental results. Carbon **33**, 941 (1995)
34. L. Langer et al., Electrical resistance of a carbon nanotube bundle. J. Mat. Res. **9**(4), 927 (1994)
35. S. Iijima, Helical microtubes of graphite carbon. Nature **56**, 354 (1991)
36. W. Kratschmer, L.D. Lamb, F. Fostiropoulos, D. Huffman, Solid C60 in the form of carbon. Nature **354**, 347 (1990)
37. S. Iijima, T. Ichihaschi, Single shell carbon nanotubes, of 1 nm diameter. Nature **603**, 363 (1993)
38. D.S. Bethune et al., Cobalt crystallized growth of carbon nanotubes with single atomic layer walls. Nature **605**, 304 (1993)
39. C. Journet et al., Large scale production of single-walled carbon nanotubes by electric arc technique. Nature **388**, 756 (1997)
40. Z. Shi et al., Mass production of single-walled carbon nanotubes by arc discharge method. Carbon **37**, 1449 (1999)
41. A.V. Krestinin et al., Perspectives of single-walled carbon nanotube production in arc discharge process. Euras. Chem. Tech. J. **5**, 7 (2003)
42. M. Yadasaka, T. Komatsu, T. Ichihashi, S. Lijima, Single wall carbon nanotube formation by laser ablation using double targets of carbon and metal. Chem. Phys. Lett. **278**, 102 (1997)
43. D.T. Colbert, R.E. Smalley, Past, present, and future of fullerene nanotubes: Bucky tubes in *Perspective of Fullerene Nanotechnoloegy*, ed. by E. Osawa (Kluwer, Dordrecht, 2002), pp. 3–10
44. W.K. Maser, A.M. Benito, M.T. Martinez, Production of carbon nanotubes: the light approach, by focused solar radiation. Carbon **40**, 1685 (2002)
45. D. Laplaze et al., Carbon nanotubes: the solar approach. Carbon **36**, 685 (1998)
46. M.S. Dresselhaus, G. Dresselhaus, P. Avouris, *Carbon Nanotubes Synthesis, Structure, Properties, and Applications* (Springer, Berlin, 2000), p. 23
47. T.W. Ebbesen, Seamless cylindrical shells of graphite carbon have novel and electronic properties that suggest new high strength, fibers, submicroscopic test tubes, and perhaps semiconductor materials. Phys. Today **49**, 26 (1996)
48. T.W. Ebbesen, P.M. Ajayan, Large scale synthesis of carbon nanotubes. Nature **358**, 220 (1992)
49. D.T. Colbert, J. Zhan, S.M. Mcclure, P. Nikolaev, Z. Chen, J.H. Hafner, D.W. Owens, P.G. Kotula, C.B. Carter, J.H. Weaver, A.G. Rinzler, R.E. Smalley, Growth and sintering of fullerene nanotubes. Science **266**, 1218 (1994)
50. S. Iijima, T. Ichihashi, Single shell carbon nanotubes, of 1 nm diameter. Nature **363**, 603 (1993)
51. D.S. Bethune, C.H. Kiang, M.S. Devries, G. Gorman, R. Savoy, J. Vazquez, R. Beyers, Cobalt catalyzed growth of carbon nanotubes with single atomic layer. Nature **363**, 605–607 (1993)
52. Y. Ando, S. Iijima, Preparation of carbon nanotubes by arc-discharge evaporation. Jpn. J. Appl. Phys., Part 2 **32**, 107 (1993)
53. Y. Ando, Preparation of carbon nanotubes. Fuller. Sci. Technol. **2**, 173–180 (1994)
54. M. Wang, X.L. Zhao, M. Ohkohchi, Y. Ando, Carbon nanotube grown on the surface of cathode depositing by arch discharge. Fuller. Sci. Technol. **4**, 1027 (1996)
55. Y. Ando, X. Zhao, M. Ohkohchi, Production of petal like graphite sheets by hydrogen arc-discharge. Carbon **35**, 153 (1997)
56. X. Zhao, M. Ohkohchi, M. Wang, S. Iijima, T. Ichihashi, Y. Ando, Preparation of high grade carbon nanotubesby hydrogen discharge. Carbon **35**, 775 (1997)
57. X.K. Wang, X.W. Lin, V.P. Dravid, J.B. Ketterson, R.P.H. Chang, Carbon nanotube synthesized in a hydrogen arc-discharge. Appl. Phys. Lett. **66**, 2430 (1995)

58. Y. Ando, X.L. Zhao, M. Ohkohchi, Song of purified carbonnanotubes. Jpn. J. Appl. Phys., Part 2 **37**, 61 (1998)
59. Y. Tai, K. Inukai, T. Osaki, M. Tazawa, J. Murakami, S. Tanemura, Y. Ando, Identification of compounds produced through contact arc vaporization of graphite under $CH_4$ ambience. Chem. Phys. Lett. **224**, 118 (1994)
60. Y. Jiang et al., Influence of $NH_3$ atmosphere on the growth and structure of carbon nanotubes synthesized by arc-discharge method. Inorg. Mater. **45**, 1237 (2009)
61. N. Parkansky et al., Single pulse arc production of carbon nanotubes in ambient air. J. Phys. D. Appl. Phys. **37**(19), 2715 (2004)
62. J. Prasek et al., Methods of carbon nanotubes synthesis—Review. J. Mater. Chem. **21**, 15872 (2011.) Royal Soc. Chem Pub
63. T.K. Gupta, *Copper Connect Technology* (Springer, NY, 2009)
64. D. Dokin, M.K. Zuraw, *Principles of Chemical Vapor Deposition* (Kluwer, Norwell, 2003)
65. H.O. Pierson, *Hand Book of Chemical Vapor Deposition*, 2nd edn. (Elsevier, Amsterdam, 1991)
66. M. Hitchman, K. Jensen, *Chemical Vapor Deposition* (Elsevier, Amsterdam, 1993)
67. E. Flahaut, C. Laurent, A. Peigney, Catalytic CVD synthesis of double and tripple walled carbon nanotubes by the control of the catalyst preparation. Carbon **43**, 375 (2005)
68. X. Xiang, L. Zhang, H.I. Hima, F. Li, D.G. Evans, Co-based catalysts from co/Fe/al layered double hydroxides for the preparation of carbon nanotubes. Appl.Clay Sci. **42**, 405 (2009.) and also K. Hata et al., Water assisted highly efficient synthesis of impurity free sinle walled carbon nanotube synthesis, Science **306**(5700), 1362 (2004)
69. S.C. Lyu, B.C. Liu, C.J. Lee, H.K. Kang, C.W. Wang, C.Y. Park, High quality double walled carbon nanotubes produced by catalytic decomposition of benzene. Chem. Mater. **15**, 3951 (2003)
70. S.M. Kim, L. Ganglof, Growth of carbon nanotubes (CNTs) on metallic by diffusion plasma enhanced chemical vapor deposition. Phys. E (Amsterdam, Neth.) **41**, 1763 (2009)
71. H. Wang, J.J. Moore, Different growth mechanisms of vertical carbon nanotubes by r-f or dc plasma enhanced chemical vapor deposition. J. Vac. Sci. Technol., B: Nanometer Struct. Process. Meas. Phenom. **28**, 1081 (2010)
72. M. Chhowala et al., Growth process conditions of vertically aligned carbon nanotubes using plasma enhanced chemical vapor deposition. J. Appl. Phys. **90**, 5308 (2001)
73. Y. Hayashi, T. Negishi, S. Nishino, Growth of well aligned carbon nanotubes on nickel by hot filament assisted dc plasma chemical vapor deposition. J. Vac. Sci. Technol. A **19**, 1796 (2001)
74. Y. Kojima et al., Growth of high quality carbon nanotubes by grid-inserted plasma enhanced chemical vapor deposition for field emitters. Jp. J. Appl. Phys. **44**, 2600 (2005)
75. G. Sato, T. Kato, W. Oohara, R. Hatakeyama, Production and application of reactive plasmas using helicon wave discharge in very low magnetic field. Thin Solid Films **506**, 550 (2006)
76. T. Hirao et al., Formation of vertically aligned carbon nanotubes by dual-RF-plasma chemical vapor deposition. Jp. J. Appl. Phys. **40**, 1631 (2001)
77. J.D. Ferguson, G. Arikan, D.S. Dale, A.R. Woll, J.D. Brock, Measurements of surface diffusivity and coarsening during pulse laser deposition. Phys. Rev. Lett. **103**(25), 256103 (2009)
78. D.B. Chrisey, G.K. Hubler, *Pulse Laser Deposition of Thin Films* (Wiley, NY, 1994)
79. T. Guo, P. Nikolaev, A. Thess, D.T. Colbert, R.E. Smalley, Catalytic growth of single-walled nanotubes by laser vaporization. Chem. Phys. Lett. **243**, 49 (1995)
80. F. Bonaccorso, C. Bongiorno, B. Fazio, P.G. Gucciardi, et al., Pulsed laser deposition of multi-walled carbon nanotubes thin films. Appl. Sci. Surf. **254**, 1260 (2007)
81. L.L. Lebel, B. Aissa, M.A. El Khakani, D. Therriault, Preparation and mechanical characterization of laser ablated single-walled carbon nanotubes/polyurethane nanocomposite microbeams. Compos. Sci. Technol. **70**, 518 (2010)

82. J.P. Gore, A. Sane, Flame synthesis of carbon nanotubes, in *Carbon Nanotubes—Synthesis, Characterization, Application*, ed. by S. Yellampalli (INTECH, Indianapolis, 2011)
83. R.L. Vander Wall, L.J. Hall, G.M. Berger, Optimization of flame synthesis for carbon nanotubes using supported catalyst. J. Phys. Chem. B **106**(51), 13122–13132 (2002)
84. P. Gopinath, J. Gore, Chemical kinetic consideарations for post flame symthesis of carbon nanotubes in pre-mixed flames using a support catalyst. Combust. Flame **151**(3), 542–550 (2007)
85. I. Khatri et al., Synthesis and characterization of carbon nanotubes via ultrasonic spray pyrolysis method on zeolite. Thin Solid Films **518**(23), 6756 (2010)
86. I. Khatri et al., Synthesis of single walled carbon nanotubes by ultrasonic spray pyrolysis method. Diam. Relat. Mater. **18**, 319 (2009)
87. G. Kucukayan et al., An experimental and theoretical examination of the effect of sulfur on the pyrotically grown carbon nanotubes from sucrose based solid state precursors. Carbon **49**, 508 (2011)
88. J.P. Camarena et al., Molecular assembly of multiwall carbon nanotubes with amino crown ether: Synthesis and characterization. J. Nanosci. Nanotechnol. **11**, 5539 (2011)
89. Y.H. Kuang, K.Z. Li, H.J. Li, Z.W. Xu, Y.J. Wang, A solid state hybrid method to Synthsize straight carbon nanotubes by pyrolysis of two metal Phthalocyamines. Chinese J. Inorg. Chem. **26**(6), 951 (2009)
90. J. Prasek et al., Chemical vapor depositions of carbon nanotubes synthesis, Chapter-7, in *Applications of Carbon Nanotubes* ed. by A.K. Mishra (Nova, Hauppauge, 2013)
91. P. Nikolaev et al., Gas phase catalytic growth of single walled carbon nanotubes from carbon monoxide. Chem. Phys. Lett. **313**, 91 (1999)
92. M. Kumar, Y. Ando, Gigas growth of carbon nanotubes. Def. Sci. J. **58**, 496 (2008)
93. M. Endo, T. Hayashi, Y.A. Kim, M. Terrones, M.S. Dresselhaus, Applications of carbon nanotubes in the twenty first century. Phil. Trans. R. Soc. Lon. A **352**, 2223 (2004)
94. H. Cheng, C. Liu, Y.Y. Fan, F. Li, H.T. Cong, L.L. He, M. Liu, Synthesis and hydrogen storage of carbon nano fibers and single-walled carbon nanotubes. Appl. Phys. Lett. **72**, 3282 (1998)
95. J. Haruyama, I. Takesue, N. Kobayashi, et al., Superconductivity in early end-bonded multiwalled carbon nanotubes. Phys. Rev. Lett. **5**, 96 (2006)
96. M. Tinkam, *Introduction to Superconductivty* (McGraw Hill, 1996)
97. E. Yasuda et al., *Carbon Alloys: Novel Concept to Develop Carbon Science and Technology* (Elsevier, NY, 2003)
98. W. Shi et al., Superconductivity in bundles of double-walled carbon nanotubes, nature nanotubes. Sci. Rep. **2**, 625 (2012)
99. P.F. Sullivan, G. Seidel, Staedy state ac-temperature calorimetry. Phys. Ther. Rev. **173**, 679 (1968)
100. Y. Wang, C. Senatore, V. Abacherli, D. Uglietti, R. Flukiger, Specific heat of $Nb_3Sn$ wires. Supercond. Sci. Technol. **19**, 263 (2006)
101. P.L. McEuen, M.S. Fuhrer, H. Park, Single-walled carbon nanotube electronics. IEEE Trans. Nanotechnol. **1**, 78 (2002)
102. J. Haruyama, Superconductivity in carbon nanotubes, Chapter-33, in *Carbon Nanotubes*, ed. by J.M. Marulanda (Rijeka, 2010)
103. I. Takesue, J. Haruyama, N. Kobayashi, S. Chiashi, S. Maruyama, T. Sugai, H. Shinohara, Superconductivity in entirely end-bonded multi-walled carbon nanotubes. Phys. Rev. Lett. **96**, 057001 (2006)
104. R. Saito, M. Fujita, G. Dresselhaus, M.S. Dresselhaus, Electronic structure of chiral graphene tubules. Appl. Phys. Lett. **60**, 1579 (1992)
105. H. Katura et al., Optical properties of single walled carbon nanotubes. Synth. Met. **108**, 2555 (1999)
106. M. Bachilo et al., Structure-assigned optical spectra of single walled carbon nanotubes. Science **298**, 2361 (2002)

107. S. Botti, R. Ciardi, L. de Dominics, L.S. Asilyan, R. Fantoni, T. Marolo, DFWM measurements of third order succeptibility of single-wall carbon nanotubes grown without catalyst. Chem. Phys. Lett. **378**, 117 (2003)
108. R. Saito, G. Dresselhaus, M.S. Dresselhaus, *Physical Properties of Carbon Nanotubes* (Imerial College Press, London, 1998)
109. Y-C. Chen, Raman spectroscopy studies on Single-Walled Carbon Nanotubes (SWNTs) and SWNT/Ag Nanostructure, Ph.D. Thesis, The University of Manchester, Engineering and Physical Science, London, (2008)
110. B. Bendjemll, Electronic and optical properties of the excess purified SWCNTs produced by HIPCO. J. Nanoelectron. mater. **2**, 173 (2009)
111. H. Kataura et al., Optical properties of single-wall carbon nanotube. Synth. Met. **103**, 2555 (1999)
112. M.S. Bell et al., Carbon nanotube deposition br plasma enhanced CVD. Pure Appl. Chem. **76** (6), 1117 (2006)
113. L. Zhang et al., Four-probe charge transport measurements on individual vertically aligned carbon nano-fibers. Appl. Phys. Lett. **84**, 3972 (2004)
114. E.J.H. Lee, L. Zhi, M. Burghard, K. Mullen, K. Kern, Electrical properties and photoconductivity of stacked graphene carbon nanotubes. Adv. Mater. **22**, 1854 (2010)
115. T.K. Gupta, Effect of temperature on electrical conduction of carbon-black filled polyimide. IEEE Trans. On Comp. Hybrids. Manuf. Technol. **12**, 696 (1989)
116. N.F. Mott, Conduction in non-crystalline materials. Phil. Mag. **19**, 835 (1969)
117. R.J. Chen et al., Molecular desorption from single-walled carbon nanotubes. J. Appl. Phys. Lett. **79**, 2258 (2001)
118. Q. Zhao, Z. Gan, Q. Zhuang, Electrochemical sensors based on carbon nanotubes. Electroanalysis **14**, 1609 (2002)
119. M. Musameh, J. Wang, A. Merkoci, Y. Lin, Low potential stable NADH detection at carbon nanotube modified carbon electrodes. Electrochem. Commun. **4**, 743 (2002)
120. J.J. Goodling et al., Protein electrochemistry using aligned carbon nanotube arrays. J. Am. Chem. Soc. **125**, 9006 (2003)
121. X. Yu, D. Chattopadhyay, F. Papadimitrakopoulos, J.F. Rusling, Peroxidase activity of enzymes bound to the ends of single-wall carbonnano tube forest electrodes. Electrochem. Commun. **5**, 408 (2003)
122. J. Wang, Carbon nanotube based electrochemical biosensors: a review. Electroanalysis **17**(1), 7 (2005)
123. F. Patolsky, Y. Weizmann, I. Willner, Long range electrical contacting of redox enzymes by SWCNT connectors. Angew. Chem. Int. Ed. **43**, 2113 (2004)
124. V. Ambegaokar, B.L. Harpin, J.S. Langer, Hopping conductivity in disordered systems. Phys. Ther. Rev. **B4**, 2612 (1971)
125. J. Li et al., Carbon nanotubes sensors for gas and organic vapor detection. Nano Lett. **3**(7), 929 (2003)
126. Y. Wang, T.W. Yeow, A review of carbon nanotubes-based sensors, J. Sens. **2009**, 24, article ID 493904, (2009)
127. T. Someya, J. Small, P. Kim, J.C. Nuckkolls, J.T. Yardley, Alcohol vapor sensors based on single-walled carbon nanotube field effect transistors. Nano Lett. **3**(7), 877 (2003)
128. J. Cong et al., Nanotube molecular wires as chemical sensors. Science **287**, 5453 (2000)
129. S.M. Khamis, R.A. Jones, A.T.C. Johnson, G. Preti, J. Kwak, A. Gelperin, DNA-decorated carbon nanotube (DNA-NT) -based FETs as ultrasensitive chemical sensors: discrimination of homologues, structural isomers, and optical isomers. AIP Adv. **2**(2), 022110 (2012)
130. A. Star et al., Label-free detection of DNA hybridization using carbon nanotube network field effect transistors. Proc. Natl. Acad. Sci. **103**(4), 921 (2005)
131. A.A. Bhirde et al., Targeted killing cancer cells in vivo and in vitro with EGF-directed carbon nanotube-based drug delivery. ACS Nano **3**(2), 307 (2009)

132. N.W.S. Kam et al., Carbon nanotubes as multifunctional biological transporters and near infrared agents for selective cancer destruction. Nat. Acad. Sci. **102**(33), 11600 (2005)
133. B. Zhao et al., A bone mimic based on the self-assembly of hydroxyapatite on chemically functional single-walled carbon nanotubes. Chem. Mater. **17**, 3235 (2005)
134. M. Endo et al., Lithium secondary battery using vapor G carbon fibers as negative electrode and analysis of electrode mechanism by TEM observation. Trans. IEEJ pn.A. **115**, 349 (1995)
135. K. Takasumi et al., A modification in the preparation process of a carbon whisker for anode performance of lithium rechargable batteries. J. Power Sources **54**, 425 (1999)
136. P.J. Britto et al., Improved charge transfer at carbon nanotube electrodes. Adv. Mater. **11**, 154 (1999)
137. L. Childress, R. Hanson, Diamond NV centers for quantum computing and quantum networks. MRS Bul. **38**, 134 (2013)
138. H. Liu et al., Large scale single chirality separation of single wall carbon nanotubes by simple gel chromatography. Nat. Commun. **2**, 309 (2011)
139. R. Rao et al., In situ evidence for chirality dependent growth rates of individual carbon nanotubes. Nat. Mater. **11**, 213 (2012)
140. F. Feng et al., Templated symthesis of single-walled carbon nanotubes. ACC Chem. Res. **49** (4), 606 (2016)
141. J. Zaumseil, Single-walled carbon nanotube networks for flexible and printed electronics. Semicond. Sci. Technol. **30**(7), 074001 (2015)

# Chapter 9
# The Transparent Carbon: the Diamond

## 9.1 Introduction (Diamond)

Carbon is capable of forming many allotropes due its valency. *Diamond* is one of the well-known allotropes of carbon. The hardness and high dispersion of light of diamond make it useful for both industrial applications and jewelry. The Gemological Institute of America (GIA) classifies natural diamonds from D grade (colorless) to Z grade (light yellow). The blue diamonds fall under a different grading scale [1].

The fascinating colors, the transparency, and the physical properties have made diamond the most precious stone than any other gem stones. Diamonds feature more predominantly in the history and cultural heritage of the human race [2]. They were prized for their scarcity for centuries, and still remain a symbol of wealth and love and prestige to this day [3]. In 1971, the seventh spy film in James Bond series by Eon Productions was *Diamonds are forever*.

Indeed, the idea of connecting diamonds to romance was captured in a brilliant ad campaign begun in 1940s, causing demand for diamonds to increase. As a precious gemstone, diamonds are judged on several factors that determine their beauty. The beauty comes from 4Cs, *cut*, *clarity*, *carat*, and *color* (shown in Fig. 9.1). The other unique qualities of diamonds are transparency, luster, and dispersion of light. A diamond which contains hundred% carbons will be completely transparent. The defects in the lattice or presence of any other chemicals as impurities or dopants can turn diamond to blue, red, black, pale green, or violet. The coloration in diamonds is due to interstitial impurities or structural defects [4, 5].

Figure 9.2 shows the advertisement of the picture *Diamonds are forever*, the seventh spy film produced by Eon Productions in 1971. Diamonds have been treasured as gemstones since their use as religious icons in ancient India. Their usage in engraving tools also dates back to early human history. The most familiar



T. Gupta, *Carbon*, DOI 10.1007/978-3-319-66405-7_9

**Fig. 9.1** A well cut diamond crystal that shows *clarity*, luster, and transparency (Materials in your world, WIKI)

**Fig. 9.2** Advertisement of the picture *Diamonds are forever*, the seventh spy film produced by Eon Productions in 1971 (Courtesy: WIKI)

use of diamonds (derived from ancient Greek, *adamas*, "*proper*, unalterable, *unbreakable*, and *untamed*") today is as gemstones used for adornment. In 1772, Antoine Lavoisier proved that the origin of diamond is from carbon [6].

## 9.2 Structure of Diamond

Figure 9.3 shows the picture of cut and shaped diamonds of different sizes (left, up). A diamond, no matter what size, may be considered to be a single molecule of carbon atoms, each joined to four other carbons in regular tetrahedrons, or triangular prisms. A diamond can be cleaved, or split, along its crystal faces into smaller pieces with the sides of the cleavage remaining smooth. This property is very important to the diamond cutter and the jeweler. On the left bottom of Fig. 9.3 is shown the diamond structure. The figure also shows a chunk of graphite (right up) and its structure (right bottom). Both the diamond and the graphite are the allotropes (meaning, one of two or more existing forms of an element) of carbon. Diamond crystallizes in the cubic form, whereas graphite crystallizes in hexagonal form. Carbon can exist in different forms such as chaoite, a mineral believed to have been formed in meteorite impacts, a metallic body centered C8 carbon. Figure 9.4 shows the cubic structure of C8 carbon with eight atoms in the unit cell [7–10].

There is evidence that white dwarf stars have a core of crystallized carbon and oxygen nuclei. A news released from the Harvard Smithsonian Center for

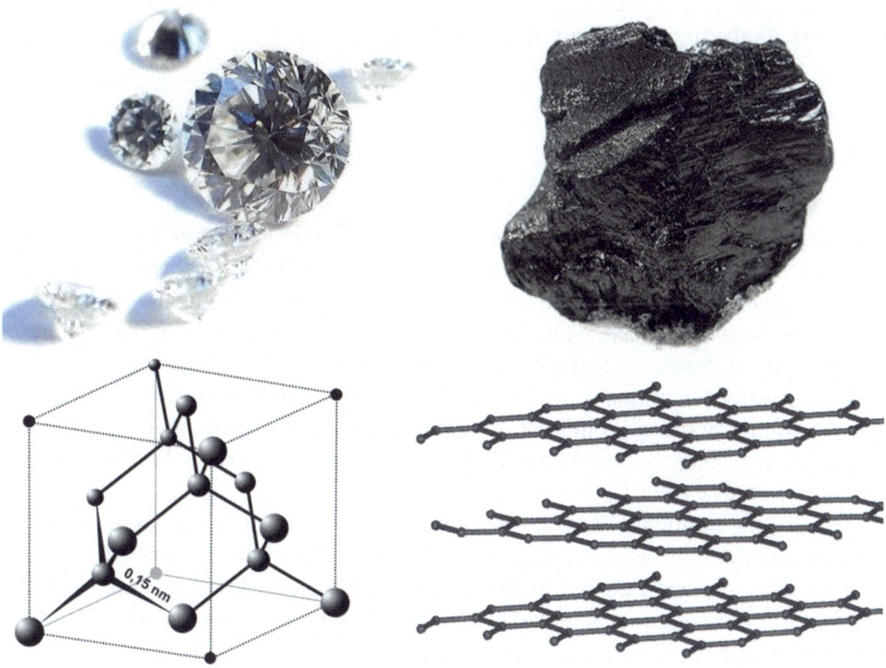


**Fig. 9.3** Picture of cut and shaped diamonds (*left*, *up*) and diamond structure (*left bottom*). The top of the right-hand side shows a chunk of graphite and at the bottom it shows its structure (Courtesy: Materials scientists, WIKI)

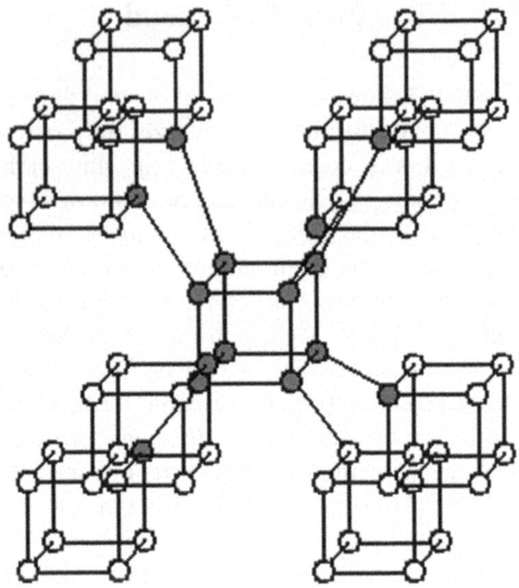

**Fig. 9.4** Cubic structure of C8 carbon (Courtesy: WIKI)

Astrophysics described the 2500-mile wide stellar core as diamond, and it was named as Lucy after the Beatles' song "Lucy in the sky with diamonds." However, it is more likely an exotic form of carbon [11].

## 9.3 History of Diamond

Diamond is the symbol of love. The *love crystal* can come either from mines or from the laboratory where it is prepared synthetically. The price for the symbol of love, the diamond, depends upon whether it is a natural crystal, cut and polished; or synthetic, cut and polished. Most natural diamonds are formed at high temperature and pressure at depths of 140–190 km in Earth's mantle (Fig. 9.5).

The growth of diamond from carbon-containing minerals occurs over periods of 1–3.5 billion years, and brought close to the Earth's surface through deep volcanic eruptions by a magma, which cools into igneous rocks known as kimberlites and lamproites (Fig. 9.6) [12].

Most of the diamonds were formed millions or billions of years ago. Powerful magma eruptions brought the diamonds to the surface creating kimberlite pipes. The magma eventually cooled inside the kimberlite pipes, leaving behind conical veins of kimberlite rock that contains diamond. Kimberlite is named after Kimberley, South Africa, where these pipes were first found. Figure 9.6 shows the formation of diamond 161 km below the surface of Earth [13–15].

Diamonds occur most often as euhedral or rounded octahedral and twinned octahedral known as *macles* (Fig. 9.7). As the structure of diamond has a cubic arrangement of atoms, they have many facets that belong to a cube, octahedron,

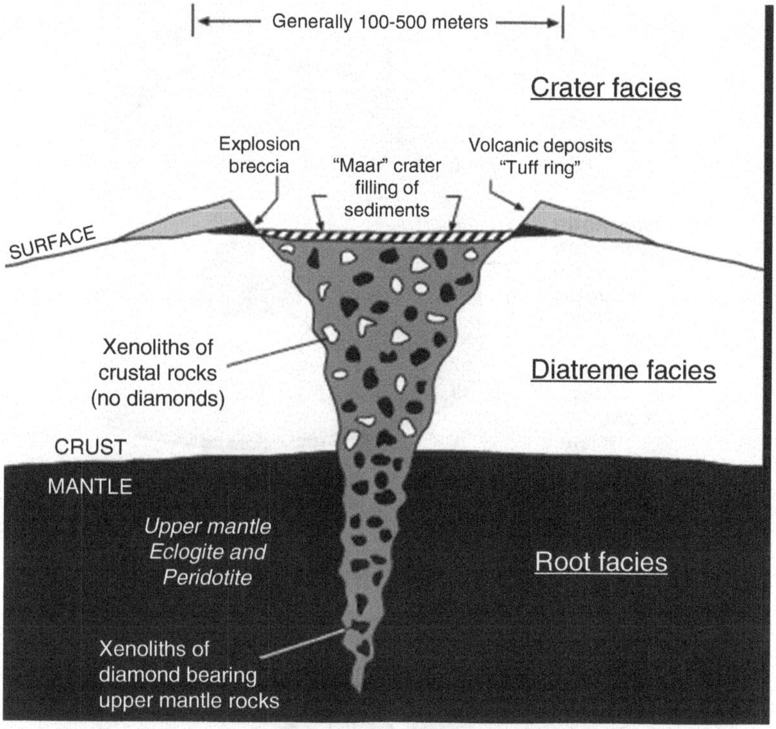


**Fig. 9.5** Formation of diamond inside Earth's surface (Courtesy: Colorado Geological Survey, CO)

rhombicosidodecahedron, tetrakis hexahedron, or disdyakis dodecahedron. The round faces of crystals are found to have a coating or gum-like skin [16]. Figure 9.7 shows a slightly misshapened octahedral shape of diamond crystal inside the ore.

Diamond is the hardest known natural material on the Mohs scale of mineral hardness. The hardness, however, depends on the purity of diamond, crystalline perfection, and orientation. Flawless pure crystal oriented along cubic diamond lattice in <111> direction shows the maximum hardness [13]. Industrial diamonds are valued mostly for their hardness and thermal conductivity, making many of the gemological characteristics of diamonds, such as 4Cs, irrelevant for most applications. The boundary between gem-quality diamonds and industrial diamonds is poorly defined.

Historically, diamonds were found only in alluvial deposits in Guntur and Krishna district of Krishna river delta in Southern India in the ninth century BC [17, 18]. In the United States, diamonds have been found in Arkansas, Colorado, Wyoming, and Montana [19]. Australia boasts the richest diamantiferous pipe, with production from the Argyl diamond mine [20].

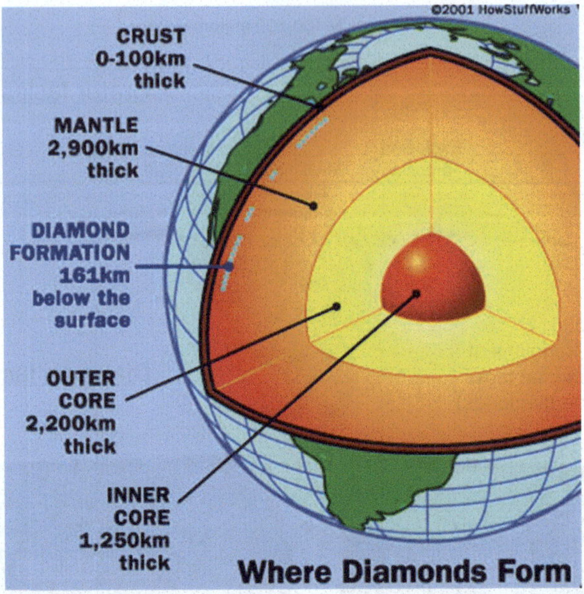


**Fig. 9.6** Formation of diamond 161 km below the surface of Earth (Courtesy: How Stuff Works, 2007)

**Fig. 9.7** Slightly misshapened octahedral shape of diamond crystal inside the ore (Courtesy: Materialscientist WIKI)

According to the report of British Geological Survey, Russia produced almost one-fifth of the global diamond output in the year 2005. On the other hand, Kimberly, the capital of Northern Province of South Africa, discovered an 83.5

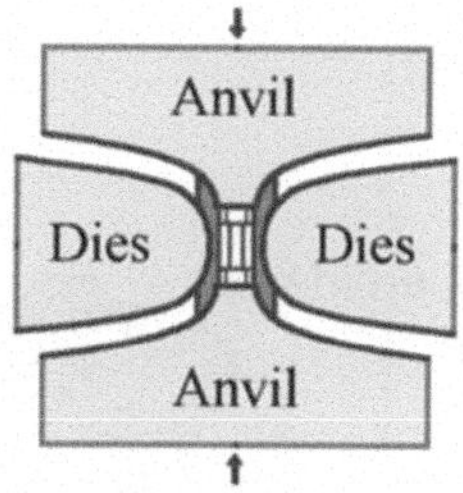


**Fig. 9.8** Schematic of a belt press (Courtesy: Hinonby, WIKI)

carat (16.70 g) diamond in 1871. The igneous rock kimberlite best known for diamond ore has been named after the town Kimberly. Many kimberlites are emplaced as carrot-shaped (Fig. 9.5), vertical intrusions termed *pipes*. The morphology of kimberlite pipes, and the classical carrot shape, is the result of explosive diatreme volcanism from very deep mantle-derived sources [21].

## 9.4 Early History and Production of Synthetic Diamonds

Synthetic diamond also called the cultured or cultivated diamond mostly produced by high-pressure high-temperature (HPHT) or chemical vapor deposition (CVD) method. The earliest reported attempt to make synthetic diamond is by James Ballantyne Hannay in the year 1879 [22]. In the year 1890, Moissan published a series of articles on the HPHT diamond. In 1941, an agreement was made between General Electric (GE), Norton and carborundum companies to produce HPHT diamonds. They were able to heat carbon to about 3000° C (5430° F) under the pressure of 3.5 gigapascals (510,000 psi). The GE group improved on the anvils (tungsten carbide) within the hydraulic press to squeeze the carbonaceous material held in a catlinite container to produce diamond [23]. Figure 9.8 shows the schematic of a belt press.

### *9.4.1 HPHT Process*

In the high-pressure high-temperature (HPHT) method, there are three main press designs, belt press, the cubic press, and the split-sphere (BARS) press used to supply the pressure and temperature necessary to produce synthetic diamond. The original invention of GE by Tracy Hall used the belt press wherein the upper and lower anvils supply the pressure load to a cylindrical inner cell. The internal pressure is confined radially by a belt of prestressed steel bands. The anvils also serve as electrodes for electric current to flow in the compressed cell. Hydraulic pressure is used to vary the belt pressure, rather than steel belts, to cofine the internal pressure. A schematic of a belt-type HPHT press is shown in Fig. 9.9.

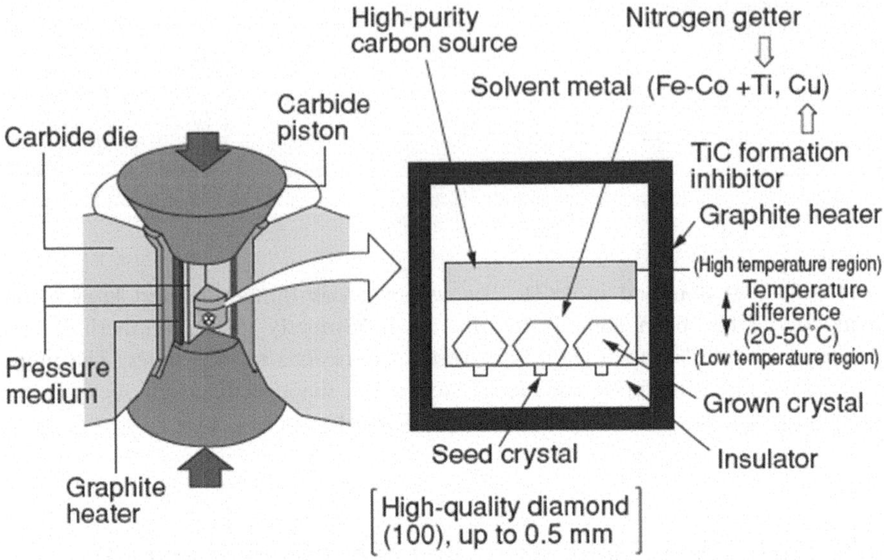


**Fig. 9.9** Schematic of a HPHT belt-type system (Courtesy: Int. diamond Labs)

Diamond seeds are placed at the bottom of the press. The internal part of the press is heated by a tubular, graphite heater which can supply heat up to 1400° C. This heater is capable of melting solvent metal which dissolves the high-purity carbon source. The dissolved carbon source is then transported to the diamond seeds and precipitated [24–26].

The cubic press design has six anvils, which provide simultaneous pressure on all faces of a cube-shaped volume. The cubic press is created to fulfill the need to increase the pressurized volume. The third type of press is the BARS, developed by the Russian Academy of Sciences in Novosibrisk. It consists of eight outer anvils with a spherical outer shape to which pressure is applied. It has six inner anvils to multiply the pressure to the sample. BARS belt press HPHT system manufactured by Kobelco is shown in Fig. 9.10.

The BARS apparatus is the most compact, efficient, and economical of all the diamond-producing presses. In the center of the BARS apparatus, there is a ceramic cylindrical *synthesis capsule* of about 2 $cm^3$ in size. The cell is placed into a cube of pressure-transmitting material, such as pyrophyllite ceramics, which is pressed by inner anvils made from cemented carbide or VK10 alloy [27]. The outer octahedral cavity is pressed by eight steel outer anvils. After mounting, the whole assembly is locked in a disk-type barrel with a diameter about 1 m. The barrel is filled with oil, which pressurizes upon heating, and the oil pressure is transferred to the central cell. The synthesis capsule is heated up by a coaxial graphite heater, and the temperature is measured with a thermocouple [28]. Figure 9.11 shows a BARS system, where the size of the outer barrel is reduced for presentation purpose.

**Fig. 9.10** A HPHT press manufactured by Kobelco, Tokyo, Japan (Courtesy: Kobelco)

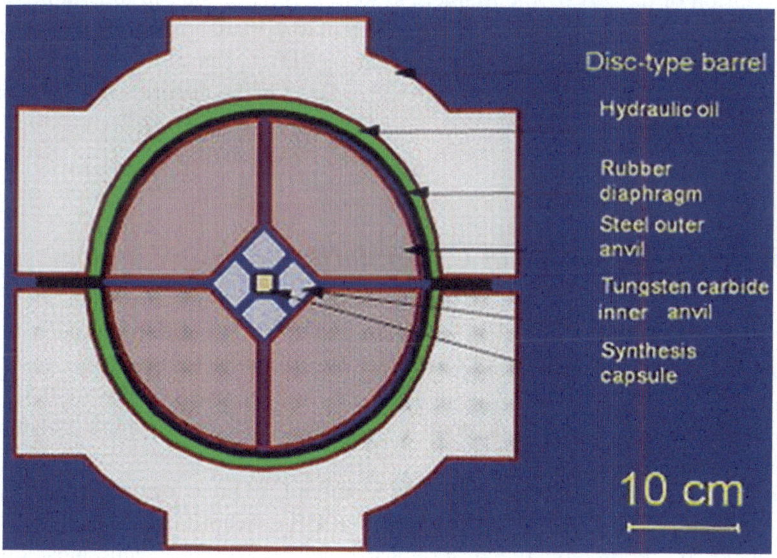


**Fig. 9.11** A BARS system, where the size of the outer barrel is reduced for presentation purpose (Courtesy: WIKI)

## 9.4.2 Chemical Vapor Deposition Method (CVD)

Diamond can be grown from the mixture of hydrocarbon gas by using chemical vapor deposition method. The method has been in use since 1950s, and quite matured. Compared to HPHT method, it does not require high temperature or high pressure, and growth occurs under low pressure (1–27 kPa) and low temperature. The popularity of growing CVD diamond by CVD methods has been gaining momentum because of its ability to grow diamond over a large area and on various substrates [29–31].

The CVD process is conducted under nonequilibrium conditions, and in presence of hydrogen radicals (atomic hydrogen) to remove carbon from the hydrocarbon gas or graphite. The hydrogen radicals are produced either by thermal dissociation on a hot tungsten (W) or tantalum (Ta) filament or in plasma by electron impact.

In hot filament CVD reactor, a mixture of hydrogen ($H_2$) and methane ($CH_4$) is activated by the filament through thermal heating and dissociation of hydrogen on the surface [32]:

$$H + CH_4 \rightarrow CH_3 + H_2 \tag{9.1}$$

$$CH_3 + CH_3 + M \rightarrow C_2H_6 + M \tag{9.2}$$

$$C_2H_y + H + M \leftrightarrow C_2H_{y\text{-}1} + H_2 + M \tag{9.3}$$

Activation of the gas by the filament system is the first step in the diamond thin film growth. The methyl radical has fundamental importance for studies of diamond film growth by chemical vapor deposition (CVD) because it is regarded as a gas-phase precursor to diamond growth [33, 34]. M is the generic collision partner in pressure-dependent reactions.

### 9.4.3 *Characterization of CVD Films*

Figure 9.12 shows a scanning electron micrograph (SEM) and Raman spectrum of a film grown by hot filament CVD method. The filament temperature was set at 2200 K, substrate temperature was set at 1273 K, and the gases used were 0.5% $CH_4$ in $H_2$ [32]. The Raman peak was found to be around 1332 $cm^{-1}$, which is also found with natural diamond.

Figure 9.13 shows a dual chamber hot filament CVD reactor (model 665 series, from $sp^3$ Diamond Technologies, Santa Clara, CA). According to the information released by the company, the proprietary filament assembly generates uniform energy distribution, operating at filament temperatures up to 2550° C and power levels up to 20 $W/cm^2$.

Methane ($CH_4$) is commonly used as a carbon source, and in the presence of hydrogen radicals (atomic hydrogen) methane is broken into methyl radicals ($C_2H_{y\text{-}1}$). The CHx and $C_2$Hy radicals are regarded as the precursors for diamond growth [35, 36].

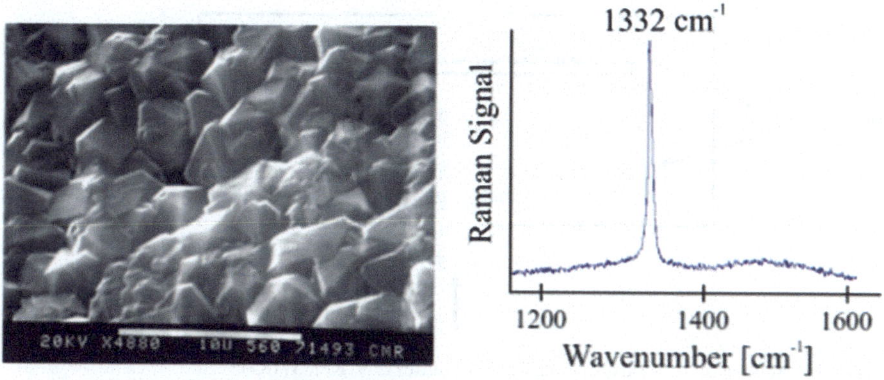


**Fig. 9.12** Scanning electron micrograph (SEM) and Raman spectrum of a film grown by hot filament CVD method (Reproduced with permission from E.H. Wahl)

**Fig. 9.13** Dual chamber hot filament CVD reactor (Courtesy: sp[3] Diamond Technologies, Santa Clara, CA)

## *9.4.4 Plasma-Enhanced CVD*

Among the entire chemical vapor deposition (CVD) methods, plasma-enhanced chemical vapor deposition (PECVD) has been used mostly to produce diamond films at low temperatures. In this method, plasma of the gas enhanced the chemical reaction rates of the precursors. A homotaxial growth of diamond by PECVD method at low concentration of methane (0.005–0.025 % $CH_4/(H_2 + CH_4)$ ratio)

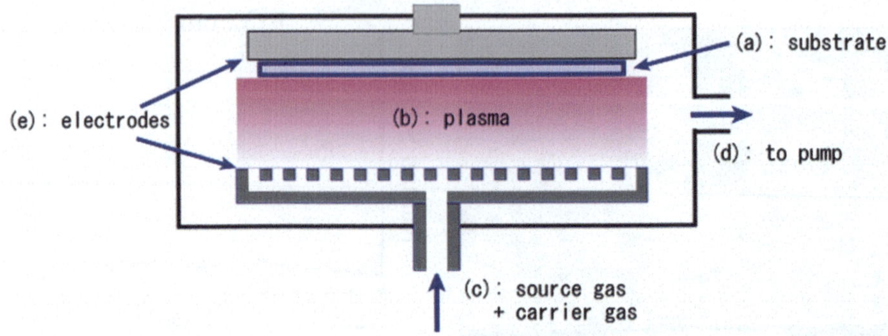


**Fig. 9.14** Schematic of a PECVD apparatus (Courtesy: S. Kei, WIKI)

**Fig. 9.15** Hydrogen-terminated diamond film with {111} orientation (Courtesy: Springer, Germany)

shows a step-free surface {111} oriented film (Fig. 9.15) [35]. Figure 9.14 shows the schematic of a PECVD apparatus.

The growth mode for the homoepitaxial diamond {111} film is three-dimensional (3D). Figure 9.15 shows hydrogen-terminated diamond film with {111} orientation. Variations in methane ($CH_4$) concentrations give rise to different fluxes of hydrocarbon precursors arriving at the surface of the substrate, and the growth rate increases with methane concentration ($CH_4$). As the methane concentration reached a higher level, adatoms form clusters landed on the surface of the substrate giving rise to a rough surface due to 3D growth.

For diamond-based electronic devices, one has to dope diamond with acceptor and donor impurities. Nitrogen is the common type of impurity in diamond that forms several types of complexes with vacancies. During homoepitaxial growth of diamond by PECVD, boron and phosphorus are introduced for p-type and n-type dopants, respectively. The highest Hall hole and electron mobility of PECVD diamond films reported are 1860 $cm^2/V$ s at 290 K and 660 $cm^2/V$ s at 300 K, respectively [37, 38]. Figure 9.16 shows a gem-cut diamond grown by CVD method.

The optical-grade diamond film growth on a single crystal of silicon heated to 1000°C is very slow (~1 μm/h) because a much smaller fraction of the precursor

**Fig. 9.16** Colorless gem-cut from diamond grown by CVD method (Courtesy: WIKI)

material is deposited as diamond since it resists atomic hydrogen etch due to strong $sp^3$-hybridized carbon bonds. The need for a powerful source for atomic hydrogen has been found with three different CVD processes, such as (i) hot wire CVD, (ii) DC arc CVD, and (iii) microwave plasma-enhanced CVD.

### *9.4.5 Hot Wire CVD*

Hot wire CVD technology does not depend on ionic species of plasma to activate the precursor. The refractory material wire such as tungsten is heated to well above 2000° C which ultimately dissociates hydrogen molecules (H–H bond energy: 4.53 eV). In hot filament assisted CVD process, incorporation of filament materials combined with limited deposition area make hot filament assisted CVD method useless for high-purity commercial applications of diamond. It is also difficult to make a 3D coating by this process.

The DC arc CVD technology and the microwave plasma-enhanced CVD depend on plasma discharge to produce atomic hydrogen. DC arc is capable of producing highest plasma densities, but it lacks 3D coating.

### *9.4.6 Microwave Plasma-Assisted CVD (MWPECVD)*

MWPECVD process can produce very dense and hot discharge. As a result, it can provide high deposition rate in a single or dual mode cavities by using several hundred mbar pressure and several killowatts of microwave power. The high power can destroy solid insulating material like dielectric atmosphere to vacuum interfaces they come in contact with. To produce an acceptable film thickness of large area and uniformity, the microwave PECVD need to allow the plasma to be in

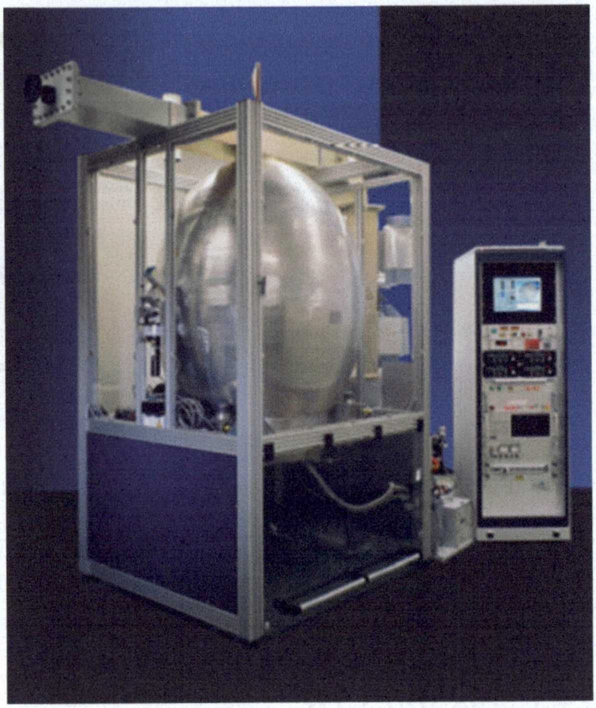

**Fig. 9.17** High-performance microwave CVD plasma reactor for diamond growth (Reproduced with permission from Christoh Nebel, Fraunhofer Inst. of Appl. Phys. Freiburg, Germany)

contact with the atmosphere to vacuum interface [39–41]. As long as the wavelength of the applied power is larger or comparable to the linear dimensions of the plasma-processing reactor, free-range wave propagation is not possible and all wave modes will have to rely on conductive components inside the chamber. The wavelengths of industrial microwave frequencies are smaller (2.45 GHz, 915 MHz) than the dimensions of the large-area plasma-processing vessels [42]. Coaxial plasma lines to feed microwave power do not reflect incident power and therefore do not need tuning devices. A high-performance microwave CVD plasma reactor for diamond growth is shown in Fig. 9.17.

The reactor consists of a multimode ellipsoidal cavity that can deliver power either between 6 and 10 kW or between 30 and 60 kW, with an operating frequency of either 2.45 GHz or 915 MHz. The film growth has been claimed between 1 and 12 μm/h (depending on the application). The reactor operations are controlled by a computer. It is suggested that repetitive pulses of microwave are better than continuous wave (cw) to deposit more uniform better-controlled diamond film with higher deposition rate. The modulated supply of power to the repetitive microwave pulse process opened up a new field of parameter setting since pulse shape, duty cycle, and pulse frequency do have strong influences on both plasma electron temperature and density as described in the Global model by Ashida et al. [43].

The mean free path of gas molecules, the electrons' mobility, and the growth of the film depend on the temperature, pressure, and the contaminants (if any) inside the chamber. The growth of the film also depends on the doping condition of the host material. All these factors change the mean free path of the gas molecules containing atomic hydrogen and carbon-containing radicals. Thus, the nuclei density, crystal quality, deposition rate, surface morphology, and texture vary with deposition conditions. It has been reported by H. Ye and coworkers that the time-dependent nucleation density reaches a maximum value at about 30 min. After that, the nucleation density decreases, and the diamond growth increases. They further reported that a two-stage growth mode is necessary for diamond film deposition [44–46]. Most of the laboratories use some kind of CVD technologies to make diamond films of high quality that are being used in nuclear medicine (radiation detectors, mostly), or cutting and as abrasive materials, cutting tools, and thermal conductors [47].

Diamond radiation detectors are able to detect deep UV photons, X-rays, gamma rays, electrons, alpha particles, charged ions, and neutrons, with a dynamic range in energies spanning from 5.5 eV up to GeV of cosmic rays. Since the bandgap of diamond is 5.5 eV, this leads into a negligible dark current noise at room temperature with no need for cooling. A 120 μm thick polycrystalline CVD-grown diamond detector has been used for monitoring 14.7 MeV neutrons emission produced with Frascate Neutron Generator (FNG). In medical science, microdiamond detectors have been used in dosimetry.

Figure 9.18 shows a diamond fly cutting machine used to generate diffraction gratings and other linear patterns with approximately contoured diamond shapes. Diamond milling can be used to generate aspheric lens arrays by annulus cutting

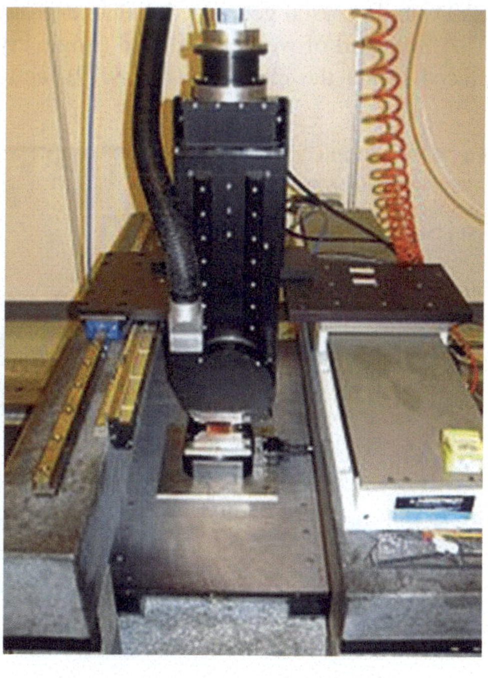

**Fig. 9.18** A diamond fly cutting machine (Courtesy: David Casale, Creative Comm., WIKI)

methods with a spherical diamond tool. The precision of these cutting tools can reach in the nanometer scale. Diamond turning machines are used in optical assemblies in telescopes, video-projectors, missile guidance systems, and lasers.

Diamond, synthetic or natural, can transmit from the ultraviolet (UV) through the visible and infrared (IR) (8–14 μm) transmission range, combined with high strength and excellent resistance to thermal shock, scratching, and erosion. Coupled with this, diamond offers good performance as a window for microwave, visible, and ultraviolet radiation. The extremely low thermal expansion coefficient and high thermal conductivity combined give diamond with the lowest thermos-optic distortion of any optical material. Moving beyond the visible band into X-rays, diamond products have a role to play in synchrotrons as windows and monochromators to deliver targeted X-rays of different wavelengths [48, 49].

The real challenge is to grow single crystal of diamond that will possess all the qualities of a natural diamond, the dream material, a symbol of love. In the beginning of 1970s, some industries took the pain to grow diamond by CVD process. Now the growth of synthetic diamond has reached a commercial phase due to the endeavor of Apollo Diamond, a US company based in Boston, MA. The company produced nearly flawless single crystal of diamond wafers and crystals for potential use in optoelectronics, nanotechnology, and consumer gem market. The company used CVD method for the production of their gem-sized synthetic diamond crystals, and obtained several patents on the process [50].

Recently, many assets of Apollo Diamond are now acquired by Scio Diamond Technology Corporation, in 2011. The company is based at South Carolina, USA, and is producing gem-sized synthetic diamonds by adopting CVD technology. The CVD single-crystal gem-sized diamond-growing machine is shown in Fig. 9.19.

A thin seed of well-crystallized (single crystal, without any defect) diamond is placed inside the chamber (Fig. 9.19), and there is a channel to introduce the

**Fig. 9.19** MWPECVD diamond crystal growth equipment (Photo courtesy: Scio Diamond Technology Corporation, Greenville, SC)

hydrocarbon gas ($CH_4$). There are also provisions for lasr cutting and polishing equipments. The gas is heated inside the chamber at a very high temperature to break down carbon atoms and hydrogen atoms inside the vacuum chamber. The pressure, temperature, and impurities if any will play important role in the shaping of the crystal. By managing the growth process, the shape of the diamond and its crystal features can be managed to a certain degree. There are some windows provided in the CVD equipment so that one can steer the CVD process to impact certain engineered properties to the ensuing diamond for application tailoring. Impurities like boron, nitrogen, or phosphorus during the growth process can be introduced if it is necessary. Within the parameter limits of each control knob is an increasing amount of tuning and precision that is currently being built into the process.

## 9.5 Growth Mechanism and Properties of Diamond

The nucleation and growth mechanism of MWPECVD diamond over a single crystal seed of diamond involves competitive growth by all the $C_1$ radical species that are present in the gas mixture close to the growing diamond surface. The model is based on competition between H atoms and $CH_3$ radicals and other $C_1$ radical species reacting with dangling bonds on the diamond surface [51]. For single crystal growth in high-power MWPECVD process, the concentration of H atom should be of the order of $10^{16}$ cm$^{-3}$, and thus $CH_3$ radical incorporation via monoradical sites will be a dominant mechanism. However, reliable values of the presence of H atom from both channels (mono and biradical) and the substrate temperature (here, the single crystal of diamond seed {100}) are necessary for a homogeneous growth of diamond on diamond seed. The model of Mankelevich and May further provides typical values of the plasma parameters (2D model); they are as follows: maximum gas temperature T ~ 3300 K, and mean electron temperature Te ~ 1.3 eV for Vp ~12 cm$^{-3}$ (assuming the height $H_p$ and diameter $D_p$ of the cylindrical plasma region are about 1.8 cm and 2.9 cm, respectively, making the averaged absorbed power density Q ~ 50 W cm$^{-3}$). Figure 9.20 shows MWPECVD-grown faceted diamond [52].

According to Steve Bebson, Communications Director at the World Jewelry Confedaration, *a man made reproduction of a diamond almost has reached the same chemical composition, crystal structure, and physical properties as its natural counterpart*. In fact, synthetic diamonds have a decade-old history beginning in the year 1950s at the General Electric Laboratories, USA. According to the scientists, the synthetic gems have a legitimate place in the market. In 1982, Sumitomo Electric synthesized a 1.2 carat crystal diamond, one of the world's largest man-made gem at the time. Lynette Gould, a spokesperson of De Beers sees exciting potential for synthetic diamonds in industrial applications. In an industrial park of Northern Singapore, a company produces the purest and rarest grade of rough diamonds (Type IIA) in a laboratory process called microwave plasma-

**Fig. 9.20** MWPECVD-grown faceted diamond (Photo courtesy: AFP/Getty Images)

enhanced chemical vapor deposition (MWPECVD). Carbon atoms are layered on top of an initial diamond seed fast-tracking in a natural process lasting many millennia to a matter of months. According to a critic, these lab-made gems are free of the *blood diamonds* stigma that has tainted their traditionally mined counterparts, and are good for the environment [53].

Table 9.1 Physical properties of a diamond grown by CVD method and the natural single crystal of diamond. The data are taken from different sources.

When someone asks a question about a material which shows superior transparency and luminescence besides being the hardest and highest in thermal conductivity, the answer should be diamond. The transmission spectrum for a diamond window is featureless for wavelengths longer than 225 nm apart from a moderate absorption in the range 2.6 to 6.2 μm. The Fresnel reflection at each diamond–air interface is 17% in the visible range. Using dispersion data for refractive index (RI), the transmission upper limit (no absorption) as a function of wavelength is shown in Fig. 9.21. The absorption in the midfrequency range (2.6–6.2 μm) arises due to the coupling of radiation with the movement of the nuclei, and is often referred to as *lattice* or *multiphonon absorption* [4, 54, 55].

The possibility of fabricating diamond detectors has been exciting research for decades [56]. Natural diamonds have been used as radiation detectors [57, 58], as dose distribution in radiotherapy [59]. Diamond dosimeter also has been used for X-ray and gamma-ray radiation [60].

A commercial version of single crystal synthetic diamond in the form of Schottky diode configuration has been used as the new type of microdiamond detector. This device has also been used as in a dosimeter for clinical radiation therapy [61].

The CVD-grown diamond detectors are also being used for the development of very fast, low-noise electronics operated at a very high voltage. Charged particles passing through the diamond detectors produce short pulses. The pulse rise time of less than 100 ps is short due to the high mobility of both electrons and holes. With

**Table 9.1** Properties of diamond

| Physical properties | CVD-grown diamond | Single crystal diamond |
|---|---|---|
| Density ($g/cm^3$) | 2.8–3.51 | 3.515 |
| Thermal capacity at 25° C (J/mol/K) | 6.10 | 6.190 |
| Thermal conductivity at 25° C (W/m/K) | 2100 | 2200 |
| Thermal expansion coefficient at 25–200° C ($\times 10^{-6}$ /0° C) | ~2.0 | 0.8–1.2 |
| Bandgap (eV) | 5.45 | 5.45 |
| Electrical resistivity (Ω-cm) | 1012–1016 | 1016 |
| Dielectric constant (at 45 MHz–20 GHz) | 5.6 | 5.74 |
| Dielectric strength (V/cm) | 106 | 106 |
| Loss tangent (at 45 MHz–20 GHz) | <0.0001 | |
| Carrier mobility ($cm^2$/V/s)<br>Electron (e)<br>Hole (p) | 1350–1500 (e)<br>480 (p) | 2200 (e)<br>1600 (p) |
| Young's modulus (GPa) | 820–900 | 910–1250 |
| Poisson's ratio | | 0.10–0.16 |
| Vickers hardness (GPa) | 50–100 | 57–104 |
| Index of refraction at 10 μm | 2.34–2.42 | 2.40 |
| Compression strength (GPa) | | 8.68–16.53 |

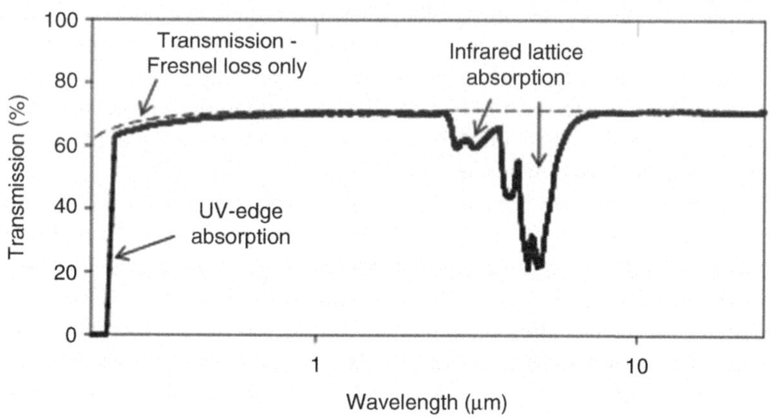


**Fig. 9.21** Transmission spectrum for a type IIa diamond, 1 mm thick (Reproduced with permission from Wiley-VCH Verlag GmbH & Co)

an impedance of 50 Ω and electrode capacitance of 10 pF, the resulting (FWHM) pulse width is reported as 1 ns or less [62].

Besides these applications, there are so many commercial equipments where diamond is the only material used. As for example, it is one of the hardest materials known and often used as an abrasive. Diamond speaker domes are very stiff, and when made into a thin dome it can vibrate rapidly without the deformation that would degrade the sound quality. Due to its highest thermal conductivity, diamond is used as heat sink to absorb or transmit heat effectively. Low-friction microbearings are needed in tiny mechanical devices. As for example, some watches have jewel bearings in their movements, and diamond is used where extreme abrasion resistance and durability are needed. Wear-resistant parts of some gadgets are produced by coating surfaces with diamond.

## 9.6 Nanodiamonds

### *9.6.1 History of Nanodiamonds*

In 1963, almost 54 years ago, in the month of July, detonation synthesis of nanodiamonds was discovered. In the year 2003, the laboratory experiments shifted to the industry as a result of the labor and success of so many scientists and technologists. However, the history of the discovery of the synthesis of ultrafine-dispersed diamonds (UDDs) is a part of the development of explosive technologies. The successful synthesis of UDDs is the history of many years of carbon condensation in a detonation wave [63].

Nanodiamonds or the nanoparticles of diamonds are found in so many materials, like in crude oil, meteorites, interstellar dust, protoplanetary nebulae, and in the earth's crust. The practical and technological applications of nanodiamonds have increased very much since its mining in India 2000 years ago. In China, it had been used as utensils as far back as 6000 years. These nanodiamonds or the nanoparticles of diamond are produced in the laboratory by chemical vapor deposition or by detonating high explosive materials [64–67].

The classification scheme of carbon nanostructures within the general hierarchy of carbon materials is shown in Fig. 9.22 [68]. The scheme is based on two major characteristics: (i) the type of carbon hybridization and (ii) the characteristic sizes of nanostructures. If we consider fullerenes, nanotubes, graphene of finite size, and the currently observed smallest nanodiamond clusters as basic structural units in the carbon nanoworld, prototype molecules can be assigned to these units at the scale of molecules (inner circle). It should be noted, however, that this scheme does not mean that these molecules are involved in the synthesis of the units; rather, we would emphasize topological similarities between organic species and inorganic materials.

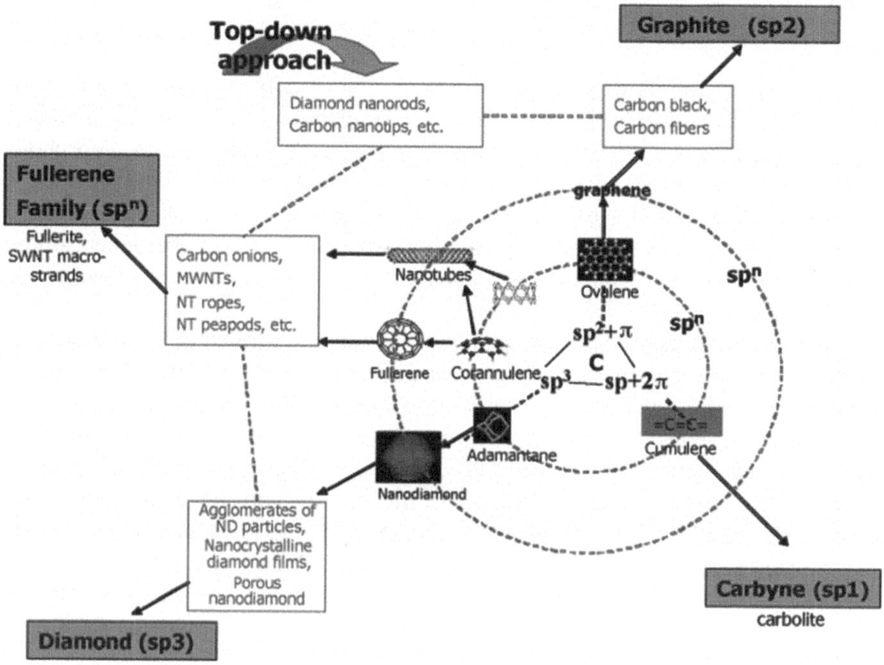


**Fig. 9.22** Classification of carbon nanostructures (Reproduced with permission from Ref. Shenderova et al. [149]

## 9.7 Synthesis of Nanodiamonds

In the laboratory, nanodiamonds are synthesized under high pressure and high temperature by using graphite or other forms of carbons, condensation from carbon-containing explosives, fullerenes, or multiwall carbon nanotubes [69–72]. Chemical vapor deposition (CVD) method is another process very much in use in the laboratory for producing nanodiamond films [73]. Besides, carbide-derived method [74], electron beam [75], ion beam [76], laser [77], and plasma [78] technologies also have been used to produce nanocrystalline form of diamonds.

Tubular form of carbon nanotubes (CNTs) under high pressure (4.5 GPa) and high temperature (1300 C) collapse, and broken graphite shells curl up and close into spheroidal networks to eliminate dangling bonds at the edges. The high curvature of the formed nested graphite shells cross-linked between the layers in the onion-like structures leads to an increased fraction of sp$^3$ bonds which facilitates the formation of diamonds [71, 79].

### 9.7.1 Low-Pressure CVD

Low-pressure chemical vapor deposition (LPCVD) of polycrystalline thin films of diamond was first reported in 1950. However, production of CVD-grown diamond films was achieved in 1980s [80]. Nanodiamond (ND) nucleation in amorphous carbon films was noticed when the substrate is negatively biased and the substrate is exposed to CVD plasma. However, formation of a dense, amorphous, hydrogenated carbon phase occurs via a sub-implantation process, in which energetic carbon, hydrocarbon, and hydrogen species bombard the substrate surface and incorporate there as surface layers.

In 1960s, Frenklach et al. studied the nucleation and growth of nanocrystalline diamonds in a low-pressure microwave-plasma CVD (LPMWPCVD). They further reported that the presence of diborane ($B_2H_6$) helps in the nucleation of nanodiamonds (NDs) at low-pressure operation, and the absence of diborane shows no nanodiamond formation. The method can produce diamond particles from 5 to 400 mm in diameter. Presence of linear twins and star-twins microstructures in the residue of CVD-grown ND diamonds provide clues regarding the mechanisms of interstellar dust formation. Spherical NDs having diameters from 150 to 600 mm have been synthesized in a gas-phase multicathode direct current plasma-assisted CVD. The grain size of the NDs is approximately 30 nm in size [79].

In 1990s, Iijima successfully used to convert the cores of concentric shell graphite onions into ND by using high-energy electron irradiation. L.A. Curtiss et al., reported ion-beam irradiated ND formation. They also reported the formation of Ne + irradiated ND from graphitic carbon soot [81]. There are several reports of ND formation from different carbonaceous material including carbon black by using intense laser radiation [82, 83].

### 9.7.2 Characteristics of Nanodiamonds

The structure of nanodiamonds has been found to be almost spherical with twinning defects. The presence of diamond cage has been found in the core of nanodiamonds and in the absence of hydrogen on the surface; both computer simulations and experimental data suggest that graphitic-like reconstructions and onion-like structures (Fig. 9.23) appear at the surface of the nanodiamonds [84–86].

J.Y. Raty et al. reported the size of the nanodiamond particles to be 2–4 nm in size having fullerene-like hybrid structures on reconstruction. During reconstruction, tensile stress is induced in the core shell, and it is contrary to the reconstruction of silicon and germanium quantum dots where reconstruction induces compression on the nanoparticle core [87].

The stability of nanodiamonds has been established by computing the binding energy of diamond-like and graphite-like carbon clusters using fixed energy values

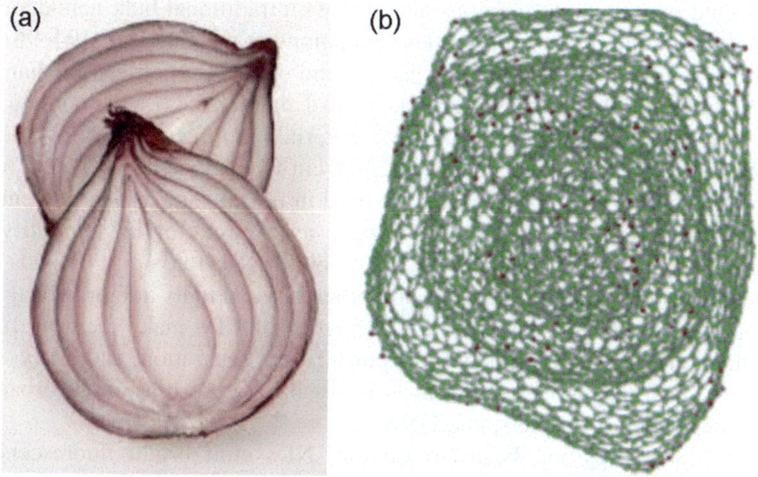


**Fig. 9.23** (**a**) Cross-section of onion showing the scales; (**b**) Carbon nano-onions

for carbon–carbon and carbon–hydrogen bonds [88]. As a matter of fact, the stability of a nanodiamond is dependent on the size of the diamond [89, 90].

The CVD-grown nanodiamond showed a persistence of quantum confinement effects up to 27 nm, a diameter surprisingly larger than in silicon (Si) or germanium (Ge) nanoparticles where quantum confinement effects disappear above 5–7 nm. On the other hand, the nanodiamond produced by hot filament CVD and high explosive detonation waves showed that quantum confinement effects disappear in particles larger than 4 nm. The nanodiamonds produced by detonation wave showed very similar properties of the bulk diamonds [86]. Applying both time-dependent and time-dependent functional theory, Raty et al. and Runge et al. concluded that quantum confinement effects disappear in nanodiamonds when the particle sizes are larger than 1 nm [91, 92].

Fokin and Schreiner applied density functional theory to study changes in bandgap effects of nanodiamonds. They also found that strong quantum confinement effects prevail when the particle sizes vary between 0.2 and at least 2 nm, and the bandgap is reduced to 6.7 eV. They further noticed that the octahedral and tetrahedral nanodiamonds show the same trends in bandgap narrowing, and it is the dimension rather than the shape or morphology of the nanodiamonds that alters the bandgap. Bandgap tuning through external (by C-H bond substitution) or internal (by replacing CH or $CH_2$ moieties) doping is nonadditive for the same dopant [93].

They also have talked about *push (electrons donating)–pull (electron accepting) doping*, (substitutional doping), where electron donating and electron accepting plays an important role in reducing the bandgap of nanodiamonds and comparable to bulk diamonds. It is hoped that *push–pull doping* might be a valuable technique to make nanodiamonds the valuable candidates for semiconductor applications. Thus, nanodiamonds (or diamondoids) offer unique opportunities to study the relationship between the substitution pattern and the properties of nanodiamond particles, and in contrast to previous attempts with traditional bulk nanodiamond, allow the fine-tuning of these properties of diamond at the nanolevel [94–96].

Nanodiamonds show excellent mechanical and optical properties. The hardness of the diamond is 50 times stronger than that of titanium (Ti) and stainless steel (SS). As a result, it can be used in abrasive materials and in cutting tools. Thermal conductivity of nanodiamonds is highest (20 W/cm-C) compared to 0.21 W/cmC of Ti and 0.16 W/cmC of SS. Thus, NDs are used in the coolant, which prevents the hot zones inside the coolant. It is also used as a paste for its reliability and durability. The hardness, chemical inertness, and low cytotoxicity of NDs have been made possible to use in implants and other surgical tools in biomedical field. The chemical modification and surface properties of NDs make them useful for immobilization of protein and DNA for purification, separation, and analysis. Its exceptional absorbing property has made it possible to use it as an absorbant for amino acids, proteins, platelets, and DNA.

During high-energy ion beam irradiation, NDs emit bright fluorescence at 550–800 nm due to its nitrogen ($N_2$) vacancies [97]. The emission together with its noncytotoxicity make NDs a promising fluorescent probe for single-particle tracking in heterogeneous environments as well as in the field of biomarkers and biolabeling [98, 99]. NDs are nontoxic and are used as drug carriers and delivery vehicles [100]. NDs can also be used as an effective gene carrier in the administration of growth homone such as insulin. Furthermore, drug ND complex does not affect the white blood cells, making them useful in cancer treatments [101]. Nanodiamond composites are also used in bone tissue engineering [102].

Figure 9.23 (a) shows the cross-section of onion showing the scales and (b) carbon nano-onions. Carbon nano-onions (CNO), also called carbon-like onions (or OLC), were discovered in the year 1960, and they are multishelled nanoparticles consisting of concentric graphitic layers and a hollow core comparable in some respect to the scale of an onion. The central shell is a C6 molecule, and is regarded as belonging to fullerene family. The typical diameter of these carbon nano-onions are from 3 to 5 nm. Their physical and chemical properties depend on the number of carbon atoms present in a particular carbon nano-onion. The number of carbon atoms may vary from one nano-onion to other.

Carbon nano-onion (CNO or OLC) can also be synthesized by high-energy irradiation of electrons (1.2 MeV $>10^{24}$ e/cm$^2$) or ion bombardment. Carbon arc or low-pressure microwave plasma CVD technique can produce a small quantity of carbon nano-onions with a mass or by-products. In the year 1994, Vladimir et al. used vacuum annealing of nanodiamond precursor to produce carbon nanodiamond (CNO or OLC) [103, 104].

The carbon nano-onion (CNO or OLC) was fabricated by annealing detonation of nanodiamond (average grain size was about 5 nm) at 1700° C. Tomita et al. obtained diamond film by annealing OLC at temperature of 500°C in air [105, 106]. The grain size of OLC at 900°C is 3.4 nm, and it increases with the increase of temperatures. Most OLC are ellipsoid, and a small portion is spherical. OLC is a member of the fullerene family.

## 9.8 Diamond-Like Carbon (DLC)

Diamond-like carbon or DLC is an amorphous material having two types of carbon bonds, i.e., hexagonal graphite bonds ($sp^2$) and tetrahedral diamond bonds ($sp^3$). The name diamond-like carbon has been attributed to this amorphous material because it possesses some characteristics like diamond, such as it is extremely hard, has low electrical resistivity, low coefficient of friction, and optical transparency over a wide range of wavelengths.

Figure 9.24 shows the ternary phase diagram of a carbon (C): hydrogen (H), ta-(C-H), and ta-C. The hardest, the strongest, and the slickest mixture is known as tetrahedral amorphous (ta) carbon, or ta-C. Coating of a thin layer of ta-C, a pure form of DLC, a $sp^3$-bonded carbon atom can increase the resistance of a stainless steel (type 304) against abrasive wear.

The detailed structure of bonding of hydrogen and carbon in DLC is not fully understood. Different models have been put forward by different investigators from time to time. McKenzie et al. proposed a two-phase nanocrystalline model consisting of polycylic aromatic hydrocarbon regions interconnected by tetrahedral carbon [107].

Robertson, [108] in the year 1986, proposed a hypothesis based on the network of covalently bonded carbon atoms in different hybridization. In the year 1988, Angus and Jansen proposed a model on the basis of theories on random covalent network (RCN) put forward by Philips [109] and Thorpe [110]. Experimental observations of Angus and Wang supported covalent network of DLC containing $sp^2$ and $sp^3$ carbon sites, having no long-range order [111].

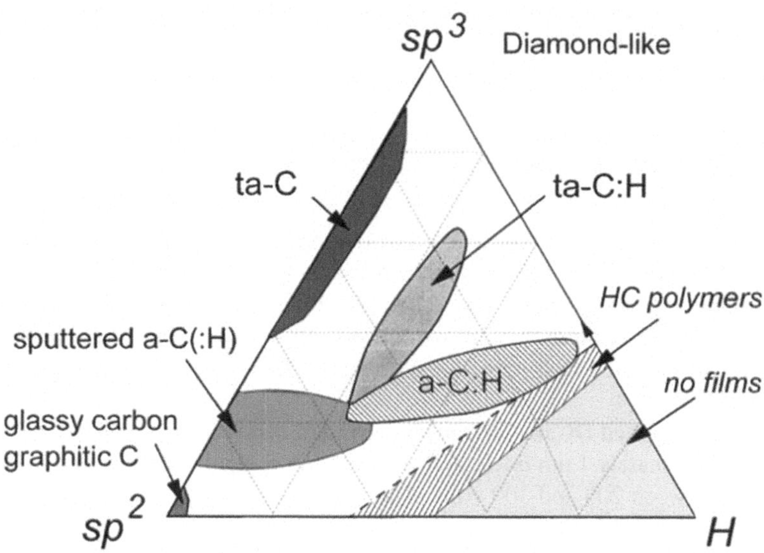


**Fig. 9.24** Ternary phase diagram of a carbon (C): hydrogen (H), ta-C-H, and ta-C.

### 9.8.1 *Synthesis of DLC*

Diamond-like carbon (DLC) can be hydrogenated or nonhydrogenated. Hydrogenated DLCs are produced from hydrocarbon gas mixture using various energy sources, such as DC-discharge, microwave and radio frequency (RF) energy, oxy-acetylene flame, and hot filaments. Nonhydrogenated DLCs are produced by (i) ion beam evaporation, (ii) magnetron sputtering, (iii) pulsed laser ablation, (iv) filtered cathodic arc deposition, (v) RF plasma-enhanced chemical vapor deposition (PECVD) techniques. These nonhydrogenated films when coated over a substrate can have both the $sp^2$ and $sp^3$ bonding with different ratios. As a result, these films differ in structural and physical properties from each other.

### 9.8.2 *Ion Beam Evaporation*

The principle of this method is to bombard a carbonaceous material (such as methane, $CH_4$) by energetic ions (of the order of 100–750 eV) with a current density of about 2.5 A/cm$^2$. Upon impact, the $CH_4$ molecules dissociate into carbon (C) and hydrogen ($H_2$). The arrival rate of the dissociated carbon atoms has been calculated to be approximately $1.5 \times 10^{16}$ atoms/cm$^2$/s, while the hydrogen atoms escape. The approximate rate of film growth is 1 μm/h.

The other ion beam evaporation which is called ion beam conversion coatings can deposit the film at a rate of 10 μm/h. The ion energies used here are high and can go up to 100 keV. The introduction of some impurities like sulfur, fluorine, or nitrogen can provide a low-friction coefficient film. Figure 9.25 shows a schematic of an Ion Beam Deposition System.

### 9.8.3 *Magnetron Sputtering*

Magnetron sputtering is a physical vapor deposition (PVD) process, where magnetic field is used to confine the highly dense plasma (of the gas) in front of the target. Figure 9.26 (a) shows the schematic of a sputtering system, where the sputtered ions are scattered and move toward the substrate. Fig. 9.26 (b) shows a magnetic field applied in front of the energetic ions to confine the gaseous ions in front of the target.

Peng et al. applied DC magnetron sputtering to deposit a nonhydrogenated DLC film of approximately 1 μm thick. A bias voltage is applied to the substrate which is generally between 200 and 300 volts. The argon ($Ar^+$) gas pressure is kept around 1 Pa.

However, the low sputtering coefficient of carbon in magnetic sputtering system is not an efficient means for depositing thick (2–3 μm) coatings. As a result,

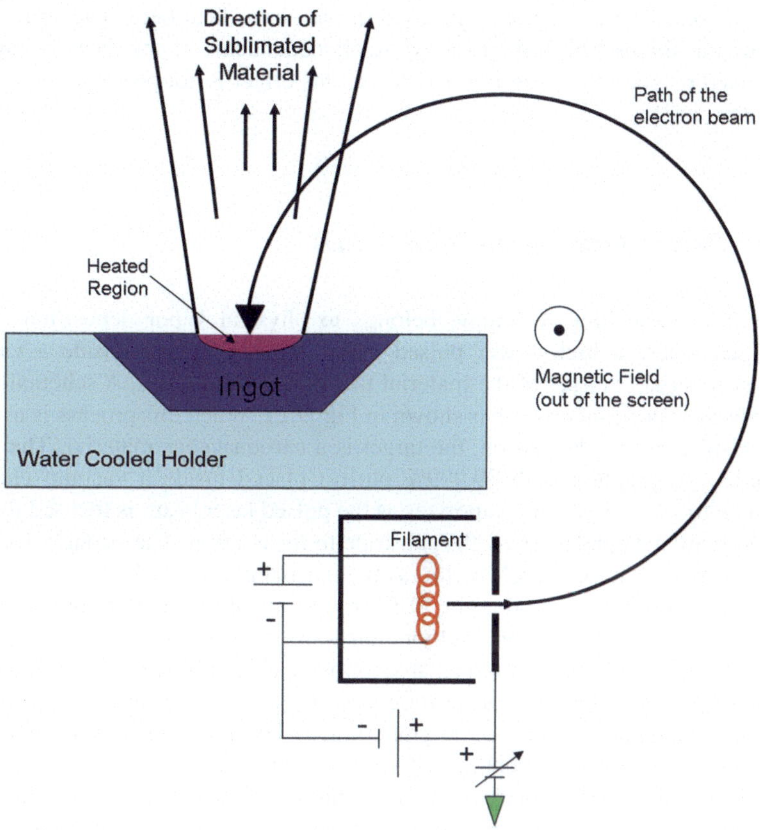


**Fig. 9.25** Schematic of an Ion-Beam Deposition System (Courtesy: WIKI)

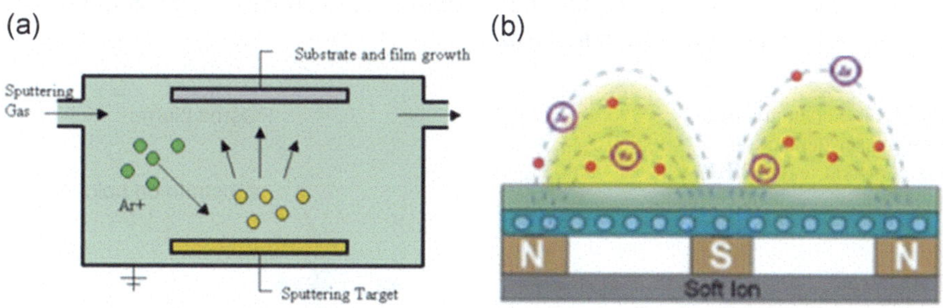


**Fig. 9.26** (**a**) Schematic of a sputtering system (Courtesy: Mattopia, WIKI), and (**b**) confinement of the ions in a magnetic field

sputtered coated DLC are not being used in the biomedical field. The structure of the film on silicon substrate is very much dependent on the bias voltage by unbalanced magnetron sputtering. However, this effect is not profound with metal substrates [112].

### *9.8.4 Pulsed Laser Deposition (PLD)*

Pulsed laser deposition technique belongs to physical vapor deposition (PVD) technique, where a high-power pulsed laser beam is focused inside a vacuum chamber to strike a target of the material that is to be deposited. A schematic of a pulsed laser deposition process is shown in Fig. 9.27. When this process is used for diamond-like carbon deposition, the target is a carbonaceous material. The target material (e.g., graphite with 99.999% purity) placed inside a vacuum chamber evacuated to $10^{-3}$ to $10^{-6}$ Pa vaporizes as the pulsed laser beam is focused through a lens system and condenses on the substrate to form a film. The strongly forward-directed flux of material is referred to as the plasma plume [113].

The several steps involved in the PLD process are absorption of photon energy, surface melting, vaporization, multiphoton ionization, plasma emission, plasma heating by inverse-Bremsstrahlung absorption, and finally supersonic expansion of plumes owing to the pressure differences. Target melting may lead to droplet emission from metal, ceramic, or polymer targets. This however is prevented due to the high melting point and conductivity of a graphite target.

In PLD system for producing DLC films, ultraviolet nanosecond pulse lasers, or excimers, or Nd-YAG lasers have been conventionally used employing a graphite target and have been reported by many workers [114, 115]. However, due to the powerful laser fluence, the graphite fragments from the solid target are found to deposit along with the film. Figure 9.28 shows a commercial PLD System used to deposit DLC films over different substrates.

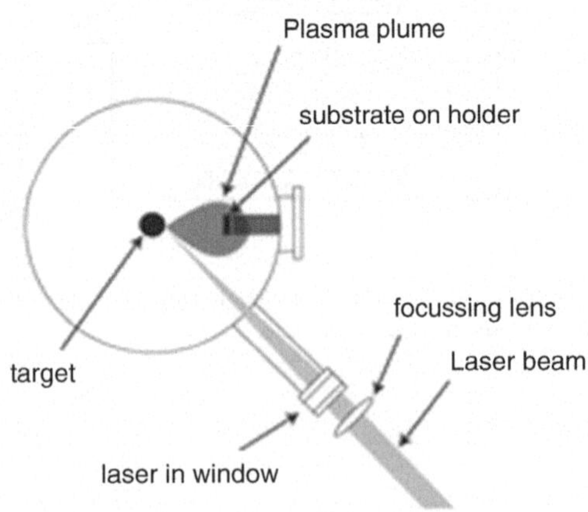


**Fig. 9.27** Schematic of a pulsed laser deposition process (Courtesy: WIKI)

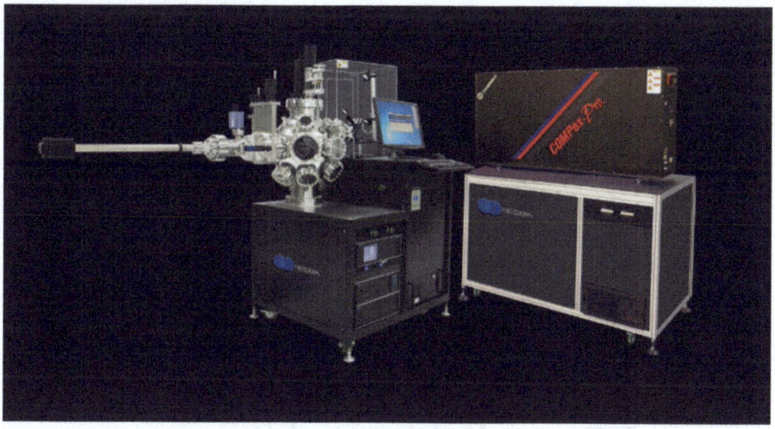

**Fig. 9.28** Commercial PLD System used to deposit DLC films over different substrates (Courtesy: Neocera company, MD, USA)

Hanabusa et al. successfully prevented such ejection of the carbon fragments from the solid target by using a frozen hydrocarbon gas target [116]. Recently, Yoshida et al. have produced DLC films onto silicone ($(SiO(CH_3)_2)n$) rubber on silicon wafer by using femto-second (fs) pulsed laser deposition technique with a frozen hydrocarbon of pentanol ($C_5H_{11}OH$). They claimed that they not only prevent the fragments of carbon inside the film but also improved the frictional property of the film which prevented the ordinary PLD deposited film to be used in medical engineering fields [117].

### *9.8.5 Filtered Cathodic Arc Deposition*

Cathodic arc deposition is a coating technology with great potential because cathodic arc plasmas are fully ionized with very energetic ions, promoting adhesion, and the formation of dense films. The origin of the vacuum arc deposition goes back as early as 1877 [118]. The subsequent first patent on the cathodic arc deposition was reported in 1892 [119]. A deposition energy between 25 and 100 eV is desirable to give a momentum to the condensing species to promote adhesion, good crystallinity, and to disrupt the columnar growth [120, 121].

Filtered cathodic arc deposition is a physical vapor deposition technique (Arc-PVD) where an electric arc is produced to vaporize the material to be condensed. It is the high-current, low-voltage arc on the surface of the cathode that gives rise to a highly energetic emitting area known as cathode spot. The localized temperature at the cathode spot is very high (~1500 C) which results in a high-velocity (~10 km/s) jet of vaporized cathode material, leaving a crater behind on the cathode surface.

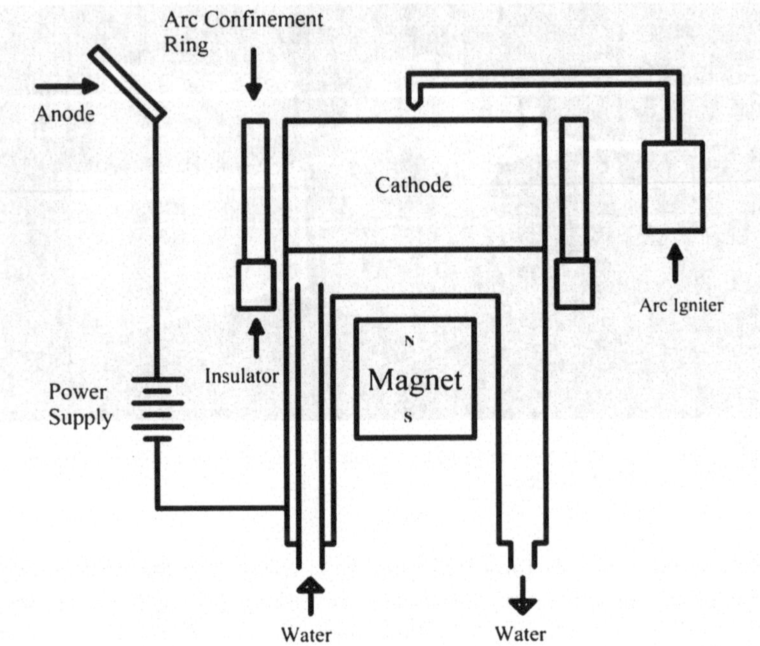


**Fig. 9.29** Sablev-type cathodic arc source with magnet (Courtesy: Sunyataburus, Creative Comm.WIKI)

The most widely used arc source which is used in the West is Sablev-type cathodic arc source (Fig. 9.29). The target is surrounded by an electrically floating metal ring working as an arc confinement ring. The anode can be either the metal vacuum chamber wall or a discrete anode. Arc spots are created by mechanical trigger and can be steered by magnetic field.

The plasma beam from the cathode source is a powerful jet, and the condensing particles sometimes contain cluster (macroparticles) of the source (target) material. This results in a very nonuniform, nonsmooth films of DLC. To deposit workable smooth films, the macroparticles ejected from the target is filtered. Macroparticle filters do not only remove macroparticles, as their name suggest, but also neutral atoms that may be present from the evaporating macroparticles. There are many designs for macroparticle filters, and Aksenov quarter-torus duct macroparticle filter using plasma optical principles is the most popular. Figure 9.30 shows the Aksenov quarter-torus duct microparticle filter developed by A.I. Morozov [122–125].

The Aksenov torroidal type filter is typically of the form of a 45–90$^0$ curved magnetic duct. More complex combinations such as the in-plane "S" and twisted "S"-type filters have also been used [126, 127]. Several modeling studies show that transport efficiencies were maximized by focusing the plasma as a jet into the duct and biasing the duct to a potential of +20 V [128].

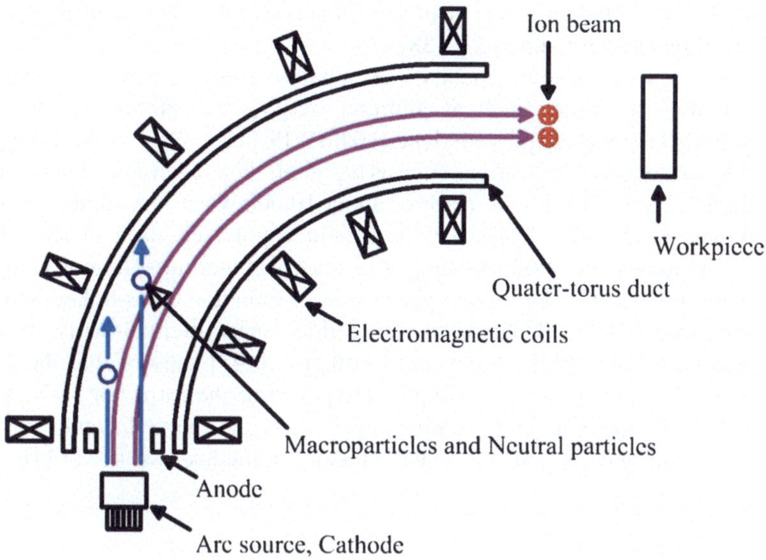


**Fig. 9.30** Aksenov quarter-torus duct microparticle filter (Courtesy: Sunyataburus, Creative Comm.WIKI)

The plasma source can be operated either in pulsed mode fed by pulse-forming network (PFN) or by a transistor-switched DC supply. System coefficients up to 3.5% were demonstrated with *pulsed twist filter* for carbon [127]. It has been shown that bias of pulsed filters can be realized simply by using voltage drop across a self-resistor, which was electrically placed between the anode of the arc plasma source and the filter coil. This approach eliminates the need for a separate bias circuit and avoids the undesirable coupling of bias capacitor and pulse-forming network, if the latter is used to drive the pulsed discharge [129].

Pulsed twist filter, a compact, open structure that operates with pulsed arcs and high magnetic field has been found to fulfill the requirements of a macroparticle-free, ultrathin (<5 nm) DLC film on magnetic storage gadget. The tribological properties of the magnetic head disk interface are not only of mechanical but also of chemical nature. The overcoat of DLC ultrathin film has been successful against wear and corrosion [130, 131]. With the recent progress in the development of the filter and the equipment, ultrathin DLC films are showing outstanding combination of advantageous properties: they can be very hard, tough, super-smooth, chemically inert, well adherent to the substrate, and compatible with lubricants [132, 133].

Besides the processes described above, diamond-like carbon films have been synthesized by argon sputter deposition using graphite target [134, 135], electron cyclotron resonance plasma chemical vapor deposition [136], and radiofrequency CVD [133]. The most outstanding achievement of the filtered cathodic arc deposition is the development of filters to minimize the macroparticles reaching the substrate. The real challenge has also been in successfully implementing to grow DLC films thinner than 5 nm [137, 138].

DLC coatings on different metals like cobalt–chromium (Co-Cr), titanium-6 aluminum alloy, ceramics such as alumina ($Al_2O_3$) and zirconia ($ZrO_2$), and ultrahigh molecular weight polyethylene (UHMWPE) have been in use in medical science like orthopedic applications such as hip joint. It has been seen that corrosion rate of the DLC is reduced by a factor of about 10,000 when it is applied to alloys and polymers [139, 140]. UHMWPE is a unique polymer, used in the clinical (medical) industry since 1962, because of its chemical inertness, lubricity, impact, and abrasion resistance. The second generation of radiation cross-linked and vitamin E stabilized UHMWPE has been emerged as a new internationally standardized biomaterial [141, 142]. Butter and Lettington [143] showed that the needle coated with DLC reduces the force required to penetrate the cornea of pig's eyes by about 30%. It is hoped that DLC coatings in a variety of orthopedic and cardiovascular instruments and gadgets will make a history in medical science [144].

### *9.8.6 Special Applications of Diamond*

The exotic features of quantum mechanics have the potential to revolutionize information technologies. The nitrogen vacancy (NV) center in diamond has recently emerged as one of the leading candidates for such quantum information technologies [145]. The electron spin coherence times of single NV centers in isotropically purified CVD diamond have been measured near 1 s for $T_1$ in the 4–80 k regime, 90 μs for $T_2$*, 2–15 ms for $T_2$ ($T_2$ ~ 0.53 for ensemble using dynamical decoupling techniques), while the communication time for 100 km through a fiber is approximately 500 μs [146].

The nitrogen vacancy center is a point defect in the diamond lattice. Two charge states of this defect, namely neutral $N\text{-}V^0$ and negative $N\text{-}V^-$ have been detected from optical measurements. The paramagnetic $N\text{-}V^0$ has one unpaired electron, and can be converted to $N\text{-}V^-$ by changing the Fermi-level position. This can be achieved by applying external voltage to a p–n junction made from doped diamond, e.g., in a Schottky mode. In the negative charge state $N\text{-}V^-$ an extra electron is located at the vacancy site forming a spin pair with one vacancy electron. The N-V centers are randomly oriented within a diamond crystal, and the quantum state of electron at the N-V center can be manipulated by optical excitation and magnetic resonance [147, 148].

## 9.9 Summary

The history of diamond goes back to 800–1000 BC when diamond was first discovered in India, and the first diamond was brought to Greece by Alexander, the king of the ancient Greek state Macedon in 327 BC. Whatever the story that we have heard about diamond, it is certain that it was renowned as a material with superlative physical qualities, most of which originate from the strong covalent bonding between its atoms.

It is a face-centered cubic (fcc) crystal, and because of its extreme rigidity in lattice it can be contaminated by very few types of impurities. Most natural diamonds are formed at high temperature and high pressure at depths of 140–190 km. Because of its unique optical, chemical, and physical properties, the diamond industry is separated by two distinct categories, one dealing with *gem-grade* diamonds and another with *industrial grade* diamonds.

Gem-grade diamond is the symbol of love. The *love crystal* can come either from mines or from the laboratory where it is prepared synthetically. Synthetic diamond, also called the cultured or cultivated diamond, is mostly produced by high-pressure high-temperature (HPHT) or chemical vapor deposition (CVD) method. The earliest reported attempt to make synthetic diamond is by James Ballantyne Hannay in the year 1879. In the year 1941, an agreement was made between General Electric (GE), Norton and carborundum companies to produce HPHT diamonds. In the beginning of 1970s, some industries took the pain to grow diamond by chemical vapor deposition (CVD) process. Now the growth of synthetic diamond has reached a commercial phase due to the endeavor of Apollo Diamond, a US company based in Boston, MA, by microwave plasma-enhanced CVD (MWPECVD).

In 1963, almost 54 years ago, in the month of July, detonation synthesis of nanodiamonds was discovered, and in the year 2003, the laboratory experiments shifted to the industry. The result of the labor and success of so many scientists and technologists brought successful synthesis of ultrafine dispersed diamonds (UDDs, also called nanodiamonds).

Nanodiamonds (NDs) show excellent mechanical and optical properties. The hardness of the diamond is 50 times stronger than that of titanium (Ti) and stainless steel (SS). As a result, it can be used in abrasive materials, and in cutting tools. Thermal conductivity of nanodiamonds is highest (20 W/cm-C) compared to 0.21 W/cm C of Ti, and 0.16 W/cm C of SS. Thus, NDs are used in the coolant, which prevent the hot zones inside the coolant. It is also used as a paste for its reliability and durability. The hardness, chemical inertness, and low cytotoxicity of NDs have been made possible to use in implants and other surgical tools in biomedical field.

Another member of the family of carbon or diamond is diamond-like carbon or DLC. It is an amorphous material having two types of carbon bonds, i.e., hexagonal graphite bonds ($sp^2$) and tetrahedral diamond bonds ($sp^3$). The name diamond-like carbon has been attributed to this amorphous material because it possesses some characteristics like diamond. As a matter of fact, it is extremely hard, with low electrical resistivity, low coefficient of friction, and it is optically transparent over a wide range of wavelengths.

Diamond-like carbon (DLC) can be hydrogenated or nonhydrogenated. Hydrogenated DLCs are produced from hydrocarbon gas mixture using various energy sources, such as DC-discharge, microwave and radiofrequency (RF) energy, oxy-acetylene flame, and hot filaments.

Nonhydrogenated DLCs are produced by (i) ion beam evaporation, (ii) magnetron sputtering, (iii) pulsed laser ablation, (iv) filtered cathodic arc deposition, (v) RF plasma-enhanced chemical vapor deposition (PECVD) techniques.

## References


1. G.E. Hallow, *The Nature of Diamonds* (Cambridge Univ. Press, Cambridge, London, 1998)
2. F. Ward, *Diamonds* (Gem Book Pub, Bethesda, 1998)
3. P.W. May, Diamond thin films: 21st century material. Royal Soc. Lond., A **358**, 473 (2000)
4. J. Walker, Optical absorption and luminescence in diamonds. Rep. Prog. Phys. **42**(10), 1605 (1979)
5. K. Lakoubovskii, G.J. Adriaenssens, Optical characterization of natural argyl diamonds, diamond relat. Mater **11**(1), 125 (2002)
6. A. Lavoisier, Premier Memoire sur la destruction du diamante par le feu, Histoire de l' Academie royale des sciences. Avec les Memoires de Mathematique & de Physique, Part-2, 564 (1772)
7. L.R. Johnstone, R. Hoffmann, Superdense carbon, C8: supercubane or analog of gamma-silicon? J. Am. Chem. Soc. **111**(30), 810 (1989)
8. C. Murphy, *The Atomic Difference Between Diamonds anf Graphite, Sustainable Nano* (Pub. Of Royal Soc. Chem. Piccadilly, London, 2014)
9. M. Rossi, *How can Graphite and Diamond be so Different if they are Both Composed Pure Carbon* (Sci. Am., Washington, DC, 2007)
10. M. Itoh et al., New metallic carbon crystal. Phys. Rev. Lett. **102**, 055703 (2009)
11. S. Cauchi, *Biggest Diamond Out of This World* (The Age, Melbourne, 2007)
12. E.I. Erlich, W.D. Hausel, *Diamond Deposits* (Society For Mining, Metallurgy, and Exploration, Aptos, 2002), pp. 74–94
13. A.J. Neves, M.H. Nazare (eds.) IEE publications, Glasgow, London, UK (2001)
14. W. Hershey, *Book of diamonds* (Hearthside Press, New York, 1940), p. 22
15. T. Smithson, On the nature of the diamond. Phil. Trans. Royal. Soc. Lond. **87**, 127 (1797)
16. R. Webster, P.G. Read, *Gems their Sources, Descriptions and Identifications* (Butterworth-Heinemann, London, 2000), p. 17
17. W.R. Catelle, The Diamond, John Lane Company (East London, 1911), p. 159
18. V. Ball, *Chapt-1, Diamonds, Gold and Coal of India* (Trubner and Company, London, 1881), p. 1
19. S. Cooke, *Microscopic Diamond Found in Montana* (The Montana Standard, Montana, 2004)
20. V. Lorenz, Argyl in Western Australia: the world's richest diamantiferous pipe; its Past and Future. Gemmologie, Zeitschrift der Deutschen gemmologischen Gesellschaft **56**(1–2), 35 (2007)

21. J. Ralph, *Towards Pretoria, A Record of the War Between Briton and Boer to the Relief of Kimberley* (A. Stoke Co. Pub, New York, 1900)
22. J.B. Hanny, On the artificial formation of the diamond. Proc. Royal Soc. Lond. **30**, 200 (1879)
23. M. Hazen, *The Diamond Makers* (Cambridge Univ. Press, Cambridge, London, 1999)
24. H.T. Hall, Diamond Synthesis, US Patent 2, 947608, Aug. 2, (1960)
25. H.T. Hall, Ultrahigh pressure apparatus. Rev. Sci. Instrum. **31**(2), 125 (1960)
26. H.T. Hall, Ultrahigh pressure research,: at ultrahigh pressure new and sometimes unexpected chemical and physical events occur. Science **128**(3322), 445 (1958)
27. M.G. Lohak, L.I. Alexandrova, Rise in the efficiency of the use of cemented carbide as a matrix of diamond containing studs of rock destruction tool. Int. J. Refract. Metals Hard Mat. **15**, 5 (2001)
28. N. Pal'Yanov, A.G. Sokol, M. Borzdov, A.F. Khokhryakov, Fluid bearing alkaline carbonate melts as the medium for the formation of diamonds in the earth's mantle: an experimental study. Lithos **60**(3–4), 145 (2002)
29. S. Koizumi, C.E. Nobel, M. Nesladek, *Physics and Application of CVD Diamond* (Wliey-VCH, Gernany, 2008)
30. T.K. Gupta, *Handbook of Thick and Thin Film Hybrid Microelectronics* (Wiley, New Jersey, 2003)
31. M. Werner, R. Locher, Growth and application of undoped and doped diamond films. Rep. Prog. Phys. **61**(120), 1965 (1998)
32. E.H. Wahl, Laser Based Diagnostics of Diamond Synthesis Reactor, Report No. TSD-136, DOE Sponsored Research, Mech. Eng. Dept. Stanford Univ., CA, (2001)
33. M.A. Childs, K.L. Menningen, L.W. Anderson, J.E. Lawler, Atomic and radical densities in hot filament diamond deposition system. J. Chem. Phys. **104**(22), 9111 (1996)
34. S.A. Redman, C. Chung, K.N. Roser, M.N.R. Ashfold, Resonance enhanced multiphonon ionization probing of H atoms in a hot filament chemical vapor deposition. Phys. Chem. Chem. Phys. **1**(7), 1415 (1999)
35. N. Tokuda, Homoepitaxial Diamond Growth by Plasma Enhanced Chemical Vapor Deposition, in *Novel Aspects of Diamond from Growth to Applications*, ed. by N. Yang, (Springer, Switzerland, 2015)
36. J.E. Butler, Y.A. Mankelevich, A. Cheesman, J. Ma, M.N.R. Ashfold, Understanding the chemical vapor depositionof diamond: recent progress. J. Phys. Cond. Mat. **21**(5), 364201 (2009)
37. A. Lazea, Y. Garino, T. Teraji, S. Koizumi, High quality p-type chemical vapor deposited {111}-oriented diamonds: growth, and fabrication of related electrical devices. Phys. Status Solidi **A 209**(10), 1978 (2012)
38. M. Katagiri, J. Isoya, S. Kozumi, H. Kanda, Lightly phosphorus doped homoepitaxial diamond films grown by chemical vapor deposition. Appl. Phys. Lett. **85**(26), 6365 (2004)
39. Y.-C. Chu, Y. Tzeng, O. Auceillo, Microwave plasma enhanced chemical vapor deposition of nanocrystalsdiamond films by bias enhanced nucleation and bias enhanced growth. J. Appl. Phys. **115**, 024308 (2014)
40. X.J. Guo, S.L. Sung, J.C. Lin, F.R. Chen, H.C. Shih, Hetroepitaxial diamond nucleation and growth on silicon by microwave plasma enhanced chemical vapor deposition synthesis. Diam. Relat. Mater. **9**(11), 1840 (2000)
41. H. Ye, C.O. Sun, P. Hing, H. Xie, S. Zhang, J. Wei, Nucleation and growth dynamics of diamond films by microwave plasma CVD, surface and caot. Technology **123**, 129 (2000)
42. M. Liehr, F. Fendrych, A. Taylor, M. Nesladek, Large area, low temperature, nano crystalline diamond coating technology with microwaves, MRS Proc. **1282**, 987–1282 (2011)
43. S. Ashida, C. Lee, M.A. Lieberman, Spatially averaged (global) model of time modulated high density argon plasmas. J. Vac. Sci. Technol. **A 13**(5), 2498 (1995)
44. C.A. Wolden, K.K. Gleason, On the pressure limits of diamond chemical vapor deposition. Diam. Relat. Mater. **5**(12), 1503 (1996)

45. H. Ye et al., Nucleation and growth dynamics of diamond films by Microwave Plasma – Enhanced Chemical Vapor Deposition (MWPECVD). Surface Coating Technol. **123**, 129 (2000)
46. T. Teraji, H. Wada, M. Yamamoto, K. Arima, T. Ito, Highly efficient doping of boron into high quality homoepitaxial diamond films. Diam. Relat. Mater. **15**(4–8), 602 (2006)
47. M. Sakamoto, J.G. Endriz, D.R. Scifres, 120 W CW output power from monolithic AlGaAs (800nm) LaserDiode array mounted on diamond sink. Electronics Lett. **28**(1), 197 (1992.) and also M. Angelone et al., Development and applications of CVD diamond detectors to 14 MeV NEUTRON Flux. Moni. Radiat. Prof. Dosimetry. **110**(1–4), 233 (2014)
48. P. Dore et al., Infrared properties chemical vapor deposition of polycrystalline diamond windows. Appl. Opt. **37**(24), 9 (1998)
49. C.S.J. Pickles, T.D. Madgwick, R.S. Sussmann, C.N. Hall, Industrial diamond review. IND DIAM RE **60**(4), 293 (2000)
50. Appollo Diamond, Boston, now sold to Scio Diamond, Grenville, South Carolina, USA (2011), Wikipedia, the free encyclopedia, 24th October, (2013)
51. Y.A. Mankelevich, P.W. May, New insights into the mechanism of CVD Diamond growth: single crystal diamond in MWPECVD reactors. Diam. Relat. Mat. **17**, 1021 (2008)
52. W. Wang, M.S. Hall, K.S. Moe, J. Tower, T.M. Moses, Latest-generation of CVD-grown synthetic diamond from apollow diamondinc. Gemes Gemol. **43**(4), 294 (2007)
53. S. Jegarajah, *Would you Buy a Man made Diamond?* (CNBC Rept., EDT, Manhattan, 2013)
54. R.P. Mildren, Intrinsic Optical Properties of Diamond, in *Optical Engineerinmg of Diamond*, ed. by R. P. Mildren, J. M. Rabeau (Eds), 1st edn., (Wiley-VCH verlag GmbH & Co., Germany, 2003)
55. A.M. Zaitsev, *Optical Properties of Diamond: A Data Hand Book* (Springer, Berlin, 2001)
56. T.E. Schlesinger, R.B. James, *Semiconductors for Room Temperature Nuclear Detector Applications* (Academic Press, San Diego, 1995)
57. S.F. Kozolov, R. Stuck, M. Hage-Ali, P. Siffert, Preparation and charactics of natural diamond nuclear radiation detectors. IEEE Trans. Nuclr. Sci. **NS-22**, 160 (1975)
58. T.K. Gupta, *Radiation, Ionization and Detection in Nuclear Medicine* (Springer, Hiedelberg, 2013)
59. S. Vatnisky, H. Jarvinen, Application of a natural diamond detector for the measurement of relative dose distribution in radiotherapy. Phys. Med. Biol. **38**, 173 (1993)
60. S.F. Kozolov et al., Diamond dosimeter for X-ray and gamma radiation. IEEE Trans. Nuclr. Sci. **NS-24**, 235 (1977)
61. W.U. Laub, R. Crilly, Clinical radiation therapy measurements with a new commercial synthetic single crystal diamond detector. J. Appl. Clinical Med. Phys. **15**(6), 1 (2014)
62. P. Moritz, E. Berdermann, K. Blasche, H. Stelzer, B. Voss, Broadband electronics for CVD-diamond detectors. Diam. Relat. Mater. **10**, 1765 (2001)
63. V.V. Danilenko, On the history of the discovery of nanodiamond synthesis. Phys. Solid State **46**(4), 595 (2004)
64. P.J. Lu et al., The earliest use of corundum and diamond in prehistoric China. Archaeometry **47**, 1 (2005)
65. R.S. Lewis et al., Interstellar diamonds in meteorites. Nature **326**, 160 (1987)
66. R.S. Lewis, E. Anders, B.T. Draine, Properties, detectability, and origins of interstellar diamonds in meteorites. Nature **339**, 117 (1989)
67. S. Ameari, R.S. Lewis, E. Anders, Interstellar grains in meteorites. Geochim Cosmocim Acta **58**, 459 (1994)
68. O.A. Shenderova, V. Zirmov, D. Brenner, Carbon materials and nano structures. Crit. Rev. Solid State Mater. Sci. **27**(3/4), 227 (2002)
69. S. Sattel, J. Robertson, Z. Tass, M. Scheib, D. Wiescher, E. Ehrhardt, Formation of nano-crystalline diamond by hydrocarbon plasma beam deposition. Diam. Relat. Mater. **6**, 255 (1997)

70. V.Y. Dolmatov, Detonation synthesis ultradispersed diamond: properties and applications. Russ. Chem. Rev. **70**, 607 (2010)
71. H.D. Espinosa et al., Mechanical properties of ultrananocrystalline diamond thin films relevant to MEMS/NEMS devices. Exp. Mech. **43**, 256 (2003)
72. T. Zimmerman et al., Ultracrystalline /single crystal diamind heterostructure diode. Diam. Relat. Mater. **14**, 416 (2005)
73. D.M. Gruen, Nanocrystalline diamond films. Annu. Rev. Mater. Sci. **29**, 211 (1999)
74. G. Collin, On the history of technical carbon, CFI—Ceram. Forum. Int. **77**, 28 (2000)
75. S. Iijima, Helical microtubules of graphite carbon. Nature **354**, 56 (1991)
76. H.W. Kroto, J.R. Heath, S.C. O'Brien, R.F. Curl, R.E. Smalley, C-60-Buckministerfullerine. Nature **318**, 162 (1985)
77. P. de Carli, J. Jamieson, Formation of diamond by explosive shock. Science **133**, 1821 (1961.) US Patent 3238019 (1966)
78. S. Iijima, High resolution electron microscopy of some carbonaceous materials. J. Micros. **119**, 99 (1980)
79. O. A. Shenderova, D. M. Gruen (eds.), *Ultrananocrystalline Diamond: Synthesis, Properties and Applications* (William Andew Pub, New York, 2006)
80. B.J. Derjaguin, D.V. Fedoseev, Synthesis of diamond at low pressure. Sci. Am. **223**, 102 (1975)
81. L.A. Curtiss et al., Carbon Ad-dimer defects in carbon nanotubes. Phys. Rev. Lett. **96**(7), 975506 (2006)
82. D.M. Gruen and J.W. Elam, Nanotube Daimond Composites, MRS Fall Meeting, Paper Q2.3, December 1–5 (Boston, MA, 2003)
83. K.V. Volkov, V.V. Danilenko, V.I. Elin, Diamond synthesis from carbon of detonation products. Fiz. Goren. Vzriva **26**, 123 (1990.) (in Russian)
84. L. Landt et al., Optical response of diamond nanocrystals as a function of particle size, shape, and symmetry. Phys. Rev. Letts. **103**, 047402 (2009.) and also, J.Y. Raty, G. Galli, C. Bostedt, T.W. van Buuren and L.J. Terminello, Phys. Rev. Lett. 90, 037401/1, (2003)
85. A.E. Aleksenski et al., Ultradisperse diamond cluster aggregation studied by atomic force microscopy. Technol. Phys. Lett. **26**, 819 (2000.) and also A.E. Aleksenskii, M.Baidakova, V. Osipov, and A. Vul', Nanodiamonds, 55, (2010)
86. G. Galli, Structure, Stability and Electronic Properties of nanodiamonds, in *Computer Based Modeling of Novel Carbon Systems and Their Properties, Carbon Materials*, ed. by L. Colombo, A. Fasolino, (Springer Science, Heidelberg, 2010)
87. A.J. Williamson, J.C. Grossman, R.Q. Hood, A. Puzder, G. Galli, Quantum monte carlo calculations of nanostructure optical gaps: application to silicon quantum dots. Phys. Rev. Lett. **89**, 196803 (2002)
88. P. Badziag, W.S. Verwoerd, W.P. Ellis, N.R. Greiner, Nanometer sized diamonds are more stable than graphite. Nature **343**, 244 (1990)
89. J.Y. Ratty, G. Galli, Ultradispersity of diamond at nanoscale. Nat. Mater. **2**, 792 (2003)
90. B. Wen, J. Zhao, T. Li, Relative stability of hydrogenated nanodiamond and nanographite from density function theory. Chem. Phys. Lett. **441**(4–6), 318 (2007)
91. J.Y. Raty et al., Quantum confinement and fullerene like surface reconstruction in nanodiamonds. Phys. Rev. Lett. **90**, 037401/1 (2003)
92. E. Runge, E.K.U. Gross, Density-functional theory for time dependent systems. Phys. Rev. Lett. **52**, 997 (1984)
93. A.A. Fokin, P.R. Schreiner, Band gap tuining in diamonds: first principle computational studies. Mol. Phys. **107**, 823 (2009)
94. K. Lenzke et al., Experimental determination of ionization potentials of the first five members of the nanodiamond series. J. Chem. Phys. **127**, 084320 (2007)
95. P.-H. Chung, E. Perevedentseva, C.-L. Cheng, Surface defects induced fluorescence from nanometer sized diamond particles for biological applications. Surf. Sci. **601**, 3866 (2007)

96. N. Brown, O. Hod, Controlling the electronic properties of nanodiamonds via surface chemical functionalization: a DFT study. J. Phys. Chem. **C118**, 5530 (2014)
97. J.P. Boudou et al., High yield fabrication of fluorescent nanodiamonds. Nanotechnology **20**, 235602 (2009)
98. M. Schhrand, S.A.C. Hens, O.A. Shenderova, Nanodiamond particles: properties and perspective for bioapplications. Crit. Rev. Solid State Mater. Sci. **34**, 18 (2009)
99. A.M. Schrand et al., Are diamond particles cyotoxic? J. Phys. Chem. **B111**, 2 (2007)
100. K.V. Purtov, A.I. Petunin, A.E. Burov, A.P. Puzr, V.S. Bonder, Nanodiamonds as carrier for address delivery of biologically active substances. Nano Scale Res. Lett. **5**, 631 (2010)
101. A.M. Schrand et al., From Manufacturing to Medical Applications, Nanostructure Science and Technology, in *Safety of Nanoparticles*, ed. by T. J. Webster (Ed), (Springer, Heidelberg, 2009), p. 159
102. Q. Zhang et al., Fluorescent PLLA Nanodiamond composites for bone tissue engineering. Biomaterials **32**, 87 (2011)
103. V.L. Kuznetsov, A.L. Chuvilin, Y.V. Botenko, I.Y. Malkov, V.M. Titov, Onion-like carbon from ultra-disperse diamond. Chem. Phys. Lett. **222**, 343 (1994)
104. J.K. McDonough and Y. Gogotsi, Carbon onions: synthesis and electrochemicall applications. Electrochemical Soc. Interface, p. 61, Fall (2013)
105. S. Tomita, M. Fujii, S. Hayashi, K. Yamamoto, Transformation of carbon onions to diamond by low pressure heat treatment in air. Diam. Relat. Mater. **9**(3–6), 856 (2000)
106. Q. Zou et al., Nanocrystalline diamond sintered from onion like carbon. J. Nano Mat., 271494 (2014, 2014)
107. D.R. McKenzie, R.C. McPhedran, N. Savvides, L.C. Botten, Properties and structure of amorphous hydrogenated carbon films. Phil. Mag. B **48**, 341 (1983)
108. J. Robertson, Amorphous carbon. Adv. In Phys. **35**, 317 (1986)
109. J.C. Phillips, Structure of amorphous (Ge, Si)-xyx alloys. Phys. Rev. Lett. **42**, 1151–1159 (1979)
110. M.F. Thorpe, Continuous defrormations in random networks. J. Non-Cryst. Solids **57**, 355 (1983)
111. J.C. Angus, Y. Wang, in *Diamond and Diamond lile Films and Coatings*, ed. by R. F. Clausing et al. (Eds), (Plenum, New York, 1991)
112. I. Ahmad, S.S. Roy, P.D. Maguire, P. Papakonstantinou, J.A. McLaughlin, Effect of substrate bias voltage and substrate on the structural properties of amorphous carbon films deposited by unbalanced magnetic sputtering. Thin Solid Films **482**, 45 (2005)
113. D.B. Chrisey, G.K. Hubbler, *Pulsed Laser Deposition of Thin Films* (Wiley-Intersci, New York, 1994)
114. T. Sato, S. Furino, S. Iguchi, M. Hanbusa, Diamond like carbon films prepared by pulsed laser evaporation. App. Phys., A **45**, 355 (1988)
115. A.V. Robe, B. Luther-Devies, E.G. Gamaly, Ultrafast ablation with high pulse rate lasers, part: experiments on laser deposition of amorphous carbon films. J. App. Phys. **85**, 4222 (1999)
116. M. Hanabusa, K. Tsujihara, Diamond like carbon films deposited by laser ablation using frozen acetylene target. J. Appl. Phys. **78**, 4267 (1995)
117. M. Yoshida, M. Okoshi, N. Inoue, Femto-second pulsed laser deposition of diamond like films onto silicone rubber. J. Phys. Conf. Ser. **59**, 368 (2007)
118. A.W. Wright, On the production of transparent metallic films by electrical discharge in exhausted tubes. Am. J. Sci. Arts **13**, 49 (1877)
119. T.A. Edison, Process of Duplicating Phonograms, US Patent, 484582 (1892)
120. R. L. Boxman, P. J. Martin, D. Sanders (eds.), *Hand Book of Vacuum Arc Science and Technology* (Noyes, New York, 1996)
121. Z.Q. Ma, Y. Kido, The atomic displacements on surface generated by low energy projectile. Thin Solid Films **359**, 288 (2000)

122. A.I. Morozov, Reports of the Academy of Sciences of the USSR, 163, 1363 (1965),
123. A.I. Morozov, Focussing of cold quasineutral beams in electromagnetic fields. Sov. Phys. d, Doklady **10**(8), 775 (1966)
124. M.M.M. Bilek, Y. Yin, D.R. McKenzie, A study of filter transportmechanism in filtered cathodic vacuum arcs. IEEE Trans. Plasma Sci. **24**, 1166 (1996)
125. A. Anders, S. Anders, I.G. Brown, Transport of vacuum arc plasmas through magnetic macroparticle filters. Plasma Sources Sci. Technol. **4**, 1 (1995)
126. A. Anders, S. Anders, M.R. Dickenson, R. MacGill, I.G. Brown, S-shaped magnetic macroparticle filter for cathodic arc deposition. IEEE Trans. Plasma Sci. **25**, 670 (1997)
127. A. Anders, R.A. MacGill, Twist filter for the removal of macroparticles for cathodic arc plasma. Surf. Coat. Technol. **133**, 96 (2000)
128. P.J. Martin, A. Bendavid, Review of the filtered vacuum acr process and materials deposition. Thin Solid Films **394**, 1 (2001)
129. E. Byon, A. Anders, Bias and self-bias of magnetic macroparticle filters foe cathodic arc plasmas. J. Appl. Phys. **93**(12), 1 (2003)
130. H. Tsai, D.B. Bogy, Critical review: characterization of diamond like carbon films and their application as overcoats on thin-film media for magnetic recording. J. Vac. Sci. Technol. A **5**, 3287 (1987)
131. C.S. Bhatia et al., Tribo chemistry at the head-disk interface. IEEE Trans. Magn. **35**, 910 (1999)
132. A. Anders, W. Fong, A. Kulkarni, F.W. Ryan, C. Sing Bhatia, Ultrathin diamond like carbon films deposited by filtered carbon vacuum arcs. IEEE Trans. On Plasma Sci. **29**(50), 1 (2001)
133. P. Lemoine, R.W. Lamberton, A.A. Ogwu, J.F. Zhao, P. Maguire, L. McLaughlin, Complimentary analysis techniques for the morphological study of ultrathin amorphous carbon films. J. Appl. Phys. **86**, 6564 (1999)
134. B. Wei, B. Zhang, K.E. Johnson, Nitrogen induced modifications in microstructure and wear durability of ultrathin amorphous carbon films. J. Appl. Phys. **83**, 2491 (1998)
135. S. Logothetidis, P. Patsalas, M. Gioti, A. Galdikas, L. Prenevicius, Growth kinetics of sputtered amorphous carbon films: composition studies and phenomenological model. Thin Solid Films **376**, 56 (2000)
136. M.K. Fung, K.H. Lai, C.Y. Chan, et al., Mechanical properties and corrosion studies of amorphous carbon on magnetic disks prepared by ECR plasma technique. Thin Solid Films **368**, 198 (2000)
137. S. Anders, I.G. Brown, C.S. Bhatia, D. Bogy, Wanted: hard, thin coatings for near contact recording. Data Storage **4**, 31 (1997)
138. S. Anders, C.S. Bhatia, F. Fong, R.Y. Lo, D. Bogy, Application of cathodic arc deposited ampophous hard carbon films to the head/disk tribology. Mat. Res. Symp. Proc. **517**, 371 (1998)
139. D. Sheeja, B.K. Tay, L.N. Nung, Tribological characterization of surface modified uhmwpe against diamond like carbon coated Co-Cr-Mo. Surf. Coat.Technol **190**, 231 (2005)
140. M.P. Gispert, A.P. Serro, R. Colaco, B. Saramango, fear and wear mechanisms in hip prosthesis: comparison of joint materials behavior in several lubricants. Wera **260**, 149 (2006)
141. F. Zivic, M. Babic, S. Mitrovic, D. Adamovic, S. Pelemis, Friction coefficient of uhmwpe during dry reciprocating sliding. Trib. In Indus. **36**(3), 281 (2014)
142. S.K. Young, M.A. Lotito, T.S. Keller, Friction reduction in total joint arthoplasty. Wear **222** (1), 29 (1998)
143. R.S. Butter, A.H. Lettington, Diamond like carbon for biomedical applications. J. Chem. Vapor. Dep. **3**, 182 (1995)
144. G. Dearnaley, J.H. Arps, Biomedical applications of diamond like carbon (DLC) coatings: a review. Surf. Coat. Technol. **200**, 2518 (2005)
145. L. Childress, R. Hanson, Diamond NV centers for quantum computing and quantum networks. MRS Bul. Vol **38**, 134 (2013)

146. K. Nemoto et al., Photonic quantum networks from NV centers. Scientific Repts **6**, 26284 (2016)
147. I. Aharonovich et al., Diamond photonics. Nat. Photonics **5**, 397 (2011)
148. C.S. Schreyvogel et al., Active Charge state control of single N-V centers in diamond by plane Al-Schottky junctions. Scientific Repts. **5**, 12160 (2015)
149. O.A. Shenderova, Z. Hu, D. Brenner, Carbon Family in Nanoscale, in *Synthesis Properties, and Applications of Ultrananocrystalline Diamond*, ed. by D. M. Gruen, O. A. Shenderova, A. Y. Vul (Eds), (Springer, Berlin, 2005)

# Chapter 10
# Fullerene

## 10.1 Introduction

*Fullerene* The molecule buckminsterfullerene is beautiful physically and intellectually. Its qualities and even some of its properties can be appreciated instantly and intuitively by nonscientists. Its uniqueness is bound to lead to novel applications–superconductivity is the leading contender at the moment. The commercial potential of buckminsterfullerene has heightened the excitement and controversy in recent years, while the exact nature of the discovery process in 1985 has been the subject of a heated feud between the British and American scientists involved [1]. The theoretical structure of *buckminsterfullerene*, and *fullerene (buckyball)* in general, was established in the year 1970 [2, 3]. Figure 10.1 shows the schematic representation of buckminsterfullerene C60 buckyball, one of the allotropes of carbon containing 60 equivalent carbon ($C_{60}$), formed with the pattern of a football (*buckyball*), giving it the highest symmetry (Fig. 10.2).

The elemental carbon (C) has been known from the millennia having two polymorphic forms: graphite having two-dimensional layers of $sp^2$-hybridized carbon atoms interlinked by weak van der Waals forces, and diamond having three-dimensional bonding in which each $sp^3$-hybridized carbon atom is bonded to four other similar atoms (Fig. 10.1) [4, 5].

Fullerenes, the latest allotropes of carbon having cage structure, are receiving much attention recently. Indeed, the structure, chemistry, physics, and the hidden science of this allotrope of carbon have attracted the attention of the scientific community [1, 6].

Fullerenes, the class of closed cage molecules, have in general formula $C_{20+2n6}$. $C_{60}$ has only one chemically distinct carbon atom. However, in some larger fullerenes, there is a possibility of isomers and some fullerenes are also chiral [6, 7].



T. Gupta, *Carbon*, DOI 10.1007/978-3-319-66405-7_10

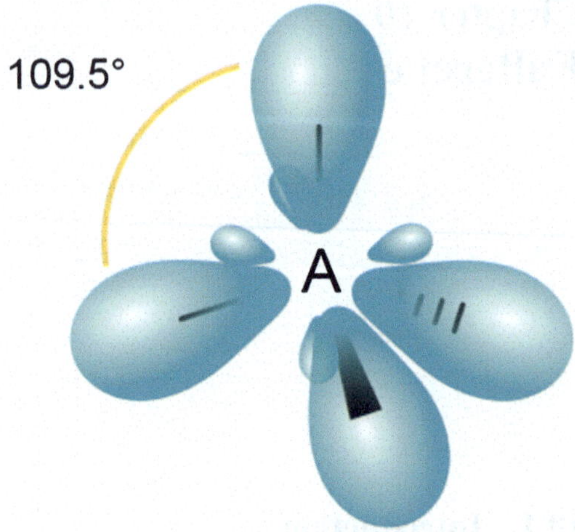


**Fig. 10.1** Four sp3-hybridized carbon atom is bonded to four other similar atoms (Courtesy: WIKI)

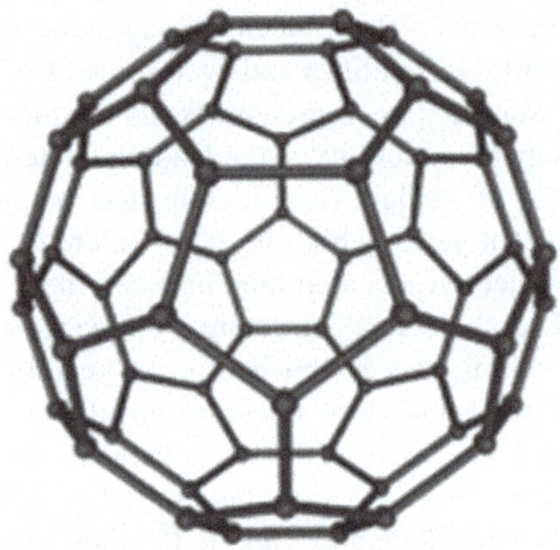

**Fig. 10.2** Schematic representation of buckministerfullerene C60 (Courtesy: Leo, WIKI)

## 10.2 Synthesis of Fullerenes

The early discovery of fullerene by arc discharge method was revolutionized due to its simplicity in producing fullerene by vaporization of resistive graphite rods. The procedure is termed as Kratschmer-Huffman method. It is a straightforward and low-cost method for generating large quantities of fullerene containing carbon shoot [8, 9].

### *10.2.1 Arc Vaporization*

Figure 10.3 shows the schematic of the arc vaporization of carbon rods for generating fullerene. Two high-purity graphite rods are clamped to the high-current

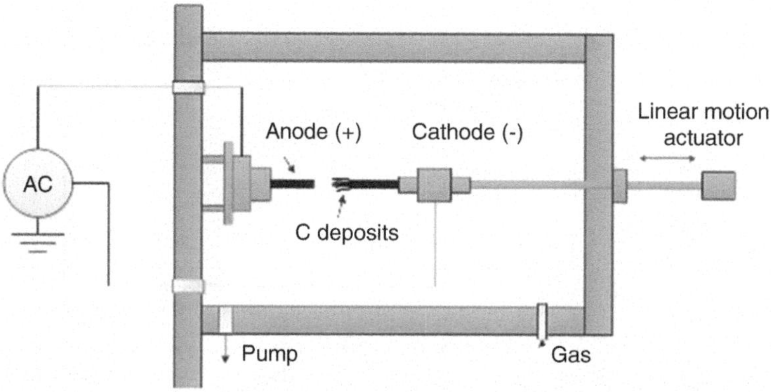


**Fig. 10.3** Schematic of the arc vaporization of carbon rods for generating fullerene (Reproduced with permission: J.P. Gore)

feedthroughs, and the chamber is pumped down to $\leq 10^{-3}$ torr and refilled with helium gas to a pressure of 150–250 torr before applying high current (100–200 A). Low voltage, either alternating current (AC) or direct current (DC), is applied for evaporation of the graphite rods. The crude carbon product or soot produced by this vaporization collects on the water-cooled inner surface of the walls and collected using a stiff brush. The soot contains a variety of carbon products including $C_{60}$ and larger fullerenes [10, 11].

The experimental Smalley's Spersonic Cluster Beam Apparatus used to discover the buckminsterfullerene ($C_{60}$) is shown in Fig. 10.4. This setup is now available commercially to study a variety of carbon clusters. A rotating disk of graphite is irradiated with a powerful laser beam to evaporate carbon (C). As the laser beam falls on the disk, a valve controlling the flow of helium (He) gas stream is opened. The gas carries the evaporated carbon species with it, and during its passage to the nozzle, the species in the vapor undergo clustering. The cluster is then analyzed by a time-of-flight mass spectrometer (shown in Fig. 10.4). Sometimes, an integration cup is introduced before the carbon cluster coming out of the chamber. This increases the percentage of $C_{60}$ and $C_{70}$ atoms.

The purification of the material is followed by the characterization of the material by a variety of spectroscopic techniques. At the same time, experiments are also done using Fourier transform ion–cyclotron resonance (FT-ICR0 apparatus) apparatus.

### 10.2.2 *Laser Ablation*

Historically, Smalley and coworkers first detected $C_{60}$ as an unusually abundant species in mass spectra recorded on carbon clusters produced by *laser ablation of graphite* in a *helium flow* [12, 13]. The laser vaporization techniques could produce

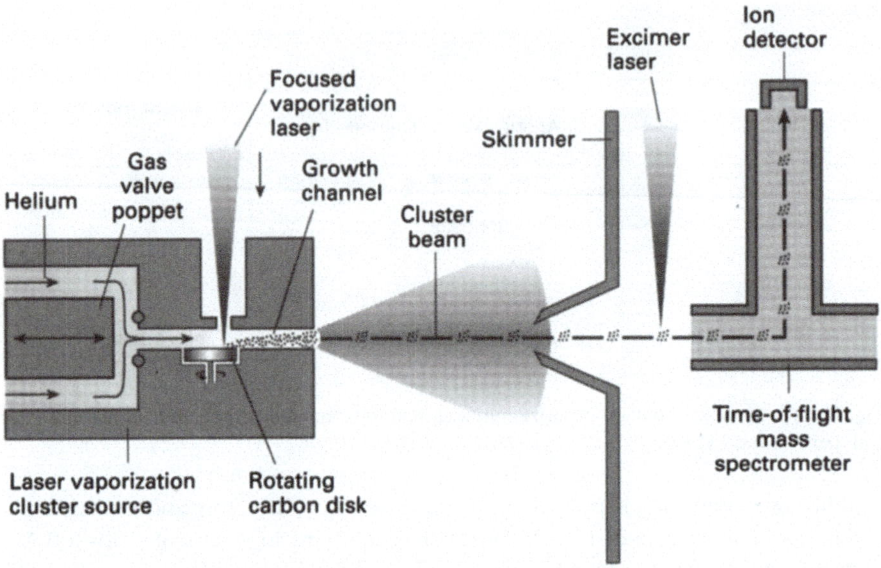


**Fig. 10.4** Schematic of Smalley's Supersonic Cluster Beam Apparatus (AP3) (Reproduced with permission from Wiley, Ref. [1])

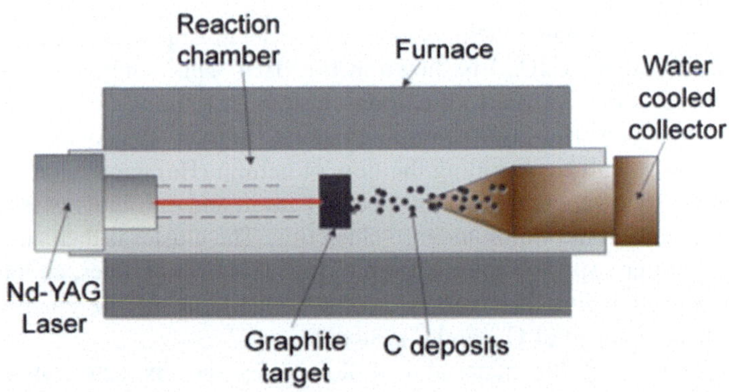


**Fig. 10.5** Schematic view of laser ablation apparatus used to prepare C60 (Reproduced with permission, J.P. Gore)

only on the order of $10^4$ clusters in the beginning of 1990. The early attempts to prepare fullerenes by laser ablation relied on ablation in an inert gas buffer at room temperature. As a result, the dense carbon plasma created during laser ablation used to cool too quickly. Smalley and coworkers improved the method by carrying out the experiment at elevated temperatures and were ultimately successful in efficient production of macroscopic quantities of $C_{60}$ and other fullerenes [14, 15]. Figure 10.5 shows the schematic view of laser ablation apparatus used to prepare C60.

The key feature of the laser ablation apparatus used to produce fullerene is that the graphite target is ablated in a high-temperature furnace, so that the plasma cools more slowly and growing carbon clusters have sufficient time to rearrange (anneal) into stable fullerenes. Lieber and Chen reported that the fullerene production is most efficient at 1200 °C [13]. They also noticed that the efficiency and selectivity of the fullerenes produced depend on the characteristics of the carbon plasma such as the laser pulse energy and wavelengths of the laser [16]. Second, the growth of the carbon cluster from the carbon plasma also depends on the buffer gas pressure and annealing of the product during growth.

Sooner, it was discovered that the Bingel-Hirsch [17] reaction (cyclopropanation reaction) is a versatile reaction to synthesize a wide range of fullerene derivatives [18]. The retro-Bingel reaction of the (alkoxycarbonyl) methanofullerenes of C60, C70, C76, and ent-C76 at the second reduction potential (upon controlled potential electrolysis (CPE)) recovered the parent fullerenes yield by 75–82%, 70%, and 5–9%, respectively, after the addition of four electrons per molecule [19]. The important retro-cyclopropanation reaction of methanofullerenes by CPE is the electrochemically induced isomerization of the addends by the migration on the C60 fullerene surface, known as the *walk on the sphere*. During walk on the sphere of bis-adducts, the major isomers formed were always the trans-2 derivative (40–50%) followed by e-(25%) isomer and the formation of trans-1 bis-adduct with 10% yield [20, 21].

The electrochemical measurements were performed by designing a cell (as shown in Fig. 10.6) which follows cyclic voltammograms (CVs) potential electrolysis process [22]. A glassy carbon working electrode, a silver (Ag) wire pseudoreference electrode, and a platinum (Pt) mesh counter-electrode were used for the experiment. To determine the potential to be applied, the CVs of each compound were recorded at the beginning of each experiment. After electrolysis under high vacuum, the solutions were exhaustively reoxidized at 0 V before purification and product analysis. An 80% yield of C60 was achieved (1 mg) after reoxidation and purification [23, 24].

Throughout the 1990s, researchers relied on the *carbon arc* method for producing fullerenes. However, the carbon arc method generated small quantities of fullerenes. Moreover, the inefficiencies of the carbon arc method and the postprocessing required to separate the fullerenes from the soot make the cost of production very high and ultimately prohibit commercial applications. Dr. Jack Howard and his team developed a commercially viable *flame combustion* method at MIT, Cambridge, MA (Fig. 10.7), to replace the carbon arc process.

### *10.2.3 Flame Combustion Method*

The experimental setup for the burning of low-pressure laminar diffusion flames of benzene vapor diluted with argon in oxygen is shown in Fig. 10.7. The vertically adjustable burner consists of a 10 mm inside diameter (i.d.) fuel outlet tube

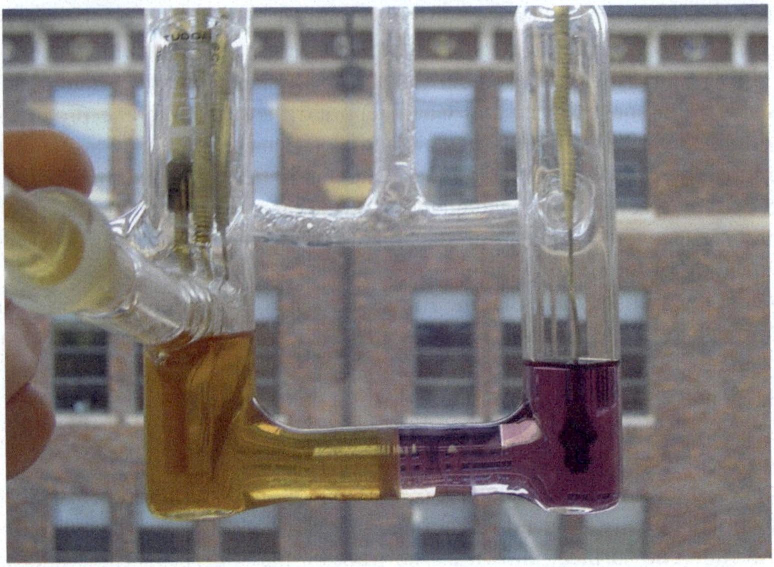

**Fig. 10.6** High-vacuum electrolysis of a C60-fullerene derivative (Photo credit: AdrianVillata, WIKI)

surrounded by a 30.5 cm diameter porous bronze plate through which oxygen is fed [25].

Recently, Sathish and Miyazawa have reported synthesis of fullerene nanowhiskers (FNW) by liquid–liquid interfacial precipitation in carbon disulphide ($CS_2$) and isopropyl alcohol system. The electron microscope images reveal the formation of nontubular FNW, with fcc and triclinic structures. It is expected that the high solubility of C60 in $CS_2$ results in partial polymerization and formation of triclinic structure [26–28].

The controlled flame combustion processing can produce higher fullerenes precisely enough to generate various mixtures of C60, C70 and the higher fullerenes by ton, and can tailor the product to a particular customer's demand without the need for expensive solvent-based postprocessing. The method also has an advantage to dope fullerene cluster by introducing the dopant into the benzene–oxygen mixture prior to combustion [29–32].

Fullerene ($C_{60}$) has 60 carbons held together by covalent bonds on the surface of a truncated icosahedron, and the molecules interact with each other by van der Waals forces. Furthermore, the molecules rotate rather freely and are expected to show some interesting phenomena. One of the many interesting phenomena is that it shows some interesting dissolution behavior (endothermic below room temperature and exothermic above room temperature), which has been interpreted earlier as due to phase change in solid $C_{60}$.

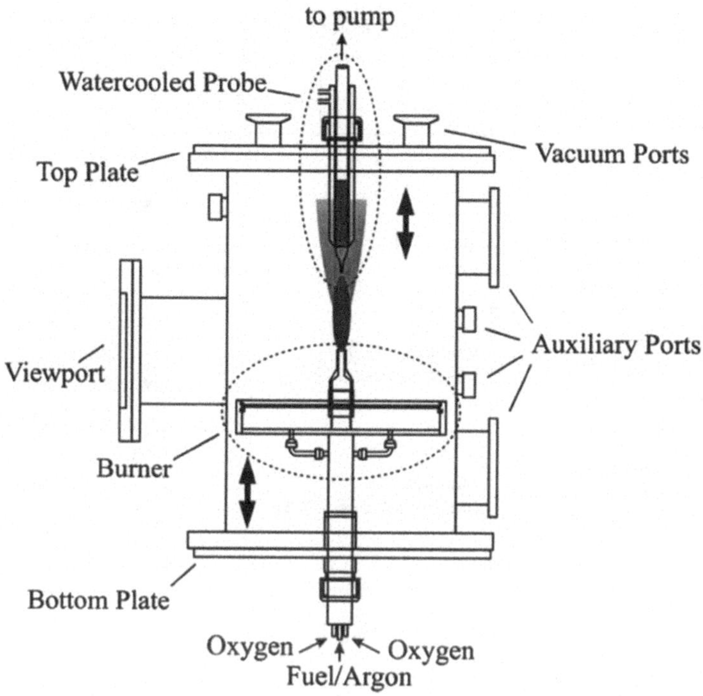


**Fig. 10.7** Experimental setup for the flame-combustion technique (Reproduced with permission from J.B. Vander Sande)

## 10.3 Separation and Purification of Fullerenes

In almost all the methods described above, fullerenes are produced as a crude mixture of C60 along with other Cn ($n \leq 60$) and conventional hydrocarbon species. Therefore, individual fullerene clusters must be isolated from the crude soot. Kratschmer and Huffman produced fullerene using benzene. The present researchers used toluene as a better alternative to benzene, as benzene is not a very friendly chemical. The extraction of fullerene is efficiently carried out using several methods. Soot is first extracted continuously by boiling toluene in a Soxhlet solid extraction apparatus with any of the solvents such as chloroform, toluene, benzene, n-hexane, 1, 2-dichlorobenzene, etc. [33, 34]. Figure 10.8 illustrates the basic separation and purification of fullerenes.

The second method adopted to extract fullerenes from soot involves sonication of toluene–soot mixture at room temperature. The second basic purification procedure of fullerenes [35] is an efficient alternative sonication method of dispersing carbon soot in tetrahydrofuran (THF) at a concentration of 0.1 $gmL^{-1}$. The experiment is performed at room temperature, and it takes 20 min to complete the sonication [36]. The solution is filtered and THF is evaporated by rotary evaporator

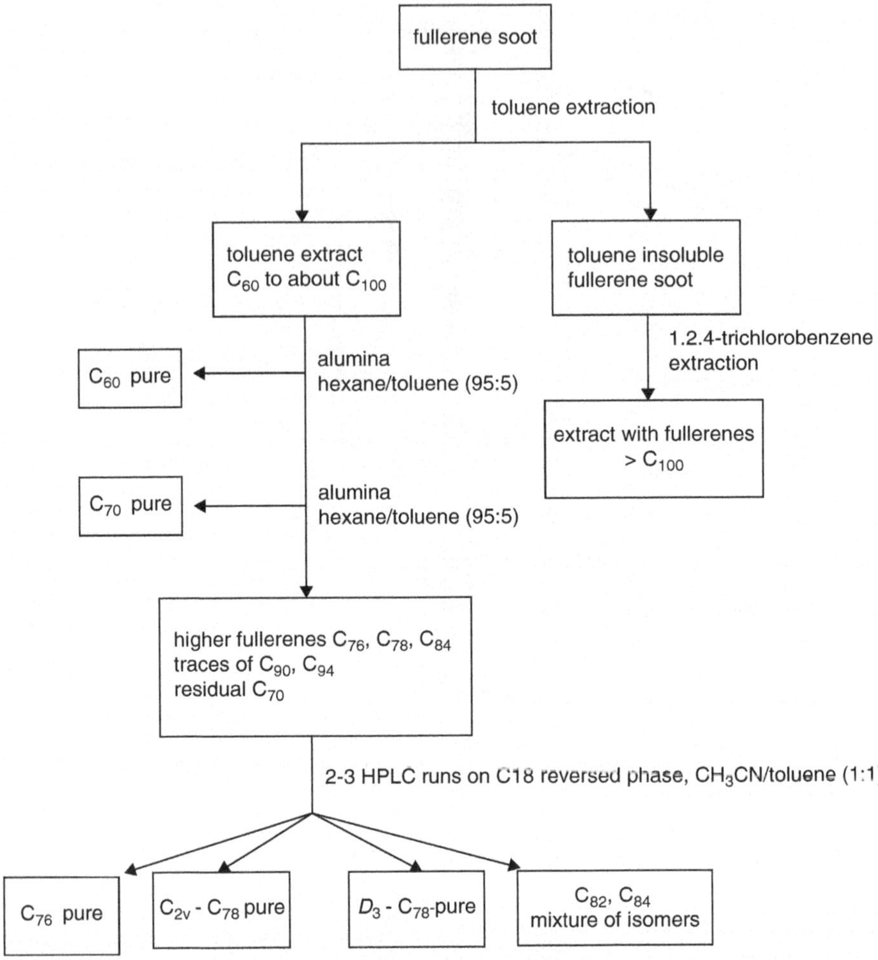


**Fig. 10.8** Illustration of the basic separation and purification of fullerenes (Reproduced with permission from Wiley-VCH, Weinheim, Germany, 2010)

at 50 °C. The remaining impurities are mostly polyaromatic hydrocarbons and are washed in diethyl ether before being purified.

A sublimation method called the nonsolvent-based method has been used successfully to extract fullerenes from the soot. C60 and C70 powders are known to sublime in vacuum at relatively low temperatures (between 350 and 450 °C). The advantage of producing fullerenes by sublimation process is that it does not contain any solvent [37, 38].

To isolate pure C60 and other individual fullerenes, liquid chromatography is used. The most widely used chromatographic separation employs a neutral alumina ($Al_2O_3$) stationary phase and simple gravity feed [39, 40]. Chromatography on alumina provides a straightforward technology for isolating C60 and C70.

However, this method is very time-consuming, and produces large quantities of organic wastes. A modified and more attractive method of separation of C60 and other fullerenes is the use of a special type of chromatogram where, the activated charcoal supported on silica gel and the mobile phase is pure toluene [41, 42]. Recently, many solid phases have been developed for high-performance liquid chromatography of fullerenes [43, 44].

According to some theoretical calculations, there exists no liquid phase below the triple point. Recently, Monte Carlo simulation study of six model fluids interacting through short-range attractive potentials has been applied to study the vapor–liquid diagrams of $C_{60}$ and $C_{70}$. It has been found that binodal curves of these systems correctly reproduce the $C_{60}$ and $C_{70}$ Girifalco potentials [45, 46].

It is important to verify the purity of individual fullerene samples, since the presence of impurities can significantly affect the solid-state properties of the sample. Among the several techniques that are applied to verify and to ascertain the purity of fullerene samples are (a) *nuclear magnetic resonance* (NMR) *spectroscopy*; (b) *mass spectroscopy*; (c) *optical spectroscopy*, and (d) *high-pressure liquid chromatography* (HPLC).

## 10.4 NMR Spectroscopy

Solution phase carbon-13 ($^{13}C$) nuclear magnetic resonance spectroscopy (NMR) combined with theoretical calculation is a powerful tool for fullerene identification [47–49]. The most convincing $^{13}C$ NMR spectrum of the purified C60 reported by Kroto and coworkers has a single peak at ~142.7 PPM, indicating a highly symmetrical truncated icosahedron structure in which all carbons are identical (as shown in Fig. 10.9). The $^{13}C$ NMR spectrum of C70 shows five sets of distinct resonance at 150.8 (e), 148.3 (d), 147.8 (c), 144.4 (b), and 130.8 (a), with inequivalent carbon atoms in the ratio of 10:10:20:20:10. These five resonances reflect the lower symmetry of C70 versus C60 [50]. This is precisely the ratio of the line intensities observed in the $^{13}C$ NMR spectrum. Figure 10.9 shows 13C NMR spectra and structural drawings of C60 and C70.

## 10.5 Mass Spectroscopy

Mass spectrometry is an analytical chemistry technique that helps identify the amount and type of chemicals present in a sample by measuring the mass to charge ratio and the abundance of gas-phase ions [51]. In the year 1990 and 1991, indirect evidence of the structures and properties of fullerenes was obtained almost entirely from mass spectrometric experiments. The reviews by Kroto, Smalley, and their coworkers [52, 53] provided an excellent firsthand perspective on the discovery, postulation of structures, and characterization of fullerenes.

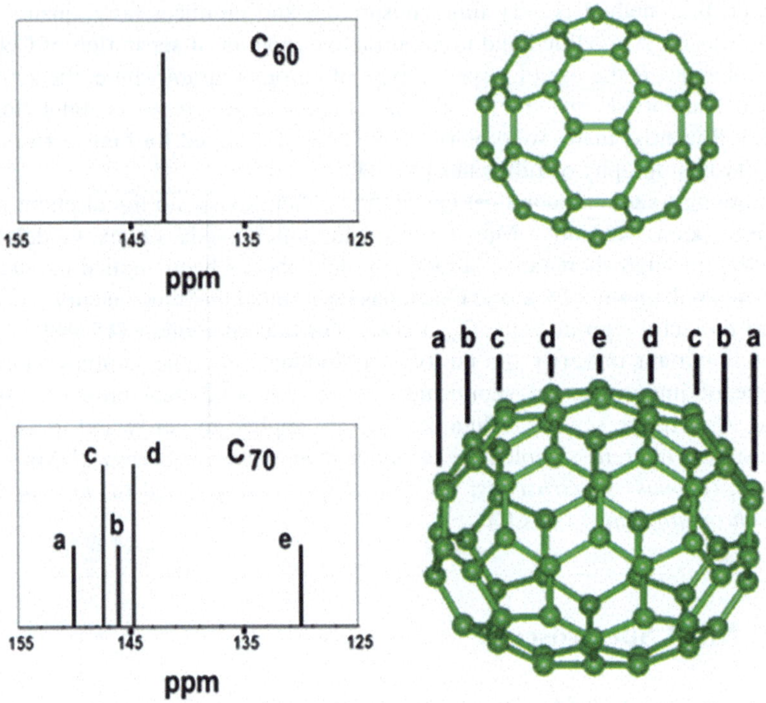


**Fig. 10.9** 13C NMR spectra and structural drawings of C60 and C70 (Photo courtesy: J.R. Bleeke and R.F. Frey)

Figure 10.10 shows electron ionization mass spectrum of a fullerene mixture. The intensity (relative abundance) is plotted along the Y-axis against the mass (m) to charge (z) ratio. The 70-eI mass spectra of the vapor (thermally desorbed) at ~350 °C yield, in addition to C60 and C70, multiply charged molecular ions and species corresponding to C56. The C56 is shown to be the fragmentation product of C60. One of the unique aspects of mass spectrometry, which is commonly used as an analytical method for bulk samples, is its critical role in discovery and early characterization of these species (C60, C70, and others) in the gas phase, which in turn resulted in their synthesis in bulk quantities [54]. In 1985, a group of researchers at Bell laboratories reported first tandem mass spectrometry (time-of-flight mass spectrometry) measurement for C60 [55]. Tandem mass spectrometry played an important role in probing the physical properties of fullerenes. The first systematic photodissociation study by Smalley and coworkers [56] used tandem time-of-flight mass spectrometry (TOFMS), which detected C60 and C70 carbon cluster ions [57].

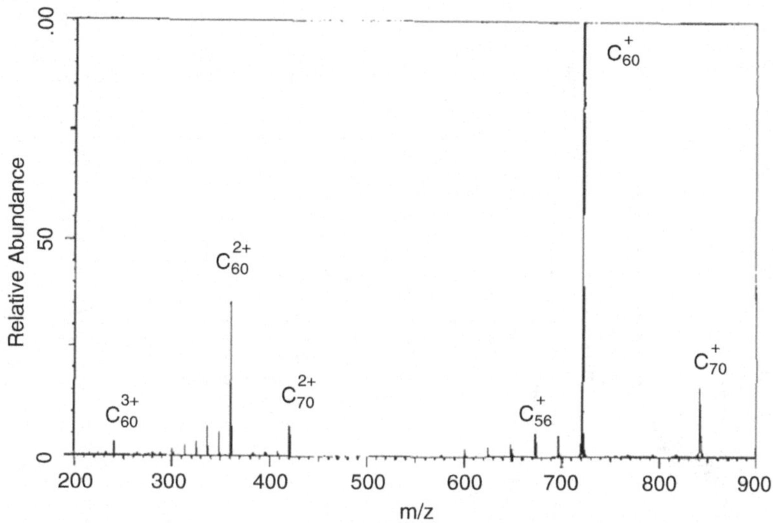


**Fig. 10.10** Electron ionization mass spectrum of a fullerene mixture (Reproduced with permission from M.M. Ross)

## 10.6 Optical Spectroscopy

Optical spectroscopy covers the ultraviolet (UV) through the infrared (IR) regions of the electromagnetic spectrum, and is considered to be another important tool for characterizing fullerene family. Due to the electronic transitions within the fullerene clusters, fullerenes exhibit size and structure dependent UV–visible (VIS) absorption spectra. Quantitatively, pure C60 solutions exhibit a deep purple color, while C70 solutions show an intense wine red color. The UV–VIS absorption spectra of C60 show strong peaks at ~257 nm and ~329 nm, and weak broad absorption at ~450 nm and 700 nm. On the other hand, C70 shows well-defined absorption peaks at ~331 nm, 360 nm, 378 nm, and 468 nm.

A calculated amount of highly oriented polycrystalline C60 material grown by molecular beam epitaxy [58, 59] is dissolved in toluene and sonicated for 5 min Toluene is evaporated under a stream of $N_2$ at 40–50 °C, after which a required volume of water is added to form a straw yellow dispersion of C60. The UV–VIS absorption spectrum of C60 in toluene solution is presented in Fig. 10.11. Expanded UV–VIS absorption spectrum shows a peak near 404 nm, which is the characteristic of monomeric C60. The broad visible reduced absorptions for blue around 450 nm and red around 700 nm have been identified as the characteristic of $(C60)_n$.

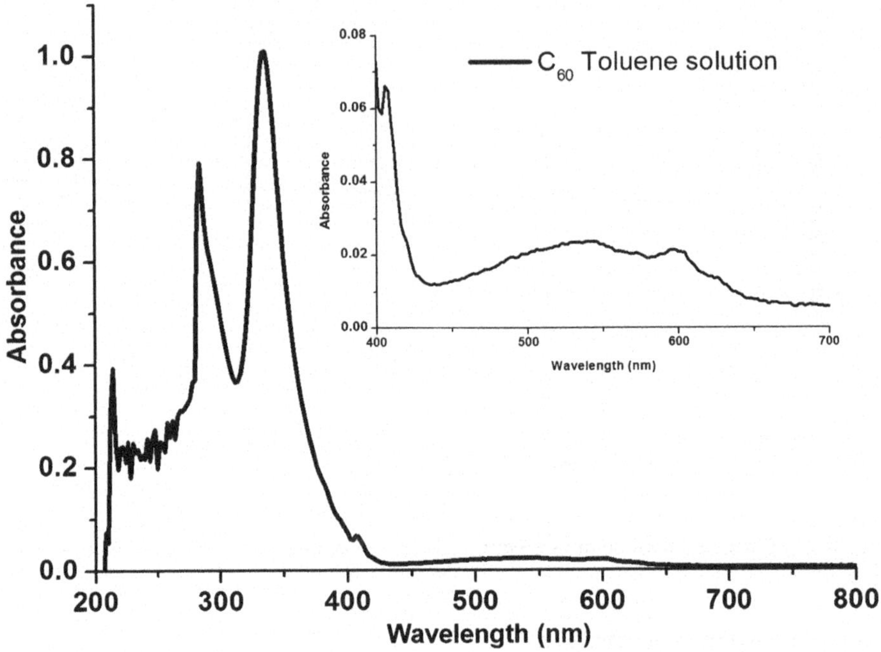


**Fig. 10.11** UV—visible absorption spectrum profiles of C60 in toluene (Courtesy: WIKI)

## 10.7 High-Pressure Liquid Chromatography (HPLC)

Since 1990, high-pressure liquid chromatography (HPLC) has played an important role in isolating fullerene clusters obtained from laser vaporization of graphite, electric arc discharge, hydrocarbon flame combustion process, and pyrolysis of carbon materials. HPLC has allowed macroscopic quantities of fullerenes (particularly C60) to be isolated and purified from the processing products [60, 61].

High-pressure liquid chromatography (HPLC) (Fig. 10.12) is also referred to as high-performance liquid chromatography. Figure 10.12 shows a high-pressure liquid chromatographic machine. HPLC relies on pumps to pass a pressurized liquid and a sample mixture through a column filled with a sorbent, leading to the separation of the sample components. HPLC is distinguished from traditional (low-pressure) liquid chromatography because operational pressures are significantly higher (50–350 bar), while ordinary liquid chromatography typically relies on the force of gravity to pass the mobile phase through the column.

Low-pressure liquid chromatography (LPLC) rather than HPLC has been used for separation based on adsorption chromatography and performed with conventional normal stationary phases (e.g. silica gel or alumina). C60 is first separated from C70 using open column liquid chromatography (LC) on alumina as stationary phase, and n-hexane or n-hexane/toluene (95:5) as mobile phase. The method presents several disadvantages, since n-hexane is not a good solvent for fullerenes.

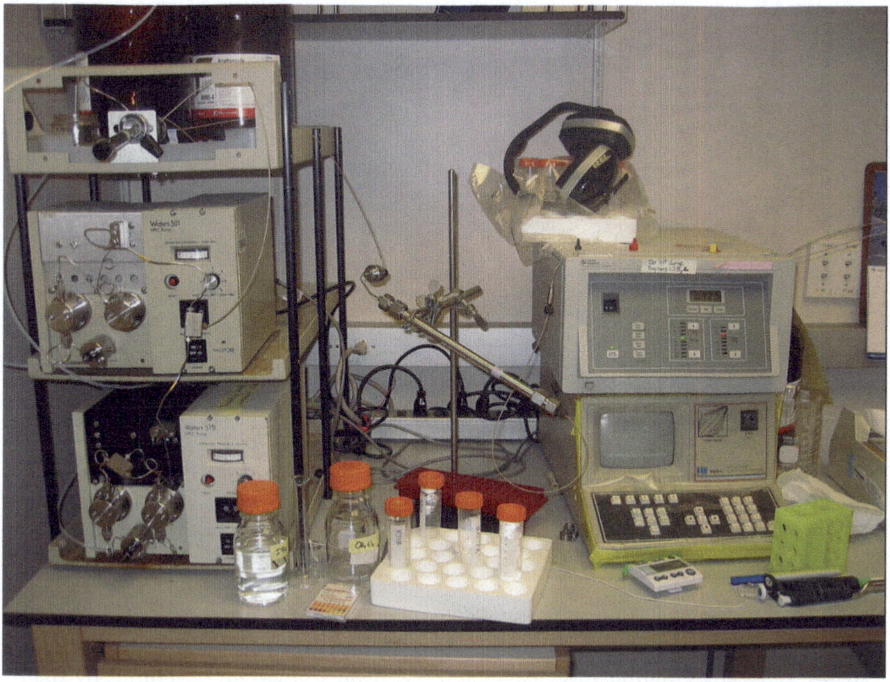

**Fig. 10.12** High-pressure liquid chromatographic machine (Courtesy: Kjaergaard, WIKI)

## 10.8 Characteristics of Fullerenes

A modified method for the characterization of fullerenes is the combination of Soxhlet extraction and LC on alumina, in which hot solvent is continuously recycled. However, the most efficient and inexpensive method involved using activated charcoal and silica gel mixture (1:2) as the stationary phase. Another promising method for fullerene separation is charge transfer chromatography which is based on the donation or acceptance of electrons between the stationary phase.

Saito and Oshiyama using a one electron model calculated the highest occupied molecular orbital and lowest unoccupied molecular orbital (HOMO–LUMO) gap of 1.92 eV for the free C60 molecule and a direct bandgap of 1.5 eV for the crystal [62]. However, the experimental value of the direct gap from the optical absorption spectra and photoconductive spectra ascertains the value to be ~1.6 eV [63–65].

Figure 10.13 shows crystalline C60 solid with brownish color. Both C60 and C70 are soft solids, and their compressibilities are comparable to the c-axis compressibility of graphite. At fairly low pressures (below the amorphization pressure), the photoluminescence band originally shows at 1.6 eV and undergoes a redshift [66, 67]. For nonlinear optics, especially for third harmonic generation, C60 is an interesting material which shows photoconduction and photo-induced electron

**Fig. 10.13** C60 solid with brownish color (Photo courtesy: J. Gschnaller, GNU,WIKI)

transfer, optical switching properties, and has been studied and reported by many scientists [68].

Figure 10.14 shows the model of a fcc lattice C60 solid showing the tetrahedral (upper arrow) and the octahedral sites (lower arrow). Extensive structural studies of fullerene C60, by X-ray, neutron diffraction, and electron diffraction studies, have established that upon crystallization, it forms a face-centered cubic (fcc) lattice, with lattice constant of 14.17 Å at room temperature. The distance between nearest neighbors of C60 clusters is 10 Å and the intercluster separation is 2.9 Å [68–71]. Further studies indicated that there exists two tetrahedral holes and one octahedral hole with radii 1.12 and 2.06 Å in C60 molecule. Both the octahedral and the tetrahedral sites can in principle accommodate metal dopants. The process of introducing dopants to these lattice holes in solid C60 fullerene is similar in many respects to intercalation of metals and other species into the interlayer region of graphite.

## 10.9 Doped Fullerenes

Earlier studies at Bell Laboratories showed that exposure of C60 to alkali metal vapor resulted in the uptake of alkali metal into the lattice with a concomitant increase in conductivity [72, 73]. When C60 is doped with potassium (K), it showed superconducting properties at 18 K [76]. However, cesium (Cs)-doped C60 showed superconductivity at a higher temperature (30 K) [74]. The probable reason for the

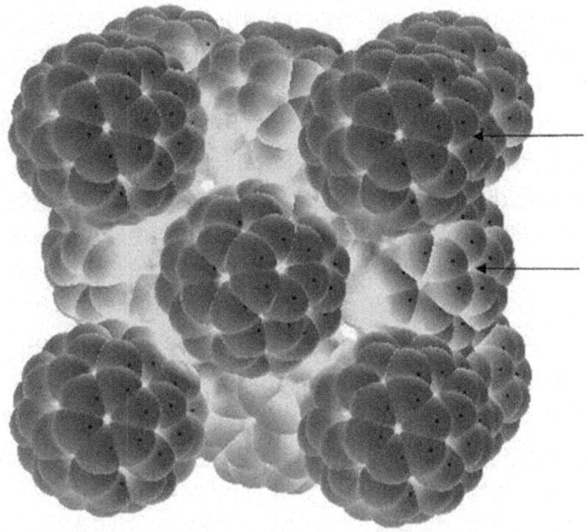

**Fig. 10.14** Model illustrating the packing of individual clusters in fcc lattice solid C60 fullerene. The *upper arrow* represents the tetrahedral and the *lower one* represents the octohedral sites (Courtesy: Thesaurus.rusnana, Creative Comm. WIKI)

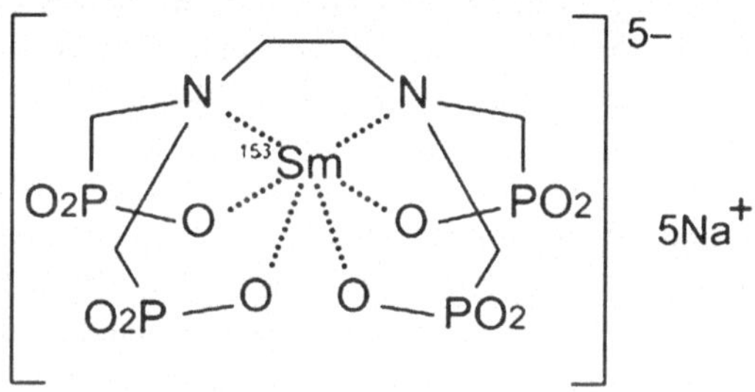


**Fig. 10.15** Model of C60 doped with samarium (Sm) (Photo courtesy: Anypodetos, WIKI)

superconductivity with doped C60 was thought to be intercalation of the dopant in octahedral and/or tetrahedral holes in the lattice or due to the change of the intercage distance through intercalation of the metal.

The rare earth metals doped with C60 fullerides have been reported to show superconductivity with superlattice crystal phase of $Sm_{2.75}C60$ [75, 76]. Figure 10.15 shows a model of C60 doped with samarium (Sm).

Recently, Takeda and coworkers studied the wealth and novel properties and functions produced by the capture of different atoms inside the fullerene cage. They found that doped with argon in a hot isostatic-pressed (HIP) equipment in the presence of argon (argon-doped) under a magnetic field of 10 Oe, C60 shows a superconducting behavior near 17.8 K (Tc), while the same material, without argon and doped with potassium (K), shows a superconducting behavior at 19.2 K (Tc). The hysteresis between the zero-field cooled and the field cooled data indicates type II superconductivity [77].

Since its discovery in 1970s and 45 years after its discovery, fullerene is still a topic of interest for its potential application in medicinal chemistry, especially with the advent of nanoscience, nanomedicine, and nanotechnologies where fullerene can be positioned properly to play a significant role in medicine. The more promising applications, among the ones so far explored, are the use of fullerene as a new platform for drug delivery, as a neuroprotector and as a contrast agent in MRI [78].

Antioxidant property of fullerene has made this allotrope to be used in the treatment of alcohol-induced encephalopathy and in alcoholism prophylaxis. Fullerene derivatives can also be used in the protection of neurodegenerative diseases mediated by β-amyloid. Doxorubicin (DOX) is widely used in cancer therapy but has adverse effects, especially on heart due to overproduction of superoxide. It has been demonstrated that pretreatment with C60 $(OH)_{24}$ as an antioxidant protects cardiomyocytes from the oxidative damage of a single dose of DOX, by means of hydroxyl radical scavenging [79].

Low-toxicity, and a photosensitizer, fullerene seems to be the ideal scaffold for the design of a photodynamic therapy to selectively destroy cancerous cells. The direct interaction between photoactivated C60 and DNA can lead to the oxidation of guanine; hence weakened bond is hydrolyzed resulting in the cleavage of DNA strand [80].

One of the fullerene C60 buckyball derivatives is PCBM (phenyl-C61-butyric acid methyl ester), which has been used as an organic photovoltaic device. Its chemical formula is $C_{72}H_{14}O_2$. PCBM is found to form red-brown platelet-shaped crystals, with monoclinic unit cell [81].

Unique chemistry and physics of fullerene (C60) materials and their derivatives continue to stimulate advances in applied and fundamental science. PCBM is known as an excellent electron acceptor and can be chemically modified to improve solubility in organic solvents. Such soluble fullerene derivatives are known as some of the best n-type organic semiconductors [82].

PCBM is soluble in the same organic solvents as excellent p-type semiconductors such as MDMO-PPV, MEH-PPV, and P3HT [81]. This simplifies the preparation of blends and solution-processing of heterojunction PV cells and OFETs. The high affinity of PCBM results in efficient photo-induced electron transfer from p-type polymers as well as from metal electrodes in biased thin-film OFETs [83].

Fullerene derivatives have been adopted widely as n-type semiconductors in solution-processed bulk heterojunction (BH) organic photovoltaic (OPV) devices. Various performance improvements have been made for OPV devices by modifying the p-type material-processing conditions leading to the present promising power conversion efficiency ($\eta$) of about 5% [82]. Recently, 11% power conversion efficiency of organic photovoltaic cell has been reported [83–85]. Figure 10.16 shows the architecture of an organic photovoltaic cell. In the figure, **a** represents the molecular formula, **b** represents the structure of the organic cell, and **c** represents the J–V characteristics of the cell under special conditions [86].

Another important application of fullerene derivatives is molecular heterojunctions (MJ). These MJs are made by covalently linking fullerenes with electron-donating or photoactive macromolecules. These MJs have shown promise as intrinsic p/n-type semiconductors and even as artificial mimics of biological photosynthesis.

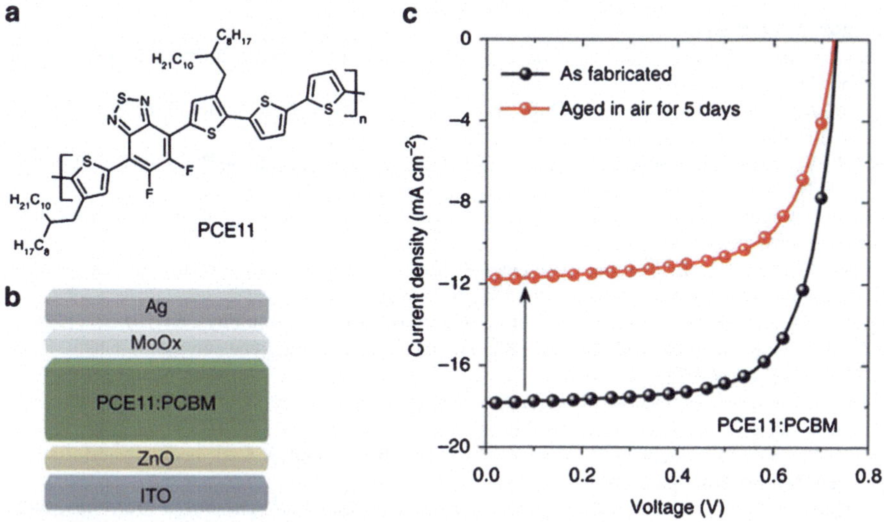


**Fig. 10.16** Architecture of an organic photovoltaic cell. (**a**) represents the molecular formula, (**b**) represents structure of the organic cell, and (**c**) represents the J–V characteristics of the cell under special conditions (Adapted from Ref. [86] with permission from Springer-*Nature*)

## 10.10 Summary

Fullerene C60, or the buckminster or bucky-ball C70, or others produced by any one of the methods previously stated, needs to be isolated or separated from the other fullerenes present in the product. There are several methods, such as NMR spectroscopy, mass spectroscopy, high-pressure/low-pressure liquid chromatography, or optical spectroscopy, that are available to identify and characterize a particular fullerene. On a per-C60 basis, there is one octahedral hole and two tetrahedral holes. These pores are available to accommodate a large variety of species (including alkali metal or rare-earth dopants) without a change in the basic structure of C60. Much of the studies on conductivity and superconductivity in potassium and rubidium doped C60 were initiated from the reports of Bell Laboratories. Some scientists believe that metallofullerenes could in the future bring a revolution in advanced materials research like the discovery of high-temperature superconductors, and can be used in drug delivery to destroy the cancerous cells.

## References

1. H. Aldersey-Williams, *The Most Beautiful Molecule: The Discovery of the Buckyball* (Wiley, New York, 1995)
2. E. Osawa, The original conjecture of a stable C60 molecule. Chem. Abstr. **74**, 75698v (1971)

3. D.A. Bochvar, E.G. Gal'pern, Electronic structure of C20 and C60. Proc. Acad. Sci. USSR **209**, 239 (1973)
4. L. Pauling, *The Nature of Chemical Bond*, 3rd edn. (Oxford University Press, London, 1960), p. 111
5. W. Kutzelnigg, Orthogonal and non-orthogonal hybrids. J. Mol. Struct. THEOCHEM **169**, 403 (1988)
6. H. W. Kroto, J. E. Fischer, D. E. Cox (eds.), *The Fullerene* (Pergamon Press, Oxford, London, 1993)
7. W. Andreoni, *The Physics of Fullerene-Based and Fullerene Related Materials* (Springer, Heidelberg, 2000), p. 24
8. W. Kratschmer, K. Fostiropoulos, D.R. Huffman, The success in synthesis of macroscopic quantities of C60. Chem. Phys. Lett. **170**, 167 (1990)
9. D.V. Afanas'ev et al., Fullerene creation in arc discharge. J. Tech. Phys. **64**(10), 76 (1994)
10. T.W. Ebbesen, J. Tabuchi, K. Tanigaki, The mechanics of fullerene formation. Chem. Phys. Lett. **191**, 336 (1992)
11. G.N. Churilov, Synthesis of fullerenes and other nanomaterials in arc discharge. Fullerenes Nanotubes Carbon Nanostruct **16**, 395 (2008)
12. H.W. Krot, J.R. Heath, S.C. O'Brien, R.F. Curl, R.E. Smalley, C60: buckminsterfullerene. Lett. Nature **318**, 162 (1985)
13. C.M. Lieber, C.-C. Chen, *Preparation of Fullerenes and Fullerene-Based Materials*, Solid state physics, vol 48 (Academic, New York, 1994), pp. 109–148
14. H. Kroto, Space, stars, C60, and soot. Science **242**, 1139 (1988)
15. R.E. Haufler, Y. Chai, L.P.F. Chibante, J. Conceicao, C. Jin, L.S. Wang, S. Maruyama, R.E. Smalley, Carbon arc generation of C60. Mater. Res. Soc. Symp. Proc. **206**, 627 (1991)
16. C.-C. Chen, C.M. Lieber, Synthesis of pure 13C60 and determination of the isotope effect for fullerene superconductors. J. Am. Chem. Soc. **114**, 3141 (1992)
17. I. Lamparth, A. Hirsch, Water soluble melonic acid derivatives of C60 with a defined three-dimensional structure. J. Chem. Soc. Chem. Commun. **116**, 1727 (1994)
18. A.L. Ortiz-Hernandez, Design and regioselective synthesis of two and three pronged C60 fullerene derivatives and their applications in molecular electronics, Ph.D. Dissertation, Clemson University, South Carolina, 2010
19. R. Kessinger et al., Preparation of Enantiometrically pure C76 with a general electrochemical method for the removable of Di (alkoxycarbonyl) methano bridges from methanofullerene the reto-Bingel reaction. Angew. Chem. Int. Ed. **37**, 1919 (1998)
20. R. Kessinger et al., Walk on the sphere: electrochemically induced isomerization of C60 Bis-adducts by migration of Di (alkoxycarbonyl) methano bridges. J. Am. Chem. Soc. **120**, 8545 (1998)
21. F. Diederich, R. Kessinger, Templated regioselective and stereoselective synthesis in fullerene chemistry. Acc. Chem. Res **32**(6), 537 (1999)
22. S. Fukuzumi, F. D'Souza, D. M. Guldi (eds.), *Fullerenes 2000*, vol 8 (Electrochemical Society, Pennington, 2000)
23. O. Lukoyanova, Studies of the stability and potential applications of pyrrolidinofullerenes and ather fullerene derivatives, Ph.D. Thesis, Clemson University, Clemson, SC, USA 2007
24. S. Zhang, O. Lukoyanova, I. Echegoyen, Synthesis of fullerene adducts with terpyridyl groups in trans-1 positions. Chem. Eur. J. Org. Chem., **12**, 3396 (2009)
25. P. Hebgen, A. Goel, J.B. Howard, L.C. Rainey, J.B. Vander Sande, Synthesis of fullerenes and fullerenic nanostructures in a low-pressure benzene/oxygen diffusion flame. Proc. Combust. Inst. **28**, 1397 (2000)
26. M. Sathish, K. Mityazawa, Synthesis and characterization of fullerenes Nanowhiskers by liquid-liquid interfacial precipitation: Influence of C60 solubility. Molecules **17**, 3858 (2012)
27. R. Ceolin et al., Solid state studies of the C60. 2 $(CH_3)CCl_3$ solvate. Carbon **43**, 417 (2005)
28. T. Nakanishi, W. Schmitt, T. Michinobu, D.G. Kurth, K. Ariga, Hierarchical supramolecular fullerene architectures with controlled dimensionality. Chem. Commun., 47, 5892 (2005)

29. J.B. Howard, K. Das Chowdhuri, J.B. Vander Sande, Carbon shells in flame. Nature **370**, 603 (1994)
30. J.B. Howard, J.T. McKinnon, Y. Markarovsky, A.L. Lafleur, M.E. Johnson, Fullerenes C60 and C70 in flames. Nature **352**, 139 (1991)
31. J.B. Howard et al., Fullerenes synthesis in combustion, Carbon. **30**, 1183 (1992) + P.W. Stephens, *Physics and Chemistry of Fullerenes* (World Scientific, River Edge, 1993)
32. Z.A. Mansurov, Soot formation combustion process, (review). Combust. Explot. Shock Waves **41**(6), 727 (2005)
33. K.C. Khemani, M. Prato, F. Wudl, A simple Soxhelt chromatographic method for the isolation of pure fullerenes C60 and C70. J. Org. Chem. **57**, 3254 (1992)
34. M.S. Amer, *Raman-Spectroscopy and Nanotechnology* (Royal Society Chemistry, London, 2010)
35. C. Thilgen, F. Diederich, R.L. Whetten, Buckministerfullerene, 1993, 59, in *Fullerenes, Chemistry and Reactions*, ed. by A. Hersch, M. Brettreich (Wiley, Weinheim, 2010)
36. M. Diack, R.L. Hettich, R.N. Compton, G. Guiochon, Contribution to the isolation and characterization of buckminsterfullerenes. Anal. Chem. **64**, 2143 (1992)
37. J. Abrefah, D.R. Olander, M. Balooch, W.J. Siekhaus, Vapor pressure of buckminsterfullerene. Appl. Phys. Lett. **60**, 1313 (1991)
38. C. Pan, M.P. Sampson, Y. Chai, R.H. Hauge, J.L. Margrave, Heats of sublimation from a polycrystalline mixture of carbon clusters (C60 and C70). J. Phys. Chem. **95**, 2944 (1991)
39. F. Diederich et al., The higher fullerenes isolation and characterization of C76, C84, C90, C94 and C70O an oxide of D5h-C70. Science **252**, 548 (1991)
40. P. Bhyrappa, A. Penicaud, M. Kawamoto, C.A. Reed, Improved chromatographic separation and purification of C60 and C70 fullerenes. J. Chem. Soc. Chem. Commun. **13**, 936 (1992)
41. W.A. Scrivens, P.V. Bedworth, J.M. Tour, Purification of gram quantities of C60. A new inexpansive and facile method. J. Am. Chem. Soc. **114**, 7917 (1992)
42. N. Manova et al., Separation of C60/C70 mixture on activated carbon and activated carbon fibers. Carbon **33**(2), 209 (1994)
43. X.R. Xia, N. Monteiro-Riviere, J. Riviere, Trace analysis of fullerenes in biological samples by simplified liquid liquid extraction and high performance liquid chromatography. J. Chromatogr. A **1129**, 216 (2006)
44. D. Bouchard, X. Ma, Extraction of high performance liquid chromatographic analysis of C60, C70, and [6,6]-penyl C61-butyric acid methyl ester in systematic and natural waters. J. Chromatogr. A **1203**(2), 153 (2008)
45. P. Orea, Phase diagrams of model $C_{60}$ and $C_{70}$ fullerenes from short-range attractive potentials. J. Chem. Phys. **130**, 104703 (2009)
46. L.A. Girifalco, Molecular properties of fullerene in the gas and solid phase. J. Phys. Chem. **95**, 868 (1992)
47. Kaminsky, M. Budesinsky, S. Taubert, P. Bour, M. Straka, Fullerene C70 characterization by $^{13}$ C NMR and the importance of the solvent and dynamics in spectral simulation. Phys. Chem. Chem. Phys. **15**, 9223 (2013)
48. R. Taylor, J.P. Hare, A.K. Abdul-sada, H.W. Kroto, Isolation, separation, and characteristics of the fullerene C60, C70, the third form of carbon. J. Chem. Soc. Chem. Commun., 20, 1423 (1990)
49. G. Sun, M. Kertesz, Theoretical 13NMR spectra of IPR isomers of fullerenes C60, C70, C72, C74, C76, and C78 studied by density functional theory. J. Phys. Chem. A **104**(31), 7398 (2000)
50. R.D. Johnson, G. Meijer, J.R. Salem, D.S. Bethune, 2D nuclear magnetic resonance study of the structure of the fullerene C70. J. Am. Chem. Soc. **113**, 3619 (1991)
51. D.O. Sparkman, *Mass Spectrometry Desk Reference* (Global View, Pittsburgh, 2000)
52. R.E. Smalley, The great balls of carbon—the buckminsterfullerene. Sciences **31**, 22 (1991)
53. H.W. Kroto, A.W. Allaf, S.P. Balm, C60, buckminsterfullerene. Chem. Rev. **91**, 1213 (1991)

54. S.W. McElvany, M.M. Ross, Mass spectrometry and fullerenes. J. Am. Soc. Mass Spectrom. **3**, 268 (1992)
55. L.A. Bloomfield, M.E. Geusic, R.R. Freeman, W.L. Brown, Negative and positive cluster ions of carbon. Chem. Phys. Lett. **121**, 33 (1985)
56. S.C. O'Brien, J.R. Heath, R.F. Curl, R.E. Stanley, Buckminsterfullerene and other carbon cluster ions. J. Chem. Phys. **88**, 220 (1998)
57. S.W. McElvany, M.M. Ross, J.H. Callahan, Characterization of fullerenes by mass spectrometry. Acc. Chem. Res. **25**, 162 (1992)
58. Z. Dai, H. Naramoto, K. Narumi, S. Yamamoto, A. Miyashita, A three step process for epitaxial growth of (111) oriented C60 films on alkali halide substrates. Thin Solid Films **360**, 28 (2000)
59. H. Takashima, M. Nakaya, A. Yamamoto, A. Hashimoto, Solid C60 growth on hexagonal GaN (0001) surface. J. Cryst. Growth **227**, 829 (2001)
60. V.L. Cebola, L. Membrado, J. Vela, Fullerenes: liquid chromatography, in *Analytical Supercritical Fluid Chromatography and Extraction*, ed. by M. L. Lee, K. E. Markides (Academic, San Diego, 2000), p. 2901
61. J. Theobald, M. Perrut, J.V. Weber, E. Millon, J.F. Muller, Extraction and purification of fullerenes: a comprehensive review. Sep. Sci. Technol **30**(14), 2783 (1995)
62. E.A. Katz, Chapt-13, in *Nanostructured Materials for Solar Energy Conversion*, ed. by T. Soga (Elsevier, Amsterdam, 2006), p. 377
63. W. Zhou, S-hen Xie, S-fa Oian, G. Wang, L-xi Qian, Photothermal deflection spectra of solid C60, J. Phys. Condens. Matter **8**, 5793 (1996) and also S. Saito, A. Oshiyama, Phys. Rev. Lett. **186**, 281 (1991)
64. M.B. Jost et al., Inverse photoemission and theory. Phys. Rev. B **44**, 1966 (1991)
65. R.W. Lof et al., Bandgap, excitons, and Columb interaction in solid C60. Phys. Rev. Lett. **68**, 3924 (1992)
66. D.W. Snoke, K. Syassen, A. Mittelbach, Optical absorption of $C_{60}$ at high pressure. Phys. Rev. B **47**(7), 4146 (1993)
67. T. Pradeep, A glimpse into fascinating world of fullerenes. Curr. Sci **72**(2), 124 (1997)
68. J. Shinar, Z.V. Vardeny, Z.H. Kafafi, *Optical & Electronic Properties of Fullerenes and Fullerene-Based Materials* (Mercel Dekker AG, New York, 2000)
69. D.A. Neumann et al., Coherent quasielastic neutron scattering study of rotational dynamics of C60 in orientationally disordered phase. Phys. Rev. Lett. **67**, 3808 (1991)
70. J.E. Fischer, P.A. Heiney, A.B. Smith, Solid state chemistry of fullerene-based materials. Acc. Chem. Res. **25**, 115 (1992)
71. C.M. Lieber, Z. Zhang, Physical properties of meta-doped fullerene superconductors. Solid State Phys. **48**, 349 (1994)
72. R.C. Haddon et al., Conducting films of C60 and C70 by alkali metal doping. Nature **350**, 320 (1991)
73. A.F. Habard et al., Superconductivity at 18K in potasium doped C60. Nature **350**, 600 (1991)
74. S.P. Kelty, C.C. Chen, C.M. Lieber, Superconductivity at 30K in cesium doped C60. Nature **352**, 223 (1991)
75. M. Akada et al., Superconductivity appearing from C60 doped with rare earth metals. Sci. Technol. Adv. Mater. **7**, S83 (2006)
76. X.H. Chen, G. Roth, Superconductivity in samarium doped C60. Phys. Rev. B. Condens. Matter **52**(21), 15534 (1995)
77. A. Takeda et al., Superconductivity of Doped Ar C60, Chem. Commun., 912, Royal Society of Chemistry, London, (2006)
78. M. Carini, L. Dordevic, T. Da Ros, Fullerene in Biology and Medicine, Chapt-1, Vol-3, in *Handbook of Carbon Nano Materials*, (World Scientific, Singapore, 2015)
79. V.M. Torres et al., Fullerene C60 $(OH)_{24}$ prevents doxorubicin induced accute cadiotoxicity in rats. Pharmacol. Rep. **62**, 707 (2010)

80. R. Bernstein, F. Prat, C.S. Foote, On the mechanism of DNA cleavage by fullerenes: Electron transfer from guanosine and 8 Oxo-guanocine to C60. J. Am. Chem. Soc. **121**, 464 (1999)
81. Y. Lin, Y. Li, X. Zen, Small molecule semiconductors for high efficiency organic photovoltaics. Chem. Soc. Rev. **41**, 4245 (2012)
82. N. Martin, F. Giacalone (eds.) *Fullerene Polymers, Synthesis, Properties, and Applications* (Wiley, Weinheim, 2010) and also http://www.Sigmaaldrich.com/content/dam/sigmaaldrich/docs/Aldrich/General_information/1/material-matters-basics,vol.14.pdf
83. C. R. Newman et al., Introduction to organic thin film transistors and design of n-channel semiconductors, Chem. Mater. **16**(23), 4436 (2004) and also C. Brabec, V. Dyakonov, U. scherf, *Organic Phovoltaics* (Wiley, Germany, 2008)
84. J. Luis Delgado et al., Organic photovoltaics a chemical approach. J. Chem. Soc. Chem. Commun. R.S.C. **46**, 4853 (2010)
85. M. Paterno et al., Micro focussed X-ray diffraction characterization of high quality [6,6]-phenyl-C61-butyric acid methyl ester single crystals without solvent impurities. J. Mater. Chem. **C1**, 5619 (2013)
86. N. Li et al., Abnormal strong burn-in degradation of highly efficient polymer solar cells caused by spinodal donor-acceptor demixing. Nat. Commun. **8**, 14541 (2017)

The manufacturer's authorised representative in the EU is Springer Nature Customer Service Centre GmbH, Europaplatz 3, 69115 Heidelberg, Germany. If you have any concerns regarding our products, please contact ProductSafety@springernature.com

Printed and bound by CPI Group (UK) Ltd, Croydon, CR0 4YY
25/09/2026
02222836-0015